Sustainable Waste Management and Recycling: Construction Demolition Waste

Proceedings of the International Conference organised by the
Concrete and Masonry Research Group and held at
Kingston University - London on 14-15 September 2004.

Edited by

Mukesh C. Limbachiya
Reader, School of Engineering
Kingston University

John J. Roberts
Dean, Faculty of Technology
Director, Sustainable Technology Research Centre
Kingston University

 ThomasTelford

Published by Thomas Telford Publishing, Thomas Telford Ltd, 1 Heron Quay, London E14 4JD.
www.thomastelford.com

Distributors for Thomas Telford books are
USA: ASCE Press, 1801 Alexander Bell Drive, Reston, VA 20191-4400, USA
Japan: Maruzen Co. Ltd, Book Department, 3–10 Nihonbashi 2-chome, Chuo-ku, Tokyo 103
Australia: DA Books and Journals, 648 Whitehorse Road, Mitcham 3132, Victoria

First published 2004

The full list of titles from the 2004 International Conference 'Sustainable Waste Management and Recycling' and available from Thomas Telford is as follows

- Glass waste. ISBN: 0 7277 3284 6
- Construction demolition waste. ISBN: 0 7277 3285 4
- Used/post-consumer tyres. ISBN: 0 7277 3286 2

A catalogue record for this book is available from the British Library

ISBN: 0 7277 3285 4

Printed and bound in Great Britain by MPG Books, Bodmin, Cornwall

PREFACE

With the introduction of waste legislation, in the form of regulations and directives, in many parts of the world a significant move towards sustainable waste management is becoming a legal requirement. Emphasis is now being placed on increasing recycling and promoting more sustainable waste management practices, and greater co-ordination between the public, private and independent sectors, and all concerned with the management of waste and reusable materials.

However, sustainable waste management entails complex technological, environmental, social, culture and economic issues. This together with technological advances in recycling, the waste sector is facing enormous challenges in developing suitable waste management and recycling strategies. It is therefore necessary to share/ explore existing expertise and review, and discuss the challenges and identify opportunities for improving waste management and recycling as widely as possible. Thus promote sustainable resource use.

The Concrete and Masonry Research Group (CMRG) within the School of Engineering at Kingston University organised this International Conference to explore waste minimisation practices a nd r eview l atest d evelopments i n w aste m anagement a nd r ecycling. It c onsidered regulatory and other pressures which may affect industry's approach, provided a platform to meet and discuss with the leading experts on how the industry can address challenges of sustainable waste management and recycling, and review existing opportunities. This Conference dealt with key waste materials under concurrently proceeded three themes, namely *(i) Construction Demolition Waste, (ii) Used/ Post-Consumer Tyres, and (iii) Glass Waste.* Over 130 papers were presented by authors from 33 countries during this Conference and these are compiled into three theme specific proceedings.

The Opening Addresses were given by Professor Peter Scott- Vice-Chancellor of Kingston University, Professor John Roberts- Dean of Faculty of Technology at Kingston University and Mr Hugh Carr-Harris- Chief Executive of London Remade. All the papers presented under **Construction Demolition Waste** Theme and two Conference Closing Keynote papers are assembled in this volume. Professor C S Poon, Hong Kong Polytechnic University, gave the Opening Keynote paper. The Conference was closed by Keynote papers from Dr Lindon Sear, United Kingdom Quality Ash Association and Mr John Harrison, TecEco Pty Lt. Australia.

The event was organised with support from four sponsoring organisations: Credential Environmental Ltd., ReMade Kent and Medway, London ReMade and WRAP. All sponsors are gratefully acknowledged for their invaluable support. The work of Conference was an immense undertaking and help from all those involved are gratefully acknowledged, in particular, members of the Organising Committee for managing the event from start to finish; the Authors and the Chair of Technical Sessions for their invaluable contribution. Special thanks to Day Group Ltd., and Used Tyre Working Group- UK for their support.

The Proceedings have been prepared directly from the camera-ready manuscript submitted by the authors and editing has been restricted to minor changes where it was considered absolutely necessary.

Kingston upon Thames
September 2004

Mukesh C Limbachiya
John J Roberts

SPONSORING ORGANISATIONS

Credential Environmental Ltd.

ReMade Kent and Medway

London Remade

WRAP- Waste & Resources Action Programme

ORGANISING COMMITTEE

Concrete and Masonry Research Group-Kingston University

Dr M C Limbachiya *(Chairman)*

Professor J J Roberts

Professor G Sommerville OBE

Professor N J Bright MBE

Dr A N Fried

Dr K Etebar

Mr A Koulouris

Mr Y Ouchagour

Miss S Fotiadou

Dr A Ahmed

Mr O Reutter

Mr O Kanyeto

Faculty of Technology Research Office

NATIONAL ADVISORY COMMITTEE

CONTENTS

SESSION 4- WAY FORWARD AND DEVELOPING SUSTAINABLE MARKET

KEYNOTE
PAPER

PROPERTIES OF STEAM CURED RECYCLED AGGREGATE CONCRETE

C S Poon

S C Kou

The Hong Kong Polytechnic University

Hong Kong

ABSTRCT. The effects of steam curing on the hardened properties of recycled aggregate concrete were investigated. In this study, two series of concrete mixtures with water to cement ratios of 0.55 and 0.45 were prepared. Recycled aggregates were used as 20, 50, 80 and 100 % replacements of natural coarse aggregate in the concrete mixtures. The concrete specimens underwent standard water curing and steam curing regimes. The test results showed that the strength of concrete decreased as the recycled aggregate content increased. An initial steam curing regime increased the 1-day strength of the concrete but the corresponding 28 and 90-day strength decreased. Furthermore, steam curing reduced the static modulus of elasticity of the concrete compared to that of the water cured concrete. However, the negative effect of steam curing diminished as the recycled aggregate content increased. Moreover, steam curing decreased the drying shrinkage and increased the resistance against chloride ion penetration of the recycled aggregate concrete. The results demonstrate that one of the practical ways to utilize a higher percentage of recycled aggregates in concrete is "precasting" with an initial steam curing stage immediately after casting.

Keywords: Steam curing, Recycled aggregate concrete, Precast concrete, Hardened properties.

Dr C S Poon is a Professor and Leader of the *Research Group on the Reuse of Waste Materials for Construction* at the Hong Kong Polytechnic University. His research interests include construction and demolition waste management, concrete technology and recycling of waste materials for construction.

S C Kou is a PhD candidate of the *Research Group on the Reuse of Waste Materials for Construction* at the Hong Kong Polytechnic University.

INTRODUCTION

Recycled aggregate has been used as a replacement of the natural aggregate for a number of years. The potential benefits and drawbacks of using recycled aggregate in concrete have been quite extensively studied [1-7]. The use of recycled aggregate generally increases the drying shrinkage, creep and water sorptivity and decreases the compressive strength and modulus of elasticity of concrete compared to those of natural aggregate concrete. Hasaba et al. [8] reported that the drying shrinkage of concrete prepared with coarse recycled aggregate and natural sand was 50 % higher than that of conventional concrete. When both coarse and fine recycled aggregates were used, the drying shrinkage of recycled aggregate concrete was as much as 70 % higher than that of conventional concrete. Hansen and BØEGH [9] reported that recycled aggregate concrete had 15 to 30 % lower modulus of elasticity and 40 to 60 % higher shrinkage than those of conventional concrete. Olorunsogo and Padayachee [10] found that the water sorptivity of concrete prepared with 100 % recycled aggregate was about 39 % greater than that of natural aggregate concrete at the curing age of 28 days. Dhir et al. [11] showed that the compressive strength of concrete with 100 % coarse and 50 % fine recycled aggregates was between 20 – 30 % lower than that of the corresponding natural aggregate concrete.

The negative effects of recycled aggregate on concrete quality limit the use of this material in structural concrete. However, the disadvantages of using recycled aggregate can be minimized in the precast concrete industry since it is easier to ensure the quality of such products due to the presence of an existing quality assurance system. The target precast concrete products can be in the forms of partition walls, road dividers, bridge fencing, noise barriers and paving blocks which do not require very high performance standards. In Hong Kong, a development has already been made to produce bricks and blocks with recycled aggregates [7].

The curing method used for precast concrete products differs from the normal curing method where steam curing is usually employed because it accelerates the rate of strength development. However, this curing method alters the properties of produced concrete. It was found that steam curing reduced the creep of concrete by up to 50 % compared to that of normal mosit-cured concrete [12-14]. Boukendakdji et al [15] found that steam curing decreased the drying shrinkage of Portland cement and slag concretes by about 14 and 23 % respectively. Ho et al. [16] found that steam cured concrete was more porous than water cured concrete. Erdem et al. [17] indicated that steam curing increased the 1-day compressive strength of concrete compared to that of normal cured concrete. However, lower longer term strengths were observed for the steam cured concrete when compared to the normal water cured concrete.

Although plenty information is available on the effect of steam curing on conventional concrete properties, there is limited data relating to steam cured recycled aggregate concrete. There is a need to gain more information on the effect of steam curing on the strength and durability of concrete produced with high percentages of recycled aggregates.

EXPERIMENTAL DETAILS

Cement

ASTM Type I Portland cement was used and the corresponding properties are given in Table 1.

Table 1. Chemical compositions of cement

CHEMICAL COMPOSITION	(%)
SiO_2	19.61
Fe_2O_3	3.32
Al_2O_3	7.33
CaO	63.15
MgO	2.54
SO_3	2.13
Loss On Ignition (LOI)	2.97
Density (g/cm^3)	3.16
Specific Area (cm^2/g)	3519

Aggregates
Natural and recycled aggregates were used as the coarse aggregate in the concrete mixtures. In this study, a locally available crushed granite was used as the natural aggregate and recycled aggregate sourced from a recycling facility in Hong Kong was used. The nominal sizes of both the natural and recycled aggregates were 20 and 10 mm and their particle size distributions conformed to the requirements of BS 812 [18]. The corresponding physical and mechanical properties of the coarse aggregate are shown in Table 2. The porosity of the aggregates w as d etermined u sing m ercury i ntrusion p orosimetry (MIP). River s and w ith a fineness modulus of 2.11 was used as the fine aggregate in the concrete mixtures.

Table 2. Properties of natural and recycled coarse aggregates

TYPE	NOMINAL SIZE	DENSITY (kg/m^3)	WATER ABSORPTION (%)	STRENGTH (10% fines) KN	POROSITY (measured by MIP) (%)
Crushed	10mm	2.62	1.12	159	1.62
Granite	20mm	2.62	1.11		
Recycled	10mm	2.49	4.26	126	8.69
Aggregate	20mm	2.57	3.52		

Concrete Mixtures
A total of two series of concrete mixtures was prepared in the laboratory. Series I mixes were prepared with a water-to-cement ratio (W/C) and a cement content of 0.55 and 410 kg/m^3 respectively. Series II mixes were prepared with a W/C and a cement content of 0.45 and 400 kg/m^3 respectively. The recycled aggregate were used as 0%, 20%, 50%, and 100% by volume replacements of the natural aggregate. The absolute volume method was adopted to design the mix proportions of the concrete mixtures as shown in Tables 3 and 4. In each concrete mixture, the 10 and 20 mm coarse aggregates were used in a ratio of 1:2.

Table 3. Mix proportions of the series I concrete mixes

NOTATION	NATURAL GRANITE (%)	RECYCLED AGGREGATE (%)	CONSTITUTES (Kg/m³)				
			Water	Cement	Sand	Granite	Recycled Aggregate
R0	100	0	225	410	642	1048	0
R20	80	20	225	410	642	840	204
R50	50	50	225	410	642	524	506
R100	0	100	225	410	642	0	1017

Table 4. Mix proportions of the series II concrete mixes

NOTATION	NATURAL GRANITE (%)	RECYCLED AGGREGATE (%)	CONSTITUTES (Kg/m³)				
			Water	Cement	Sand	Granite	Recycled Aggregate
R0	100	0	180	400	708	1108	0
R20	80	20	180	400	708	886	215
R50	50	50	180	400	708	554	538
R100	0	100	180	400	708	0	1078

Specimens Casting and Curing

For each concrete mix, 100 mm cubes, 70 x 70 x 285 mm prisms and 200 x 100 mm diameter cylinder specimens were cast. The cubes and the prisms specimens were used to test the compressive strength and the drying shrinkage respectively. The cylinders were used to evaluate the splitting tensile strength, the static modulus of elasticity and the chloride-ion

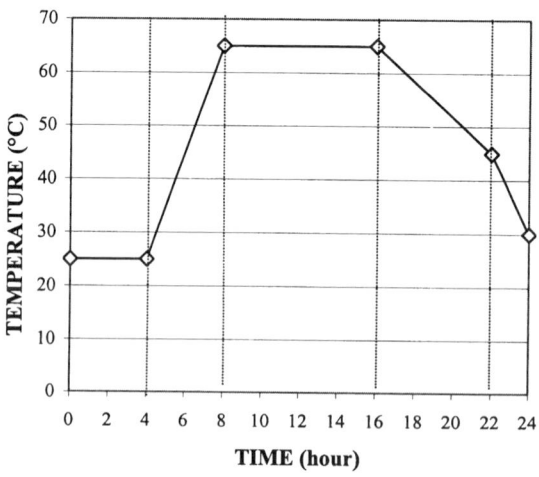

Figure1. One-day steam curing cycle

penetration of concrete. All specimens were cast in steel moulds and compacted using a vibrating table. In this study, normal water curing and initial steam curing were used.

For the water cured specimens, the specimens were cured in air for a period of 24 hours before they were demolded. After demolding, three cubes and three cylinders were immediately tested for the 1-day compressive and splitting tensile strengths, and the rest of the specimens were cured in a water tank at $27 \pm 1°C$ until other test ages were reached. For steam curing, the concrete specimens immediately after casting (without demolding) were initially cured in a steam bath at $65°C$ for 8 hours. Figure 1 shows steam curing cycle. After the steam curing stage, the specimens were demolded and three cubes and three cylinders were tested for the 1-day compressive and splitting tensile strengths. The rest of the specimens were further cured in a water tank at $27 \pm 1°C$ until other test ages were reached.

TESTS

Compressive and Splitting Tensile Strengths
The compressive and splitting tensile strengths of concrete were determined using a Denison compression machine with a loading capacity of 3000 kN. The loading rates for the compressive and splitting tensile tests were 0.2-0.4 $N/mm^2.s$ and 57 kN/min respectively. The compressive and splitting tensile strength tests were carried out at the ages of 1, 4, 7, 28 and 90 days.

Static Modulus of Elasticity
The static modulus of elasticity of concrete was determined in accordance with ASTM C 469-65. In this study, this test was carried out on the concrete specimens at the ages of 28 and 90 days.

Drying Shrinkage Test
A modified British Method (BS1881, part 5 : 1970) was used for the test. After removing the concrete prisms from the curing tank, the initial length of each specimen was measured. The specimens were then stored in an environmental chamber with a temperature of $55°C$ and a relative humidity of 95 % until the next measurements at 1, 4, 7, 28, 56, 90 and 112 days. Before each measurement was taken on the scheduled day, the specimens were first removed from the environmental chamber and conveyed to a second cooling chamber for about 4 hours at a controlled temperature of $25°C$ and a relative humidity of 75 %.

The length of each specimen was then measured within 15 minutes before delivering the specimens back to the environmental chamber for the subsequence drying process. The procedure of drying, cooling and measuring continued until the final length measurement at 112 day was recorded.

Determination of chloride penetrability
The chloride ion penetrability of concrete was determined in accordance with ASTM C1202-94 using a 50 mm thick x 100 diameter concrete disc cut from the 100 x 200 mm concrete cylinders. The resistance of concrete against chloride ion penetration is represented by the total charge passed in coulombs during a test period of 6 hour. In this study, the chloride penetrability test was carried out on the concrete specimens at the ages of 28 and 90 days.

RESULTS AND DISCUSSIONS

Compressive Strength Development

The c ompressive s trengths o f c oncrete i n b oth se ries a re s hown i n T ables 5 a nd 6 . E ach presented value is the average of three measurements. The results showed that the compressive strengths of concrete at all curing ages in both series decreased with an increase in the recycled aggregate content as shown in Figure 2. The relative compressive strength indicated in Figure 2 was the ratio of the compressive strength of the concrete to that of the water cured natural aggregate concrete at 28 days. The results showed that the 1-day compressive strengths of the steam cured concrete in both series were higher than those of the water cured concrete. However, the 28 and 90-day compressive strengths of the steam cured concrete was lower than that of the water cured concrete. Moreover, the rate of gain in strength of the steam cured concrete was much lower than that of the water cured concrete after 1 day. Also, the compressive strengths of the concrete in series II were greater than those of concrete in Series I since a lower W/C ratio was used.

Table 5. Compressive strength of the series I concrete mixtures

CURING	AGE	COMPRESSIVE STRENGTH (MPa)			
		R0	R20	R50	R100
STANDARD WATER CURED	1-day	12.8	11.9	11.6	10.2
	4-day	23.3	22.4	21.8	18.6
	7-day	30.2	29.1	27.6	24.4
	28-day	48.6	45.3	42.5	38.1
	90-day	52.7	50.8	49.5	45.5
STEAM CURED	1-day	22.9	21.3	20.9	19.2
	4-day	31.1	29.1	28.6	27.8
	7-day	36.3	34.3	33.4	31.2
	28-day	44.5	42.8	40.9	37.2
	90-day	47.8	46.6	45.7	43.0

Table 6. Compressive strength of the series II concrete mixtures

CURING	AGE	COMPRESSIVE STRENGTH(MPa)			
		R0	R20	R50	R100
STANDARD WATER CURED	1-day	25.8	23.6	21.1	15.5
	4-day	45.8	43.2	40.3	26.8
	7-day	53.8	51.2	44.8	36.2
	28-day	66.8	62.4	55.8	42.0
	90-day	72.3	68.0	61.5	50.2
STEAM CURED	1-day	41.8	41.9	38.0	28.1
	4-day	49.6	47.9	41.9	32.1
	7-day	53.2	50.3	46.7	35.7
	28-day	58.1	57.4	55.1	41.4
	90-day	63.9	65.9	62.2	48.4

The lower compressive strength of the steam cured concrete at the late curing ages may be attributed to both chemical and physical effects. The chemical effect was probably due to the rapid reaction rate at the initial elevated temperature rendered an uneven distribution of hydration products in the concrete. The possible detrimental physical effect of steam curing was the formation of microcracks induced by the high temperature. Since recycled aggregates generally have higher porosity (Table 2) compared to that of natural aggregates, the detrimental effects of steam curing on the recycled aggregate concrete was less.

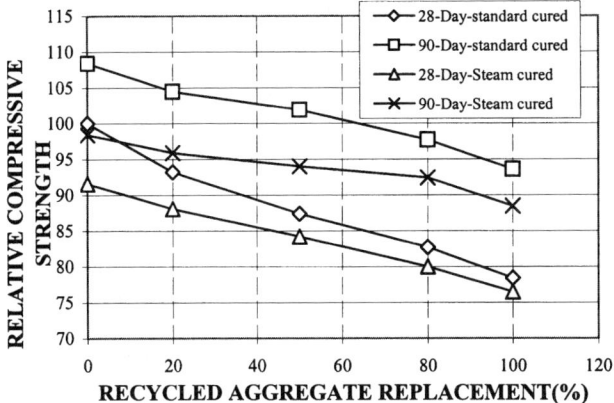

Figure 2. Relative compressive strength of series I concrete mixtures

Splitting Tensile Strength

The splitting tensile strengths of the concrete are given in Tables 7 - 8. Each presented value is the average of three specimens. The results showed that an increase in the recycled aggregate content decreased the splitting tensile strength of the concrete. Similarly, steam curing increased the 1-day tensile strength. However, after 1 day, the rate of strength gain of the steam cured concrete was much lower than that of the water cured concrete. Furthermore, a decrease in W/C from 0.55 to 0.45 increased the splitting tensile strength of concrete.

Table 7. Splitting tensile strength of the series I concrete mixtures

CURING	AGE	SPLITTING TENSILE STRENGTH (MPa)			
		R0	R20	R50	R100
STANDARD WATER CURED	1-day	1.30	1.27	1.25	1.21
	4-day	2.32	2.30	2.28	2.20
	7-day	2.62	2.53	2.51	2.38
	28-day	3.32	3.21	3.16	3.06
	90-day	3.68	3.66	3.48	3.23
STEAM CURED	1-day	2.08	2.04	1.84	1.70
	4-day	2.57	2.50	2.40	2.28
	7-day	2.84	2.78	2.70	2.52
	28-day	3.30	3.24	2.97	2.80
	90-day	3.75	3.61	3.45	3.12

Table 8. Splitting tensile strength of the series II concrete mixtures

CURING	AGE	SPLITTING TENSILE STRENGTH (MPa)			
		R0	R20	R50	R100
STANDARD WATER CURED	1-day	2.15	1.91	1.64	1.41
	4-day	2.89	2.28	2.19	2.15
	7-day	2.92	2.85	2.66	2.59
	28-day	3.43	3.16	2.97	2.84
	90-day	3.81	3.62	3.17	3.02
STEAM CURED	1-day	2.28	2.37	2.35	2.13
	4-day	2.77	2.65	2.47	2.31
	7-day	2.80	2.76	2.67	2.47
	28-day	3.06	3.58	3.24	3.08
	90-day	3.72	3.96	3.62	3.22

Static Modulus of Elasticity
The static modulus of elasticity values of the concrete in both series are presented in Tables 9 and 10. Similarly, each presented value is the average of three specimens. The results showed that the static modulus of elasticity of concrete decreased with an increase in the recycled aggregate content as shown in Figure 3.

The relative static modulus of elasticity indicated in Figure 3 was the ratio of the static modulus of elasticity of concrete to that of the water cured natural aggregate concrete at 28 days. The static modulus of elasticity of concrete prepared with 100% recycled aggregate was about 20% lower than that of the natural aggregate concrete at the curing ages of 28 and 90 days. The results also showed that the static modulus of elasticity of the steam cured concrete was lower than that of the water cured concrete. However, the detrimental effect of steam curing on the static modulus of elasticity diminished as the replacement level increased (Figure 3). At the recycled aggregate replacement level of 20 %, the static modulus of elasticity of the steam cured concrete was 7.8 % lower than that of the water cured concrete. However, at the recycled aggregate replacement level of 100 %, the static modulus of elasticity of the steam cured concrete was only 1 % lower than that of the water cured concrete.

Table 9. Static modulus of elasticity of the series I concrete mixtures

CURING	AGE	STATIC MODULUS OF ELASTICITY (GPa)			
		R0	R20	R50	R100
STANDARD WATER CURED	28-day	30.1	28.8	26.3	24.2
	90-day	31.2	29.7	27.2	24.7
STEAM CURED	28-day	27.7	26.8	25.3	23.9
	90-day	29.9	28.8	26.6	24.7

Table 10. Static modulus of elasticity of the series II concrete mixtures

CURING	AGE	STATIC MODULUS OF ELASTICITY (GPa)			
		R0	R20	R50	R100
STANDARD WATER CURED	28-day	38.7	32.1	28.6	23.4
	90-day	39.5	36.5	33.2	28.5
STEAM CURED	28-day	32.3	30.4	26.6	22.8
	90-day	34.4	34.3	31.0	27.6

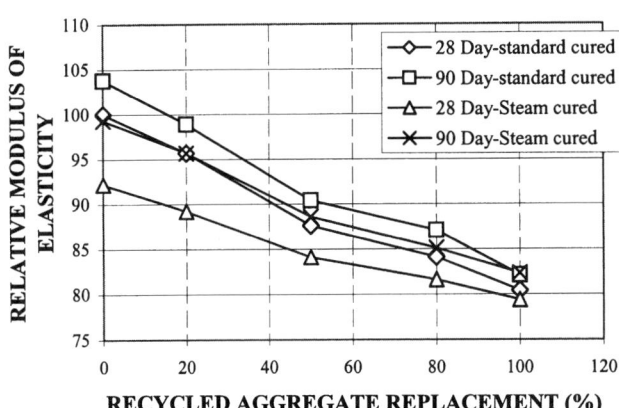

Figure 3. Relative static modulus of elasticity of series I concrete mixtures

Drying Shrinkage

The drying shrinkage values (tested at 112 days) of the concrete mixtures are shown in Tables 11-12. It can be shown that the drying shrinkage values increased with an increase in recycled aggregate content. However, steam curing reduced the drying shrinkage of the concrete. These results are similar to those reported by Brooks et al. [12]. Furthermore, the drying shrinkage of concrete in Series II was lower than that of concrete in Series I which indicated that a decrease in the W/C ratio reduced the drying shrinkage.

Table 11. Drying shrinkage of the series I concrete mixtures at 112 days

NOTATION	NATURAL GRANITE (%)	RECYCLED AGGREGATE (%)	DRYING SHRINKAGE (10^{-6})	
			STANDARD WATER CURED	STEAM CURED
R0	100	0	405	350
R20	80	20	447	382
R50	50	50	476	410
R100	0	100	540	466

Table 12. Drying shrinkage of the series II concrete mixtures at 112 days

NOTATION	NATURAL GRANITE (%)	RECYCLED AGGREGATE (%)	DRYING SHRINKAGE (10^{-6})	
			STANDARD WATER CURED	STEAM CURED
R0	100	0	389	334
R20	80	20	412	352
R50	50	50	435	370
R100	0	100	470	401

Chloride Penetrability

The resistances against chloride-ion penetration of concrete in both series are given in Tables 13 and14. The resistance decreased with an increase in the recycled aggregate content. However, steam curing increased the resistance against chloride-ion penetration of concrete. A comparison between the results of Series I and II revealed that a decrease in the W/C ratio also increased the resistance against chloride-ion penetration.

Table 13. Chloride-ion penetrations of the series I concrete mixtures

CURING	AGE	TOTAL CHARGE PASSED (COULOMBS)			
		R0	R20	R50	R100
STANDARD WATER CURED	28-day	6291	6572	6687	6905
	90-day	4940	5194	5425	5835
STEAM CURED	28-day	5297	5451	5592	5750
	90-day	4087	4328	4660	5221

Table 14. Chloride-ion penetrations of the series II concrete mixtures

CURING	AGE	TOTAL CHARGE PASSED (COULOMBS)			
		R0	R20	R50	R100
STANDARD WATER CURED	28-day	3239	3643	4620	4957
	90-day	2427	2528	2609	3562
STEAM CURED	28-day	3080	3456	3876	3940
	90-day	2285	2300	2740	2807

SUMMARY AND CONCLUSIONS

The following conclusions can be drawn from the above test results:

1. The compressive and splitting tensile strength, and static modulus of elasticity decreased as the recycled aggregate content increased. However, the reduction can be adequately compensated by the use of a lower W/C ratio.

2. Steam curing at 65 °C increased the 1-day strength of all concrete compared to that of the water cured concrete. However the 28 and 90-day strengths of the steam cured concrete were lower than those of the water cured concrete.

3. Since recycled aggregates are generally more porous than natural aggregates, the detrimental effect of steam curing on the long-term hardened properties of natural aggregate concrete are reduced in the recycled aggregate concrete.

4. The drying shrinkage of concrete increased with an increase in the recycled aggregate replacement level. The drying shrinkage value of concrete prepared with 100% recycled aggregate was about 30 % higher than that of natural aggregate concrete. Nevertheless, steam curing reduced the drying shrinkage. The reduction was about 15 % for the steam cured concrete prepared with 100 % recycled aggregate.

5. The resistance against chloride-ion penetration of the concrete decreased with an increasing recycled aggregate content where concrete prepared with 100% recycled aggregate had the highest charge passed. However, an initial steam curing regime increased the resistance of the concrete against chloride-ion penetration.

6. The results demonstrate that one of the most practical ways to utilize a higher percentage of recycled aggregates in concrete is "precasting" with an initial steam curing stage immediately after casting.

ACKNOWLEDGEMENTS

The authors would like to thank the Environment and Conservation Fund and the Hong Kong Polytechnic University for funding support.

REFERENCES

1. R.K. DHIR AND T.G. JAPPY (Editors), Proceedings of the International Conference on Exploiting Wastes in Concrete, Thomas Telford, UK, 1999.

2. HANSEN, T. C., RILEM REPORT 6: Recycling of Demolished Concrete and Masonry, published by E&FN Spon, Bodmin, UK, 1992.

3. COLLINS, R.J. The Use of Recycled Aggregates in Concrete. BRE Information Paper IP 5/94, Building Research Establishment, Watford, UK, 1994.

4. R.K. DHIR, N.A. HENDERSON AND M.C. LIMBACHIYA (Editors), Proceedings of the International Conference on the Use of Recycled Concrete Aggregates, Thomas Telford, UK, 1998.

5. RILEM, Proceedings of the 1st ETNRecy.net/RILEM workshop on Use of Recycled Materials as Aggregates in the Construction Industry, 11-12 September, 2000, Paris.

6. SAGOE-CRENTSIL, K.K. BROWN, T. AND TAYLOR, A.H.: Performance of concrete made with commercially produced coarse recycled concrete aggregate, Cem. Conr. Res. 31. 707-712, 2001.

7. POON, C. S. SHUI, Z. H. LAM, L. FOK H. AND KOU, S. C.: Influence of moisture states of natural and recycled aggregates on the properties of fresh and hardened concrete, Cem. Conr. Res. 34 (2004) 31-36.

8. HASABA,S. KAWAMURA,M. KAZUYUKI,T. AND KUNIO,T.: Drying shrinkage and durability o f c oncrete m ade f rom r ecycled c oncrete a ggregates, T ransactions o f Ja pan Concrete Institute, Tokyo, Vol 3, 1981, pp.55-60.

9. HANSEN, T. C. AND BØEGH, E.: Elasticity and drying shrinkage of recycled aggregate concrete, Journal of American Concrete Institute, 82, 5, 1985, pp. 648-652.

10. OLORUNSOGO, F.T. AND PADAYACHEE. N.: Performance of recycled aggregate concrete monitored by durability indexes, Cement and Concrete Research, 32, 2002, pp179-185.

11. DHIR, R. K. LIMBACHIYA, M. LEELAWAT, C. T.: Suitability of recycled concrete aggregate for use BS 5328 designated mixes. Proc. Inst. Civ. Engrs & Bldgs, 134, Aug. 1999, pp 257-274.

12. AMERICAN CONCRETE INSTITUTE, Pressure steam curing, ACI Journal, Vol. 60, No 8, August 1963, pp 953-986.

13. BROOKS, J.J. WAINWRIGHT, P.J. AND NEVILLE, A.M.: Time dependent behaviour of high early strength concrete containing a superplasticizer, ACI Special Publication, No 68, 1981, pp 81-100.

14. BROOKS, J.J. AND WAINWRIGHT, P.J.: Influences of steam curing on strength, shrinkage and creep of OPC and slag concretes. Proceeding of the International Conference held at the University of Dundee, Scotland, UK on 27-28 June 1996. pp 367-374.

15. BOUKENDAKDJI, M. BROOKS J. J. AND WAINWRIGHT, P.J.: Influences of steam curing on strength, shrinkage and creep of OPC and slag concrete, Proceeding of the International Conference held at the University of Dundee, Scotland, UK on 27-28 June 1996, pp 367-374.

16. HO, D. W. S. CHUA, C.W. AND TAM, C. T.: Steam-cured concrete incorporating mineral admixtures, Cement and Concrete Research, 33, 2003, pp 595-601.

17. ERDEM, T.K. TURANLI, L. ERDOGAN, T.Y.: Setting time: An important criterion to determine t he l ength o f t he d elay p eriod b efore s team c uring o f c oncrete, C ement a nd Concrete Research, 33, 2003, pp 741-745.

18. BRITISH STANDARD INSTITUTE, BS 812: Part 103.1: Methods for determination of particle size distribution, 1985.

SESSION ONE:

REGULATORY FRAMEWORK
AND GOVERNMENT POLICY

THE WRAP AGGREGATES PROGRAMME:
ACHIEVING THE POTENTIAL OF RECYCLED AGGREGATES

J Barritt

WRAP (the Waste & Resources Action Programme)

United Kingdom

ABSTRACT. The aggregate industry in the UK is going through a period of evolution towards a more sustainable resourcing of construction aggregates. WRAP is helping facilitate this process by spreading good practice, providing information on specifications and technical issues, supporting investment and addressing barriers. But what is the potential for aggregates from inert construction and demolition waste in terms of tonnage and performance? What changes are needed in the supply chain to reach this potential and what impact does poor resource us have on the total sustainability of construction aggregates? WRAP is addressing these issues and this paper considers the changes needed and the potential that can be achieved.

Keywords: Recycled and secondary aggregates, Construction demolition and excavation waste, Aggregates levy, Sustainable resource use, Procurement.

John Barritt is Aggregates Technical Advisor to the WRAP Aggregates Programme. He has over 35 years experience of construction materials in the aggregates, asphalt and ready mixed concrete industries.

INTRODUCTION

A growing economy results in a growth in the built environment and a consequent growth in the demand for construction aggregates. From the 1950's to the 1980's there was a presumed overall link between growth in GDP and aggregate sales (Figures 1 and 2) but in fact average aggregate demand growth was ahead of economic growth to the end of the 80's.

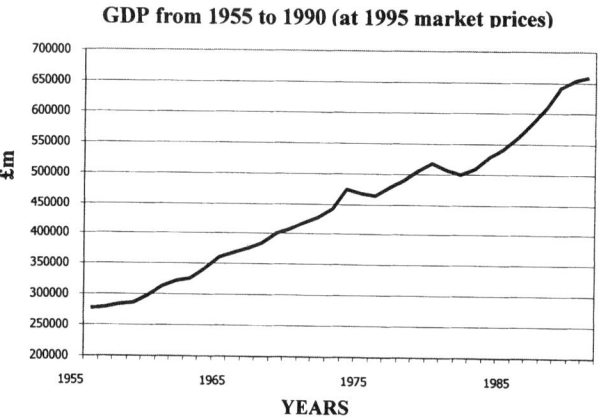

Figure1. Rise in Gross Domestic Product (GDP) Source: National Statistics office

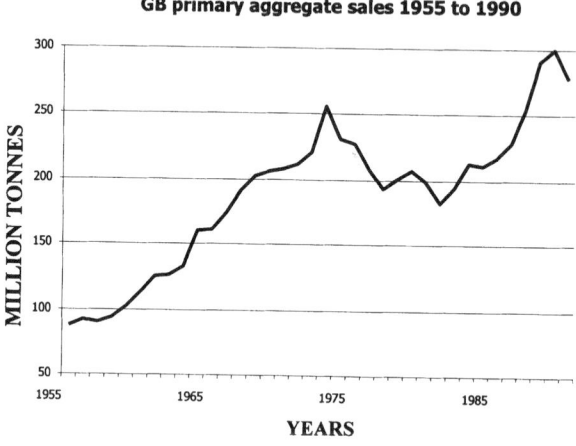

Figure 2. Primary aggregate sales for GB. Source: QPA

The economic recession from 1990 resulted in a fall in both public and private investment and as a result demand for construction reduced. As recovery developed the construction market changed, Government investment in infrastructure was low and construction practices improved with greater resource efficiency.

The level of aggregate demand per unit of construction spend reduced as did the level of aggregate demand per unit of GDP.   Aggregate statistics for recycled and secondary aggregates were included from 1989, although early figures were estimates (Figures 3 and 4).

Aggregates/£1000 construction spend

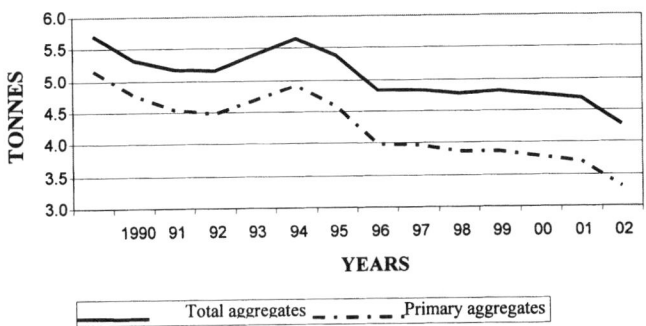

Figure 3. Aggregate demand per unit of construction spend (QPA stats)

Aggregate use/£billion GDP

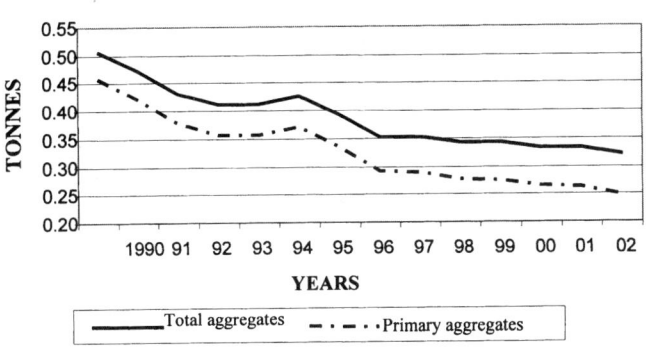

Figure 4. Aggregate use per unit of GDP (QPA stats)

This change in the market demand coincided with a change in the resourcing of aggregate demand as greater use was made recycled and secondary aggregates.

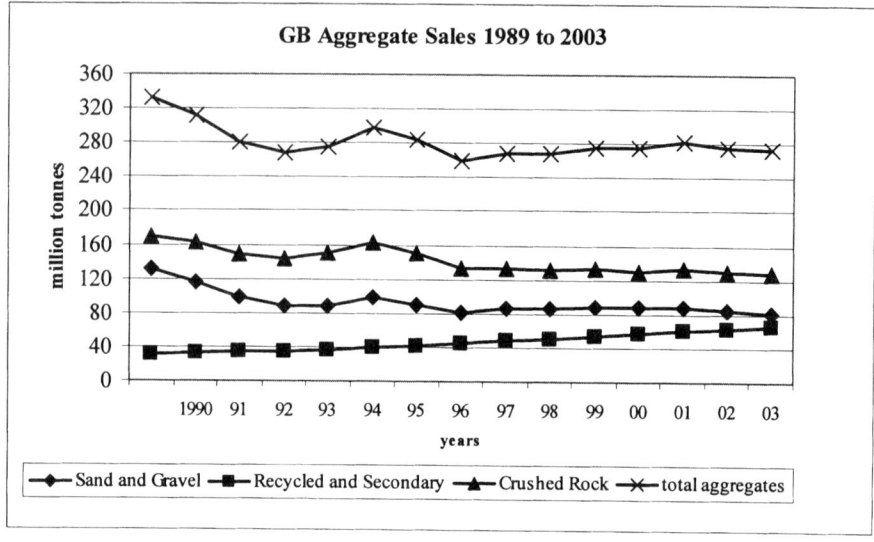

Figure 5. GB aggregate sales 1989 to 2003. (source QPA)

Growth in the use of recycled aggregates was boosted in 1996 with the introduction of the Landfill Tax and from this point to 2002 most of the growth in aggregate demand was met from increased use of recycled and secondary aggregates.

Government initiatives promoting sustainable construction continued to improve construction practices and the introduction of the Aggregates Levy in 2002 resulted in a greater decline in aggregate use per £ of construction spend and fall in total aggregate demand. However this did not reduce the continuing growth of recycled aggregates as the economics of recycling were assisted by the levy as was the competitiveness of secondary aggregates.

A small proportion, around 10%, of the Aggregates Levy was allocated to the Aggregates Levy Sustainability Fund (ALSF) and an element of that fund was designated to further reduce the demand for primary aggregates through the greater use of recycled and secondary aggregates. In April 2002 WRAP was given a brief for a two year aggregates programme to address this objective. Following the Government's decision to extend the initial two years of the ALSF by a further three years, WRAP's aggregates programme funding has been continued to March 2007.

FUTURE DEMAND

Government has a responsibility to ensure that sufficient resources are available to meet the construction aggregate demand generated by construction. This is done by the Mineral Planning section of ODPM through Mineral Planning Guidance that sets regional targets for mineral reserves and productive capacity. ODPM policy is that construction aggregate demand should be first met with recycled and secondary resources with provision being made for the balance to be provided from primary resources.

New targets for 2001 to 2016 were issued in June 2003, after a lengthy period of consultation. The total provision for the period is shown in Table 1.

Table 1. National and Regional Guidelines for aggregates provision in England, 2001 - 2016
(Million Tonnes) [source ODPM]

New Regions	GUIDELINES FOR LAND-WON PRODUCTION			ASSUMPTIONS	
	Land-won Sand & Gravel	Land-won Crushed Rock	Marine Sand & Gravel	Alternative Materials	Net Imports to England
South East England	212	35	120	118	85
London	19	0	53	82	6
East of England	256	8	32	110	8
East Midlands	165	523	0	95	0
West Midlands	162	93	0	88	16
South West	106	453	9	121	4
North West	55	167	4	101	50
Yorkshire & the Humber	73	220	3	128	0
North East	20	119	9	76	0
ENGLAND	**1068**	**1618**	**230**	**919**	**169**

The demand forecast was produced from a complex economic model that calculated the different levels of aggregate demand from different types of construction activity against projected government and private expenditure.

The level set for alternative materials (recycled and secondary aggregates) was based on the data from the two 2001 ODPM surveys on recycled and secondary aggregates respectively [1,2]and included a specific target of 60 mt by 2011. There was an assumption by ODPM that only a limited amount of the Construction Demolition and Excavation Waste remained economically viable for recycling and as such WRAP consider their target to be low.

For more simple comparison Figures 6 and 7 show the forecast demand for aggregates and the ODPM targets from Table 1 and the WRAP forecast of the greater potential from recycled and secondary resources.

The total market growth predicted by ODPM gives an overall average increase of +1.2% per annum, as a consequence the cautious ODPM forecast for recycled and secondary aggregates results in their market share showing little improvement and ending in 2016 with no improvement from 2001.

The WRAP forecast indicates a potential for growth from 23% to 28% by 2010 followed by a gradual decline as the availability of sustainable resources fails to keep pace with the growth in total market demand (Figure 8).

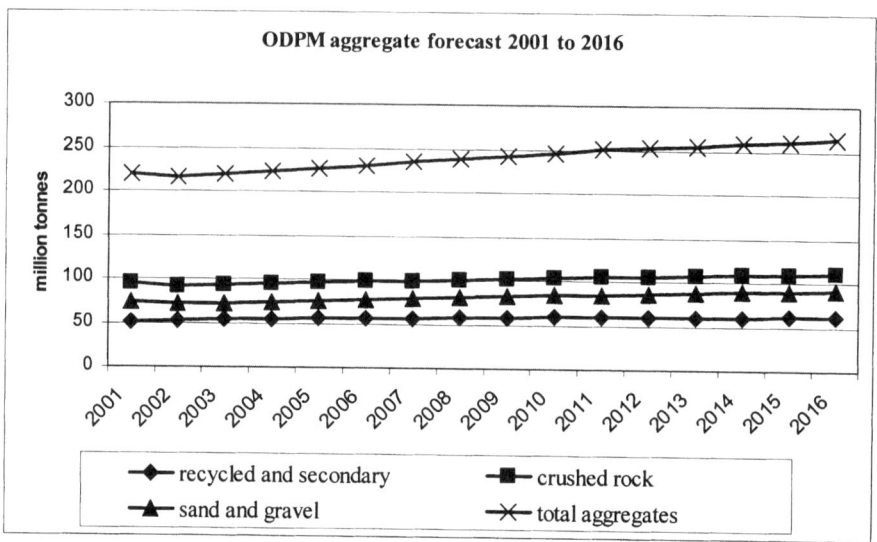

Figure 6. ODPM forecast for England

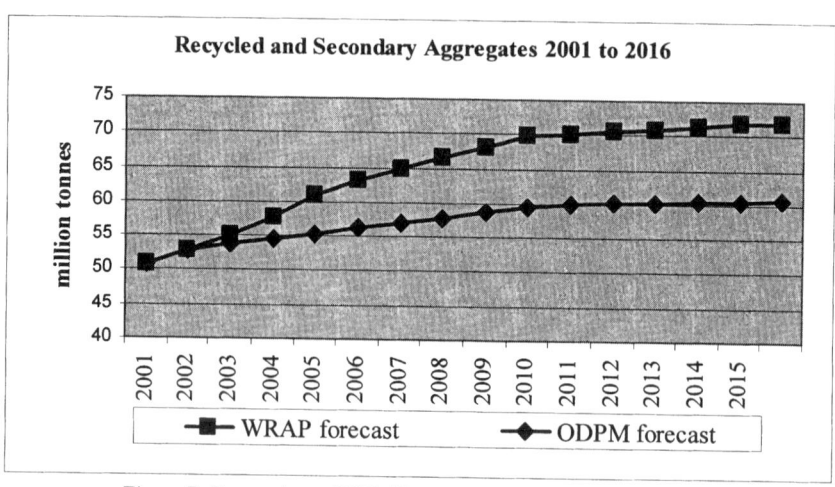

Figure 7. Comparison of WRAP and ODPM forecasts for England

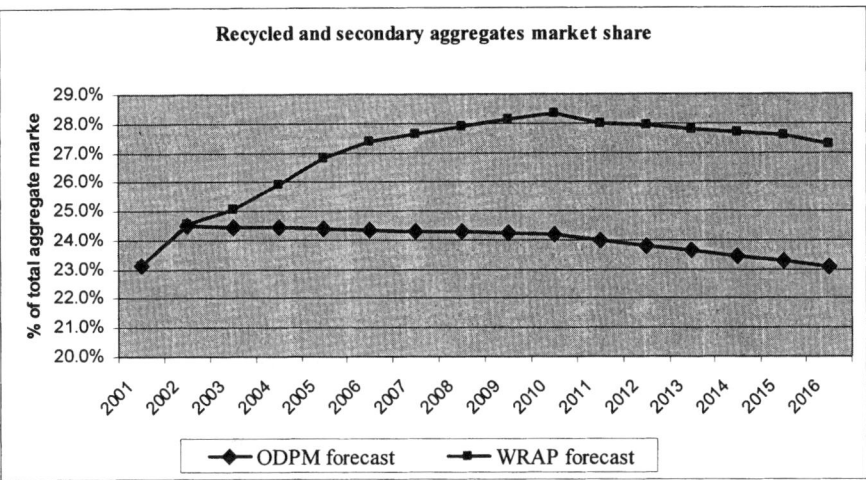

Figure 8. Comparison of forecast market shares for recycled and secondary aggregates

MEETING FUTURE DEMAND

Recycled aggregates have accounted for most of the growth in the market share of sustainable aggregates, and their use has been predominantly as fills. Secondary aggregates, such as slag, have a long track record of use in higher value applications such as asphalt whereas some only have the potential for bulk fill, such as colliery spoil.

The further growth in market share for recycled aggregates is dependant on better use of existing resources, producing aggregates that meet the broader performance needs of construction. It is also dependant on recycling being able to compete with low cost landfill options to secure more difficult to process resources and to have the processing plant needed to turn them into quality aggregates.

The barriers to the increased use of these resources were identified in detail in the original Aggregates Programme strategy. The key factors are the location of secondary resources relative to market plus the cost and availability of landfill options for recyclable resources.

There is some growth forecast for secondary aggregates with the most potential coming from China Clay waste, although this is dependant on deep water shipping facilities in Cornwall in later years. IBA growth hinges on the greater use of incineration for disposing of MSW to meet EU targets.

Recycling construction, demolition and excavation waste (CDEW) has the greatest potential for growth. The developments not considered by ODPM in their forecast were:

- DEFRA changes to waste management license exempt site regulations increasing costs and tightening regulation.
- The availability of competitive aggregate washing plant with water recirculation systems.

- The introduction of European Aggregate Standards that do not discriminate against recycled and secondary aggregates.

Further to these changes WRAP Aggregates Programme is:

- Providing capital support for processing infrastructure to process wastes being lost to landfill and to process recycled aggregates to higher value applications.
- Providing training and information across the supply chain on specifications, standards, planning, waste regulations and procurement.
- Working with the Environment Agency to promote the use of sustainable aggregates.
- Actively identifying and addressing new barriers to the growth of sustainable aggregates.

WRAP's assessment of the combined potential for recycled and secondary aggregates by 2015 is shown in Table 2. The assumed market developments to achieve these changes are:

- Increased competitiveness of recycling plants to exempt sites due to changes in regulations.
- Higher value recycling increasing economics of recycling plants and ability to compete with landfill.
- Broader market demand due to wider product range.
- Broader market demand due to improved market awareness and confidence.
- Public and private procurement policies creating demand.

Table 2. Potential for growth in recycled and secondary aggregates.

MAJOR SECONDARY AND RECYCLED RESOURCES IN ENGLAND	2001	2015	GROWTH
• Recycled Construction, Demolition and Excavation Waste	36.5	51.4	14.9
• Asphalt planings	5	5.6	0.6
• Spent Rail Ballast	1.2	1.3	0.1
• Blast Furnace and Steel Slag	1.6	2	0.4
• Colliery Spoil	0.8	1	0.2
• PFA/FBA	2.4	3	0.6
• China Clay aggregates	2.3	4.6	2.3
• Used foundry sand	0.2	0.5	0.3
• IBA	0.4	1.4	1
• Slate waste	0.3	0.5	0.2
• Tyres	0.1	0.2	0.1
• Glass	0.1	0.2	0.1
Secondary/Recycled Total	50.9	71.7	20.8

THE SUSTAINABLE USE OF RESOURCES

Recycled and secondary aggregates cannot be taken in isolation when considering the sustainable use of resources for meeting market demand for construction aggregates. It is not in keeping with sustainability to create waste products from primary aggregate extraction as a consequence of processing CDEW waste to produce recycled aggregates, but this has been a side effect of producing a disproportionate amount of recycled aggregates to low performance specifications.

The Aggregates Levy was probably the determining factor in causing this impact as previously very low priced primary aggregates could still be competitive in the market; however the levy priced them out of many contracts. The impact can be seen in Figure 9 which shows the change in the proportion of primary aggregates being used as fills and sub-base, interestingly the step change appears to commence in 2001 and therefore further data is required to understand the longer term trend.

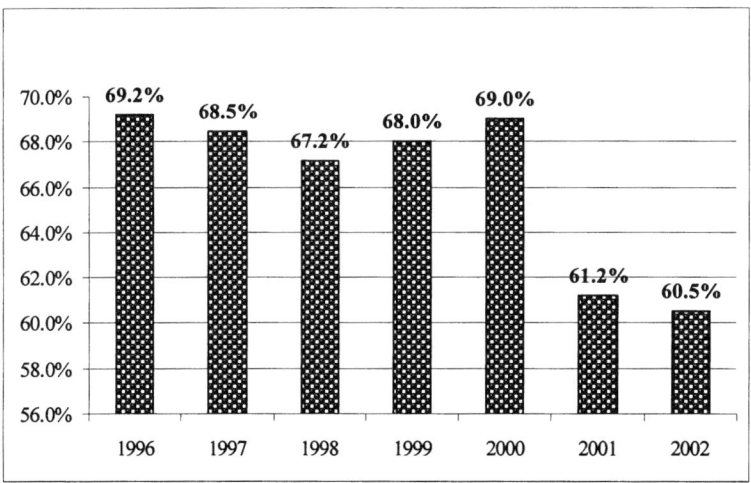

Figure 9. The proportion of crushed rock sold as fills and sub-base from AMRI survey

WRAP's assessment of the end-use of recycled and secondary aggregates for 2001 can be seen in Figure 10.

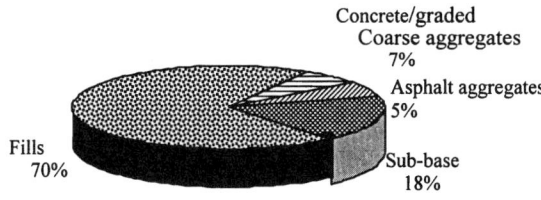

Figure10. End use of recycled and secondary aggregates 2001

An analysis of the optimum potential of these resources and their probable availability projected in 'Future Demand' above, gives a possible end-use by 2015 as shown in Figure 11.

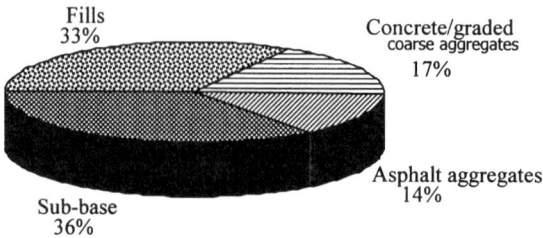

Figure 11 End use of secondary and recycled aggregates 2015

The changes in tonnage are shown below (rounding causes minor changes)

Table 3. Changes in end use of secondary/recycled aggregates

AGGREGATE GRADE	2001Sec/Recyc. mt's	2015ec/Recyc. mt's	CHANGE IN USE mt's
Concrete/graded coarse aggregates	3.3	12.1	+8.8
Asphalt aggregates	2.7	10.2	+7.5
Sub-base	9.3	25.2	+15.9
Fills	35.3	23.5	-11.8
Total	**50.6**	**71.0**	**+20.4**

The distribution in Figure 11 is very similar to that of a crushed rock quarry and considering the similarity to recycling CDEW and crushed rock processing this is not unusual.

The proportion being used in concrete is lower than some may consider but the issue here is not technical possibility but practical application of resources. Recycling economics hinge on obtaining the highest average value per tonne on all output and crushed concrete that may have the potential for use in high strength concrete has a higher value when used to raise the performance of a larger volume of lower quality resources.

However there is still a significant increase in concrete aggregates and this will be from the blending of a proportion of a general recycled aggregate with other primary aggregates in lower performance concretes.

The proportionate growth of recycled aggregates in asphalt is significantly higher than in concrete. This mainly stems from two things, recycled asphalt planings can be recycled through hot asphalt plants recovering both the high quality aggregate as well as the bitumen, generating valuable cost savings, and the wider use of cold asphalt processes. The latter has the possibility of producing asphalt with a 97% recycled/secondary aggregate content and can be used the majority of the road network.

Changes in procurement practices are also most likely to have the biggest impact on asphalt purchasing due to the large buying power of local authorities and government agencies where sustainability policies will drive procurement policy.

The increase in sub-base is easily achieved through better processing practices and a market place more confident in the use of recycled aggregates.

These changes result in a fall in the production of low performance fills enabling appropriate low performance primary aggregates to regain their lost market share. This last point illustrates the need for a broader approach to optimising the potential of all resources suitable for the production of construction aggregates. If at least 70% of market demand is always going to be produced from primary resources the greatest potential for reducing the impact of primary extraction will be achieved by ensuring the most sustainable extraction and processing of those resources, minimising waste and optimising end use.

REFERENCES

1. Survey of Arisings and Use of Secondary Materials as Aggregates in England and Wales in 2001, Planning Research by ODPM

2. Survey of Arisings and Use of Construction and Demolition waste in England and Wales in 2001, Planning Research by ODPM

ESTIMATION METHODS FOR THE GENERATION OF CONSTRUCTION AND DEMOLITION WASTE IN GREECE

D Fatta

University of Cyprus, Cyprus

A Papadopoulos, F Kourmoussis, A Mentzis, E Sgourou, K Moustakas, M Loizidou

National Technical University of Athens

N. Siouta

AKTOR

Greece

ABSTRACT. As a part of the overall EU waste management strategy, several specific waste streams have been defined to receive priority attention, the aim being to reduce the overall environmental impact of each waste stream. Construction and Demolition waste is considered as one of these 'priority' waste streams. Due to the lack of data as well as the differences in definitions between member countries, it is difficult to collect and assess data on the generation of C&D waste at EU level. The requirements for reporting these quantities derive also from the Waste Statistic Regulation of 2002. For these reasons the authors have developed a model for the estimation of the quantities of C&D waste generated, using easily accessible data.

Keywords: Construction and Demolition waste, Greece, Calculation/estimation model, Waste quantities, Priority waste streams

Dr D Fatta is a Lecturer at the Department of Civil and Environmental Engineering of the University of Cyprus. She is also an inner circle expert of the European Topic Center of Waste and Material Flows (ETC/WMF) of the European Environment Agency (EEA). Her principal research interests are in the field of Environmental Science, Technology and Management.

Dr A Papadopoulos is Research Associate the Unit of Environmental Science and Technology (UEST) of the School of Chemical Engineering of National Technical University of Athens Greece (NTUA). He has more than 10 years experience in the field of environment with 70 publications in international journals and conferences.

F Kourmoussis has an Environmental Sciences degree and is a researcher and PhD student in the UEST. He is also an outer circle expert of the ETC/WMF.

A Mentzis has a Chemical Engineering degree, an MSc in Environmental Sciences and is a researcher and PhD student in the UEST. He is also an outer circle expert of the ETC/WMF.

E Sgourou has a Chemical Engineering degree and is a researcher and PhD student in the UEST.

K Moustakas has a Chemist degree and is a researcher and PhD student in the UEST.

Professor Maria Loizidou leads the UEST. She has more than 20 years of experience in environmental issues with 180 publications in international journals and conferences.

N Siouta has a Civil Engineering degree, an MSc in Environmental Engineering and is the Environmental, Health and Safety Manager of the construction company AKTOR.

INTRODUCTION

Waste minimisation and effective waste management are considered as the basic principles of the European environmental legislation and strategy. As part of the overall waste management s trategy, s everal s pecific w aste s treams h ave b een d efined to r eceive p riority attention, the aim being to reduce the overall environmental impact of each waste stream [1]. Construction and demolition (C&D) waste according to the EU Waste Strategy is considered as one of the 'priority' waste streams. According to the Sixth Environment Action Program entitled 'Environment 2010: Our future, Our choice', recommendation action needs to be taken concerning the C&D waste stream [2].

DATA AVAILABILITY IN EU

Eurostat collects data regarding the quantities and the management methods of C&D waste in the EU member states. The available data on C&D waste for the years 1990-1999 are presented in the graph below:

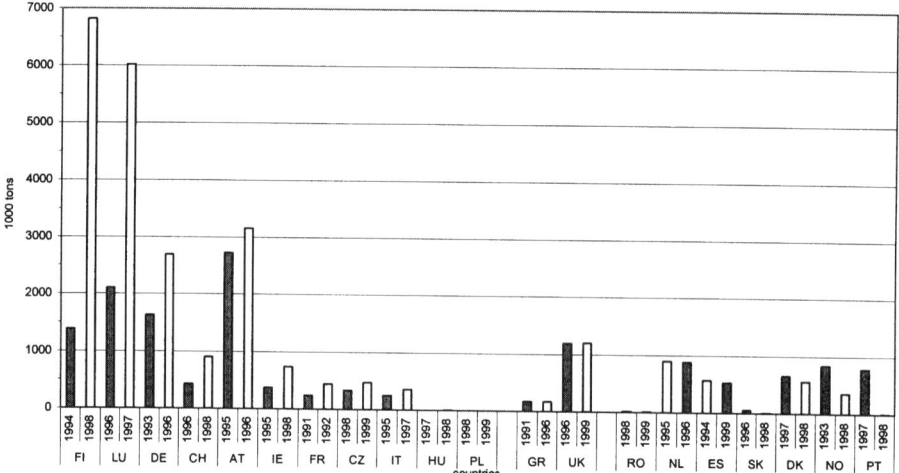

Figure 1. C&D waste generation per capita in 20 European countries, (the Figure shows the trend of the 2 most recent years during 1990-1999 for which data are available), [3]

Some information regarding Figure 1 is given in the following:
-   The figure shows only the 2 most recent years of data during 1990-1999, even though for some countries more data exist.
-   Grey colour represents first year and white colour represents last year.
-   X axis: First group is countries (FI to PL) with upward trend, followed by the group of countries (GR to UK) with stable trend and lastly the country group (RO to PT) with downward trend. Countries are sorted in descending order from left to right according to difference between the two corresponding years, i.e. FI showed more increase than LU, which showed more increase than DE and so on. Correspondingly PT showed more decrease than NO, which showed more decrease than DK and so on.

The main problem identified is the scarcity of quantified and standardized information. The EU Member countries cannot provide this kind of information for the specific waste stream. Inadequacies in the data sets make it impossible to give an accurate account of the total amount of C&D Waste generated throughout Europe as well as to assess future trends. It is estimated that the situation will improve in the coming years with the enforcement of the Waste Statistics Regulation.

RESULTS AND ASSESSMENT OF THE AVAILABLE DATA

Current Situation and Developments in the Environment
Construction and demolition waste represents a significant proportion of total waste generation. Construction and demolition waste accounts for approximately 25% of the total waste generation in the EU. The quantities of C&D waste produced show that the generation of waste is not de-linked from the economic activities. Three major concerns for the environment are related to C&D waste.

The main environmental concern is due to the existence of various hazardous substances in the C&D waste, e.g. asbestos, which is hazardous itself and PVC, which in some treatment techniques causes emissions of toxic gases (dioxin). In addition, the use of raw materials (gravel and sandpits) causes distortion of the landscape and may cause problems for the level and quality of groundwater. Finally, disposal and landfilling of this kind of waste is a problem for the limited capacity of landfills. The major part of the waste stream is inert waste; consequently the problems of leaking and emissions from landfills are limited to specific fractions of waste from construction and demolition.

Policy Relevance
Following the Council Resolution of 7 May 1990, which invited the Commission to establish proposals for action at Community level, the Priority Waste Streams Programme was initiated. The Member States identified C&D waste, as one such stream, even though at that time relatively little was known about the nature or the volumes of the flows concerned. The objectives of the priority waste stream program respond to the waste management hierarchy, which prefers waste prevention or reduction to reuse, reuse to recycling or recovery and all of these to final disposal via landfill or incineration without energy recovery.

Information Assessment
Waste amounts vary considerably from country to country. This can be partly explained by the economic, cultural and population differences that exist among countries and in some cases by various technical issues. e.g. definition of C&D waste streams varies among countries, sometimes the estimation of C&D waste includes excavated soil and miscellaneous materials such as brick and concrete, etc. Generally the amount of construction and demolition waste is increasing for most countries.

On the other hand the dominant management method is landfilling, and steps must be taken to promote other more environmental friendly options, such as recycling. The growing welfare in the EU countries leads to a demand for better accommodation and therefore to increasing renovation and demolition of old houses and buildings.

ESTIMATION MODELS

It is stipulated in the Waste Statistics Regulation that EU Member States have to provide data on specific waste streams including C&D waste beginning from the year 2004. However several countries have not yet established the mechanism to collect, assess and provide data on this waste stream. For this reason models are used to estimate the generated quantities of C&D waste.

General Estimation Model
The estimation model described below is a simple but generally relatively representative calculation model which is used by several Environmental Protection Agencies and Statistical Services in the Member States (e.g. Austria and Italy).

Construction Waste
This method is used for the estimation of construction waste: it is assumed that for every 1000 m^2 of construction approximately 50 m^3 of waste are generated. It has to be noted that the density of construction waste is in the range of 1.5 - 2.0 tn/m^3 and in order to estimate the weight of the waste an average density of 1.5 tn/m^3 was used.

Hence, the applied equation is:
$$CW = [NC + EX] * VW * D$$

    where:

        CW: construction waste in tons
        NC: new constructions in m^2 (can be provided by the Statistical Service)
        EX: extensions to existing infrastructure in m^2 (can be provided by Statistical Service)
        VW: volume of generated waste per 1000 m^2 = 50 m^3/ 1000 m^2
        D: density of waste = 1.5 tn/m^3

Demolition Waste
Furthermore for the estimation of demolition waste a similar method is used: it is assumed that each demolition refers to a building of 60 m^2 and the deriving waste is no more than 114m^3. As a result the equation applied for the calculation of demolition waste has the following form:

$$DW = ND*VD*D$$

    where:

        DW: demolition waste in tons
        ND: number of demolitions (can be provided by the Statistical Service)
        VD: volume of each demolition = 114 m^3
        D: density of waste = 1.5 tn/m^3

Results of estimated C&D waste for Greece with the general model
The results of this model for Greece are presented in the Table 1. It can be seen that there is an increase in the quantities of construction and demolition waste in the year 2000 especially in the two bigger urban centers, Attica and Thessalonica. This was an expected result due to the increase of the activities in the construction sector, in view of the Olympic Games 2004. Besides, the catastrophic earthquake of 1999 in Athens resulted in the break down of many buildings and the subsequent construction of new ones which generated more C&D waste

Table 1. Total estimated quantities of C&D waste for the years 1999-2000.

YEAR	CONSTRUCTION WASTE (tons)	DEMOLITION WASTE (tons)	TOTAL WASTE (tons)
1999	1186518	712557	1899075
2000	1276717	815670	2092387

Estimation Model Developed for Greece

The authors have developed a calculation model as a method for the estimation of the generation of construction and demolition waste in building construction activities. All the factors used in the equations are based upon observations and experience of several building constructions from construction and demolition companies. All the data that are used were provided from the National Statistical Service of Greece (NSSG). This method which takes into account the special characteristics of Greece is analyzed below.

Construction Waste

The model estimates the generation of waste that derives from the construction of new buildings, such as apartment buildings, houses, public buildings - offices, industrial buildings, hospitals, commercial buildings etc. This waste include all kinds of materials used in construction of buildings, such as leftovers of cement, bricks, aluminum and wood cavities and frames, tiles, marbles, etc.

Before the construction of a new building, the excavation phase occurs, for foundations and groundworks, such as underground spaces (e.g. parking, storage). It is noted that the waste that derives from the excavation phase are excluded.

After interviewing a large number of responsible engineers at the construction sites of various construction companies the following assumptions were made: for every 100 m^2 of construction o f a b uilding t he g enerated q uantities o f c onstruction w aste r anges f rom 4 t o 10m^3.

A representative assumption is that overall, for all kinds of buildings that are constructed in Greece; the mean value of the generated waste is 6 m^3. It has to be noted that the density of construction waste is in the range of 1.5 - 2.0 tn/m^3 and in order to estimate the weight of the waste an average density of 1.6 tn/m^3 is used. Hence, the applied equation is:

$$CW = [NC + EX] * VW * D$$

where:

NC: surface of new constructions in m^2 (provided by NSSG)

CW: construction waste in tons

EX: surface of extensions to existing infrastructure in m^2 (provided by NSSG)

VW: volume of generated waste per 100 m^2 = 6 m^3 / 100 m^2

D: density of waste = 1.6 tn/m^3

Demolition Waste

The model estimates the generation of waste that derives from the demolition of old buildings, such as apartment buildings, houses, public buildings – offices, industrial buildings, hospitals, commercial buildings etc. This waste includes all kinds of materials that these buildings contain, such as cement, bricks, aluminum and wood cavities and frames, tiles, marbles, etc.

After interviewing a large number of responsible engineers at the demolition sites of construction companies the following assumptions were made: the mean value of the number of floors that each building being demolished has, is 1.3. This value is is assumed that takes into account the following conditions:

- the buildings that are being demolished nowadays were constructed during the period of 1950-1970
- the buildings being demolished in large city centers (one city center for each prefecture) have 2 floors
- the buildings being demolished in large city centers account for 30% of the total number of buildings being demolished in Greece (according to data from the NSSG).

Furthermore it is assumed that each demolition refers to a building of 130 m^2 and the deriving waste is 0.8 m^3/m^2. Lastly it has to be noted that the density of demolition waste is approximately the same as for construction waste and an average density of 1.6 tn/m^3 is used. Hence, the applied equation is:

$$DW = ND*NF*SD*WD*D$$

where:

DW:	demolition waste in tons
ND:	number of demolitions (provided by NSSG)
NF:	mean value of number of floors that each building has = 1.3
SD:	surface of each building being demolished = 130 m^2
WD:	generated waste of each demolition = 0.8 m^3/m^2
D:	density of waste = 1.6 tn/m^3

Excavation Waste

The model also estimates the generation of excavation waste. As already mentioned, whenever a demolition of a building occurs it is mostly accompanied with an excavation phase. In several countries the quantities of this waste is either included in the construction or the demolition waste. This waste contains mostly soil and dirt. After interviewing a large number of responsible engineers the following assumptions were made.

It is assumed that each excavation has a surface equal to the building being demolished and constructed, i.e. 130 m^2. Furthermore the mean value of excavation depth is assumed to be 3 m, while the waste has an average density of 1.4 tn/m^3. Hence, the applied equation is:

$$EW = ND*ES*ED*D$$

where:

EW:	excavation waste in tons
ND:	number of licenses of new constructions (provided by NSSG)
ES:	excavation surface = 130 m^2
ED:	mean value of excavation depth = 3 m
D:	density of waste = 1.4 tn/m^3

Estimated Quantities For Greece

Table 2 presents the results obtained from the application of the model that was developed by the authors, in order to calculate the generation of construction and demolition waste in Greece, for the years 1999 and 2000. The results are presented in total and per periphery (Greece has 13 peripheries) although further data are available per prefecture (Greece has 51 prefectures) which are not presented here due to lack of space. Additionally the table contains the data of excavation waste. These data are presented in columns in the Figures 2 and 3.

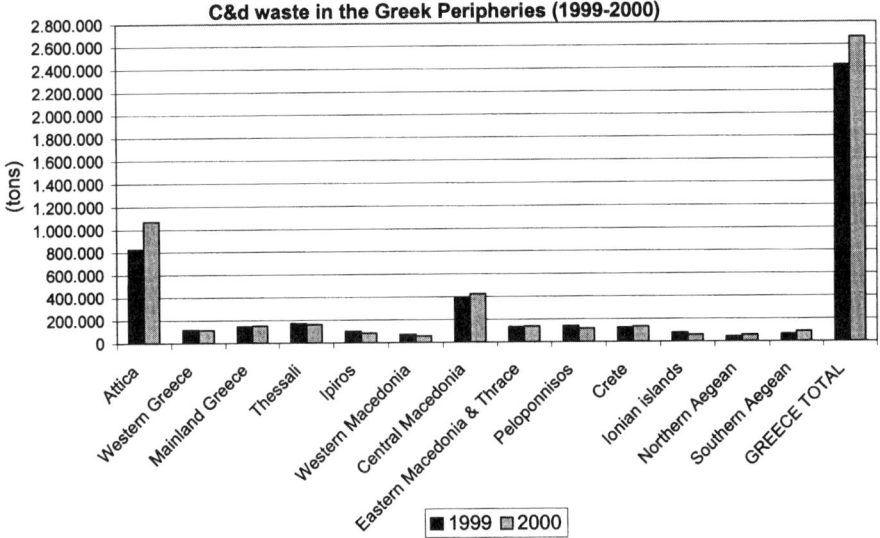

Figure 2. Construction and demolition waste in the Greek peripheries - 1999-2000

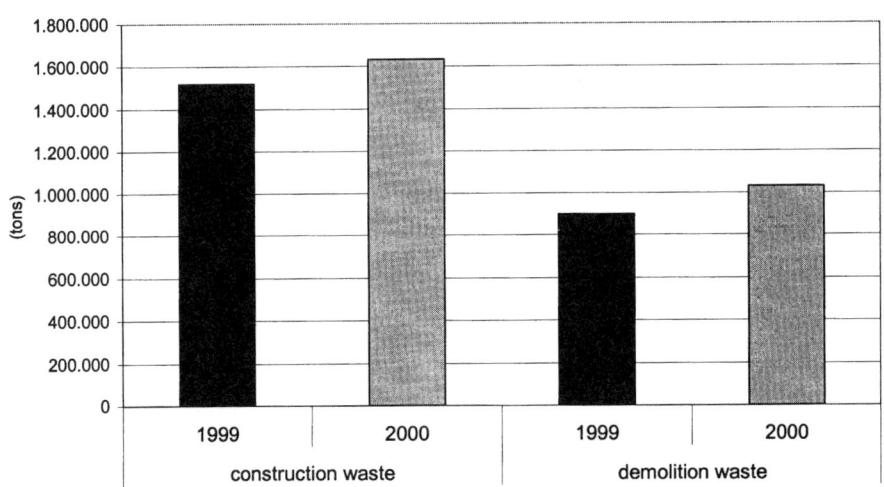

Figure 3. Construction and demolition waste in Greece - 1999-2000

Table 2. Generation of construction, demolition and excavation waste in Greece (*data generated by the estimation model developed*)

PERIPHERIES AND PREFECTURES OF GREECE	CONSTRUCTION WASTE (tons)		DEMOLITION WASTE (tons)		TOTAL C&D WASTE (tons)		EXCAVATION WASTE (tons)	
Year	1999	2000	1999	2000	1999	2000	1999	2000
Periphery of Attica	385807	487074	440860	585795	826667	1072869	3141138	4395300
Periphery of Western Greece	86493	85732	30934	27905	117426	113638	1408680	1323504
Periphery of Mainland Greece	91690	106184	54729	44995	146419	151179	1281462	1292382
Periphery of Thessalia	110808	105855	59704	56460	170513	162315	1350258	1299480
Periphery of Ipiros	70779	58011	29420	27905	100198	85917	844662	720174
Periphery of Western Macedonia	64652	53560	6922	6273	71574	59833	995904	687960
Periphery of Central Macedonia	291225	309076	105997	114217	397222	423293	3244332	3125850
Periphery of Eastern Macedonia & Thrace	91496	93985	42399	44346	133894	138330	1130766	1174992
Periphery of Peloponnisos	98847	87053	41750	31799	140597	118852	1540266	1402128
Periphery of Crete	94396	98931	32232	33313	126628	132245	1338792	1322412
Periphery of Ionian Islands	58020	44160	20334	16873	78354	61033	942942	800436
Periphery of Northern Aegean	24686	28903	21632	28771	46318	57673	546546	532350
Periphery of Southern Aegean	49843	75673	14493	13196	64337	88869	1063608	1134042
TOTAL	1518743	1634197	901405	1031846	2420148	2666044	18829356	19211010

RESULTS AND DISCUSSIONS

Figures 2 and 3 give a representative picture of Greece as far as the construction and demolition waste stream is concerned. It can be seen that there is an increase in the quantities of construction and demolition waste in the year 2000 especially in the two bigger urban centers, Athens (Periphery of Attica) and Thessalonica (Periphery of Central Macedonia).

This was an expected result due to the increase of the activities in the construction sector, in view of the 2004 Olympic Games. Besides, the catastrophic earthquake of 1999 in Athens resulted in damaging or breaking down many buildings. While new buildings for housing the population were constructed, the damaged buildings have not been demolished to date. This is the reason why there are more construction waste than demolition waste.

Construction and demolition waste represent a significant proportion of total waste generation in Europe. C&D waste accounts for approx. 25% of the total waste generation in the EU. The quantities of C&D waste produced show that the generation of waste is not de-linked from the economic activities. The main environmental concern related to C&D waste is the existence of various hazardous substances in them, e.g. asbestos, which is hazardous itself and PVC, which in some treatment techniques causes emissions of toxic gases (dioxin). In addition, the use of raw materials (gravel and sandpits) causes distortion of the landscape and may cause problems for the level and quality of groundwater. Finally, disposal and landfilling of this kind of waste is a problem for the limited capacity of landfills. The major part of the waste stream is inert waste; consequently the problems of leaking and emissions from landfills are limited to specific fractions of waste from construction and demolition.

The necessary treatment technology for the effective management of C&D waste is quite simple. In Greece the management difficulties mainly concern administrative factors and the lack of fruitful cooperation between the competent authorities. Since the main volume of this waste stream is generated in the two bigger cities, the management initiatives should focus on them.

In the framework of the development of integrated recycling schemes, most governments of EU Member States made agreements with the industries. This practice is not common in Greece, while it should be. The technique of the selective demolition, the promotion and use of materials friendly to the environment, the substitution of dangerous substances with others, the market development for the recycling products and the introduction of technical specifications are measures that could be beneficial provided that the government is in close cooperation with the industrial sector, the builders' and engineers' associations.

REFERENCES

1) Alaveras P., Papchristou E., 1999, "Methods for the determination of construction waste in Greece", HELECO '99, Proceedings-Anouncement, pg.123-131

2) Bringezu S., Schütz H., "Material Flows (MFA) – Construction materials ans Packagings", Wuppertal Institute, January 1998

3) Source of data: Wastebase: http://waste.eionet.eu.int/wastebase, European Topic Centre on Waste and Material Flows, European Environment Agency

Bibliography

1) ECA, The European Cement Association, May 2001
2) http://www.cembureau.be/News/News%202001-05.htm
3) EAPA, European Asphalt pavement Association, brochure "Asphalt in Figures 2000", http://www.eapa.org
4) GJ, Governmental Journal, Issue 723, June 9, 2000
5) HSE: Health and Safety Executive, "Managing ASBESTOS in premises", INDG223 (rev 2), September 2001
6) Tränkler J.O.V., Walker I., Dohmann M., "Environmental impact of demolition waste – an overview on 10 years of research and experience", Waste Management, Vol. 16, Nos 1-3, pp.21-26, 1996
7) U.S. Environmental Protection Agency, "Construction and Demolition Waste Landfills", Office Of Solid Waste, May 1995
8) U.S. Environmental Protection Agency, "Characterization of building related construction and demolition debris in the United States", EPA 530 R-98-010, 1995
9) Halstead et al., "A methodology for quantifying the volume of construction waste", Peter A. Yost and John M. Halstead, Waste management and research, volume 14, pages 453-461, 1996
10) Bundesministerium fur Umwelt, Abfallwirtschaft, "Grundlagen fur bundeseinheitliche regelungen fur die entsorgung von baustellenabfallen", Band 17, Schriftenreihe der Sektion III, 1995
11) European Commission, 1999. EU focus on waste management. Directorate-General Environment, Nuclear Safety and Civil Protection, Brussels
12) European Community, 2001. Sixth Environment Action Programme of the 2001-2010 (6th EAP): Environment 2010: Our future, Our choice, Brussels
13) Granados, A.J. and Peterson, P.J., Hazardous waste indicators for national decision makers, Journal of Environmental Management, (55), 249-263, 1999.
14) SCOPE, 1995. Environmental Indicators: Systematic Approach to Measuring and Reporting on the Environment in the Context of Sustainable Development, International Workshop on Indicators of Sustainable Development for Decision-Making, Ghent, Belgium.
15) WRI, 1995. Environmental Indicators: A Systematic Approach to Measuring and Reporting on Environmental Policy Performance in the Context of Sustainable Development. World Resources Institute, Washington D.C.
16) ETC/WMF, 2001. Applicable methods and tools for the assessment of information related to waste and material flows, EEA, 1st draft, Copenhagen.
17) ETC/WMF, 2002. Towards a core set of indicators on waste and material flows, prepared for the EEA, 1st draft paper, Copenhagen.
18) Donovan, C.T., Construction and demolition waste processing: new solutions to an old problem, Resource Recycling, Vol. X, No. 8: 146-155, 1991
19) Bringezu, S., Schutz, 1998. Material Flows – Construction materials and Packaging. Meeting of the Working Group 'Statistics of the environment', Joint Eurostat / EFTA Group, Luxembourg.
20) SYMONDS Group Ltd, ARGUS, Consulting Engineers and Planners and PRC Bouwcentrum, 1999. Report to DGXI: Construction and Demolition Management Practices and their Economic Impacts, Brussels, Belgium.
21) Koliopoulos, T., 1999. Sustainable Solutions for the most pressing problem within, Solid Waste Management, International Solid Waste Association Times Journal, Copenhagen, Denmark, 3, pp. 21-24.
22) European Commission, 2000. Management of Construction and Demolition Waste. Directorate-General Environment, Brussels, Belgium.
23) Poon, C.S., Yu, T.W., and Wong S.W., 2002. Minimization of building waste in Hong Kong public housing projects. Int. Conf.: Appropriate environmental and solid waste management and technologies for developing countries. International Solid Waste Association, Istanbul, 8-12 July 2002, Volume 1, pp. 515-524.
24) EEA, 'Selected waste streams: Sludge, construction & demolition waste, waste oils, waste from coal fired power plants and biodegradable municipal waste', Technical Report, January 2002

MUNICIPAL SOLID WASTE MANAGEMENT: AN OPPORTUNITY ANALYSIS AND COMPREHENSIVE BENCHMARKING OF SOLID WASTE MANAGEMENT PRACTICES

F Colangelo

R Cioffi

University of Basilicata

Italy

ABSTRACT. This paper aims at focusing on the opportunities arising by using waste materials coming from different manufacturing activities in geo-environmental applications. Three k inds o f w aste w ere i dentified: m arble s ludge, coming from m arble m anufacturing, cement kiln dust and steel slags. The use of the mentioned waste was proposed in the urban solid waste landfills cover layers according to the new directions Italian rules.

Keywords: Industrial waste, Sustainable construction, Alternative landfill, New Italian laws, Steel slags, Cement kiln dust, Marble sludge

Dr F. Colangelo is a research civil engineer at the Department of Environmental Engineering and Physics at the University of Basilicata, Italy. He is working in the area of recycling of industrial waste in concrete and in geo-environmental applications.

Professor R. Cioffi is a professor at the University of Basilicata and he teaches Materials Science and T echnology a t t he Faculty of E ngineering at t he sa me U niversity. H is m ain research activities concern the stabilization/solidification of industrial solid wastes, soils and sediments and the preparation of energy saving blended cements and concrete made with coal fly ashes, blast furnace slag, chemical gypsum, powdered tuff, lead smelter slag, MSWI ashes, etc. Another field of interest concern the inertization of asbestos wastes by mechanochemical treatments and the reuse of the obtained asbestos free products as building materials. All the above researches are proved by more than one hundred papers

INTRODUCTION

Nowadays the ecological trend aims at limiting the use of natural raw material and this increases the interest in the use of alternative materials, or waste, coming from industrial activities, which, in this way, gain a non negligible economic, energetic and environmental value [1-2]. The main geo-environmental applications can be listed as follows: road and railways building with interventions on foundation, embankments and superstructures; territory reclamation in order to create areas devoted to civil, industrial, agrarian and forest uses; restoration and improvement of areas degraded by human interventions (quarries, mines, earth moving, landfills) or by natural actions (erosion, swamps, sliding areas).

The feasibility of the waste material recovery process is especially influenced by the contemporaneous satisfaction of the economic, technical and normative aspects for each field of use. In fact, once the economic convenience has been assessed, namely that the costs related to transport to and disposal in the landfill are higher than those related to their transport and use as materials alternative to the natural ones, the experimentation must verify that the chemical physical characteristics, after possible treatments, are suitable to the specific project solutions they are devoted to. This work proposes the new utilization of industrial waste, such as marble sludge coming from stony material manufacturing, cement kiln dust and steel slags in the cover layers of Municipal Solid Waste (MSW) landfills, according to the new Italian role [3].

LEGISLATIVE FRAMEWORK

The new Italian regulations on MSW disposal [3, 4], apart from re-classifying the landfills typology (for inert waste, dangerous and not), identify new prescriptions about the cultivation modes. The decree provides for the realization of a system of waterproofing, both of the bottom and of the cover, with suitable characteristics for each kind of disposed waste, by prescribing the respect for adequate permeability values. Such limits, very restrictive indeed, represent an innovation with respect to the previous rules which did not quantify the soil's degree of impermeability. The protection of soil, phreatic and surface waters must be carried out by a combination of a geological barrier and a possible coating of the lower part which will be completed by an upper closable cover [5].

Geological Barriers

The landfill base and sides substratum must be made by a geological natural formation corresponding to requirements of waterproofing and thickness at least equal to:

- Landfill for inert waste: hydraulic conductivity $k \leq 1 \cdot 10^{-7}$ m/s and thickness ≥ 1 m;
- Landfill for non dangerous waste: hydraulic conductivity $k \leq 1 \cdot 10^{-9}$ m/s and thickness ≥ 1 m;
- Landfill for dangerous waste: hydraulic conductivity $k \leq 1 \cdot 10^{-9}$ m/s and thickness ≥ 5 m.

In the event that the geological barrier does not naturally satisfy these conditions, it can be artificially completed through a confinement barrier system, suitably made, which could supply an equivalent protection. The basement of the confinement barrier lower layer must be put above the roof of the confined water table with a clearance of 1.5 m at least and, in case o f n on c onfined w ater t able, a bove t he w ater t able m aximum e xcursion l evel w ith a clearance of 2 m at least. In inert waste landfills the artificial barrier must be thicker than 0.5 m. In case of dangerous and non dangerous waste landfills the confinement barrier system characteristics are normally guaranteed by the coupling of compressed mineral material (thickness ≥ 100 cm, $k \leq 1 \cdot 10^{-7}$ cm/s, preferably deposited in compressed uniform layers of a maximum thickness of 0.2 m) with a geo membrane. For the same kind of landfills, on the bottom, above the waterproof coating, we must always supply a layer of draining material with thickness ≥ 0.5 m. The landfill bottom, considering the foreseen settlings, must keep a suitable slope in order to favour the downward flow of the leachate to the collection systems.

Final Surface Cover

The landfill final surface cover must be able to guarantee:

- waste isolation from the external environment;
- minimization of water seepages;
- reduction to minimum of maintenance needs;
- minimization of erosion phenomena;
- resistance to settlings and localized subsidence phenomena.

The cover must be made through a multilayer system composed, from top to bottom, of:

1. Surface coating layer (S_1) with a thickness ≥ 1m favouring the development of vegetal cover species   aiming at the environmental restoration plan, supplying a suitable protection against erosion and allowing to protect the underlying barriers from thermal excursions;
2. Draining layer (S_2) with a thickness ≥ 0.5 m able to prevent the formation of a hydraulic water head on the barriers;
3. Compressed upper mineral layer (S_3) with a thickness ≥ 0.5 m with low hydraulic conductivity for inert waste landfills and compressed mineral layer with thickness >0.5 m with hydraulic conductivity $k \leq 10^{-8}$ m/s (non dangerous waste), integrated by a surface waterproof coating for the dangerous waste landfill plants;
4. Gas draining and capillary breakage layer (S_4), protected by possible blockages, with a thickness of ≥ 0.5 m only for non dangerous and dangerous waste landfill;
5. Regularization layer (S_5) for the correct positioning of the upper elements, made with draining material.

WASTE FROM EXTRACTION ACTIVITIES AND STONY MATERIAL PROCESSING

Extraction activity and stony material processing involve the production of a huge mass of rejects of different natures and characteristics. In fact, stone extraction produces a yield in blocks which usually does not overcome 60% of the extracted material, so the accumulation of production rejects represents a very important problem, causing damage to the environment and the landscape. Their possible recovery would thus offer sensible and multiple advantages. The coarsest part, the fragments and the processing scraps, obtained by the board cutting and profile turning operations, are easy to recover in the field of building, while the other kind of waste-dust coming from the block sawing and polishing operations mixed with water (cooling of the blades of the frames and of the grindstones of the polishers) - has a sludge-like composition which, once decanted in appropriate plants, is dehydrated through filter pressing. This kind of waste, even if classified as inert waste, if dispersed in the environment causes serious damage by altering, because of its extreme fineness, the soil permeability and therefore also the rain water infiltration processes.

In this work, the sludge coming from twenty Southern Italian factories dealing with marble processing was characterized; its use was proposed to build the waterproof layer (S_3) in the landfill cover. Then, the scraps waste, coarser because of their physical and chemical characteristics, could be used for the regularization layer (S_5). The results of the chemical and physical characterization made on different samples have shown the presence of $CaCO_3$ in weight percentages which are variable between 96 and 98, as well as more modest percentages of SiO_2 (0.8-1.2%), Al_2O_3 (0.1-0.3%), Fe_2O_3 (0.06-0.09%), MgO (0.6-0.8%), etc. All sludge samples analyzed is characterized by a diameter at 90% equal at 10.1 ☐m. Particle size distribution of one of the characterized samples is shown in Figure 1.

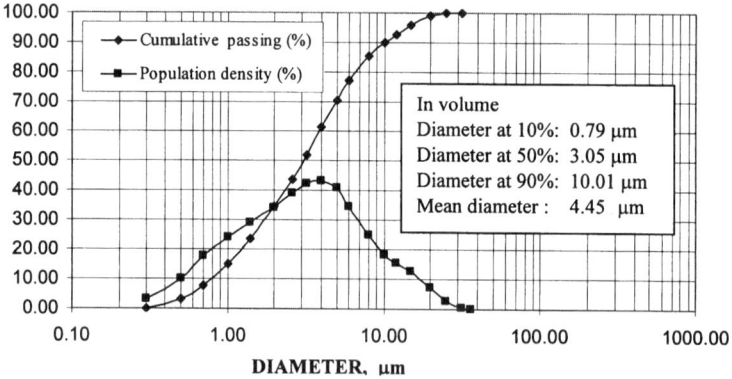

Figure 1. Particle size distribution of marble sludge

The main constraints in the reutilization of industrial waste are linked to the qualitative constancy of the following parameters: chemical composition, very small size and absence of heavy metal. In consideration of their possible reutilization, the marble sludge samples, after drying, have undergone permeability tests in triaxial cell by using compressed material 40 x 80 mm screen-tests. The measured values were very similar to each other and variable between $2 \cdot 10^{-9}$ and $8 \cdot 10^{-9}$ m/s. These values are compatible with those foreseen by the Italian Law Decree 36/03 [3] as the lower limit for the layer S_3. On the contrary, as for the percent of substitution of such a waste compared to the total of the layer, even if the Italian Law [6] provides for that the percent of the waste usable in mixture with the raw material is lower than 30%, in this experimentation we have used, with a positive outcome, samples totally made by dry compressed marble sludge.

Then, it was possible to reutilize the processing scraps, characterized by a much coarser size, in the draining layer S_5 foreseen for the regularization needed for the correct positioning. We also carried out tests to verify that such materials meet the requirements of dimensional steadiness in contact with water. Especially for marble sludge we can exclude a plastic behaviour. In fact the values of Atterberg limits are very low, (LL=0.299 , PL=0.243). Then several studies have shown that the clay permeability in contact with the landfill leachate sensibly increases [7]; marble sludge should solve this problem since it is characterized by higher chemical-physical inertia than clay. The two kinds of waste have undergone a leachate test according to the Italian regulation [6] in order to determine the possible leachate of heavy metal traces. Both kinds of calcareous waste have shown a heavy metal leachate lower than 0.001 mg/l.

CEMENT KILN DUST

Cement kiln dust (CKD) is a waste product generated by the cement industry; it is a powder mainly composed of micron-sized particles collected from electrostatic precipitators during the manufacture of cement clinker. The CKD chemical composition and particular size depend on the raw materials used to produce clinker, fuels and kiln types. For this reason, unlike the marble sludge, CKD is very heterogeneous. The Environmental Protection Agency (EPA) estimated that the amount of CKD could range from 0 to 25% of the clinker produced. This amount depends on the above-mentioned factors.

A huge quantity of CKD is produced every year in the world; for example, USA produce itself 4 to 12 million tons [8]. Only a small amount of CKD can be recycled in rotary kiln; the possibility of its recycling in other fields is thus of increasing importance. Nowadays, unfortunately, only a small quantity of this waste material is recycled even if it could be used in different ways: agricultural soil treatment, pavement base layers, blended cement, etc. The latter is surely the most important because of the quantities it can absorb. The effect of using cement kiln dust instead of ordinary Portland cement on the properties of concrete was investigated by several authors who showed that CKD decreases the mechanical properties of cement paste so it can only be used in small amounts in the cement industry.

However, the main reason for disposing of CKD is almost due to its high alkali content and this, together with the presence of a certain kind of aggregates, can cause the fearful "alkali-aggregate" reaction. The use of CKD for landfill top covers is very interesting, since it is already dry and ground. In this study the use of CKD as a material for layer S_3 is proposed. The study investigated its use for landfill top covers according to the new Italian laws [3,4]. We analyzed CKDs coming from several Italian cement works and noticed a certain variation in their chemical composition but they all showed a cumulative passing at the 40-□m sieve higher than 90% (in weight), as stated in literature. It is important to remark that the chemical composition variability is an obstacle to CKD reutilization in many production cycles while, for the use we suggest, the physical characteristics are much more important. CKD supplied by Cements of Lucania, Potenza, Italy, was used. The CKD chemical composition is shown in Table 1.

Table 1. Chemical composition of CKD used.

SiO_2	Fe_2O_3	Al_2O_3	CaO	MgO	SO_3	Na_2O	K_2O	L.O.I.
14.65 %	2.13 %	4.75 %	41.72 %	1.12 %	0.72 %	0.90 %	0.60 %	32.30 %

For CKD as well we should exclude a plastic behaviour of the material since, also in this case, the Attemberg limits were very low (LL=0.389 , PL=0.327). Like marble sludge, CKD also underwent some triaxial-cell permeability tests. The measured values were very similar to each other, ranging from $9 \cdot 10^{-9}$ m/s to $1 \cdot 10^{-10}$ m/s CKD grading curve and density population are shown in Figure 2. In this case as well we carried out some leaching tests according to the Italian rules and CKD showed low value. The above-mentioned results showed that CKD is very suitable for the proposed use.

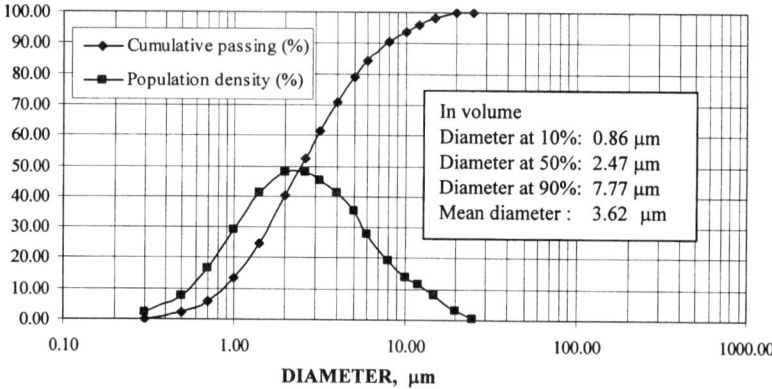

Figure 2. Particle size distribution of CKD used

STEEL SLAGS

In Italy, more than 25 million tons of steel with a resulting production of two million tons of waste are produced every year, especially from electric arc furnace. Till nowadays, more than 60% of the waste from electric arc furnace produced in Europe is placed into landfills. Such waste can be reutilized in different ways [3], for example as foundation for road embankments, as aggregates for concretes, etc. However, before reutilizing the waste in such applications, a crushing pre-treatment and the following sieving are necessary to make mixtures of suitable grading curve. The pre-treatment operations are very power expensive because of the high resistance to wear and waste crushing.

The steel slags resistance to crushing has been determined by subjecting a characteristic sample to the test used for measuring the Los Angeles coefficient [9]. This test allows determining the quantity of solid waste per weight unit which is crushed in a suitable device. Very low values are characteristic of a high resistance to crushing. The examined waste sample has shown a value of the Los Angeles coefficient equal to 16, demonstrating the clear advantage to reutilize the waste without a previous crushing treatment. In this work the use of steel slags was proposed, without a previous crushing pre-treatment, as material forming the layer S_2 (draining) and the layer S_4 (permeable to gases) of the landfill surface cover. The waste sample used is characterized by sizes between 20 and 80 mm and the outcomes of the chemical analysis and of the physical characterization are shown in Table 2.

Table 2. Chemical composition and physical properties of slags.

SiO_2	Fe_2O_3	Mn_3O_4	TiO_2	P_2O_5	Al_2O_3	CaO	MgO	SO_3
14.63 %	24.98 %	6.32 %	0.69 %	0.62 %	9.01 %	35.99 %	4.92 %	0.16 %

Specific gravity = 3.81 kg/dm³   *Apparent specific gravity = 2.01 kg/dm³*
Water absorption coefficient = 1.01

The low value of the water absorption coefficient and the average grain sizes show the very good draining capacity of this material. In addition, the very good physical-mechanical properties make it suitable for meeting the requirements of "resistance to settlings and to localized subsidence phenomena" specified by the rules. The same waste sample underwent a leaching test according to the provisions of the Italian rule [6]. The results of the tests done are reported in Table 3. The whole of results relevant to the chemical-physical characterizations allows us to propose, with high probability of success, the use of the waste in the landfill surface cover layers.

Table 3. Leaching tests results.

CONTAMINANT	UNIT	VALUE RECORDERED (V.R)	LAW LIMITS (L.L.)	V.R./L.L. (%)
Sulfates	mg/l	21.30	250.00	8.50
Chlorides	mg/l	5.30	200.00	2.65
Barium	mg/l	0.27	1.00	27.00
Branch	mg/l	0.03	0.05	60.00
Zinc	mg/l	0.30	1.30	23.00
Nichel	$\square$g/l	3.03	10.00	30.30
Cadmium	$\square$g/l	0.10	5.00	2.00
Lead	$\square$g/l	0.40	50.00	0.80

PREPARATION OF THE COVER LAYER

The following picture (Figure3) shows the scheme of the cover layer prepared using the two fractions of residues from marble processing and steel slags without pre-crushing. The great quantity of residues usable in this system, able to meet the requirements imposed by the recent rules, is clear.

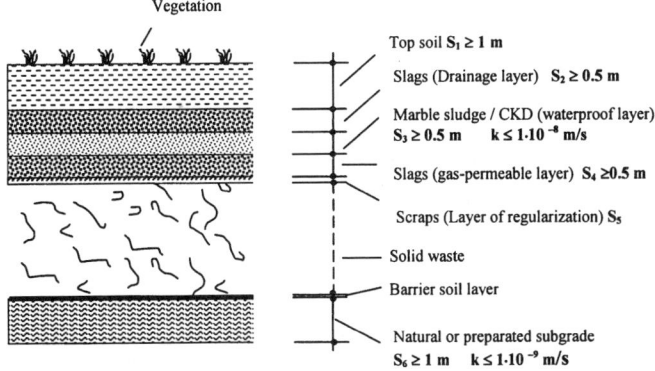

Vegetation

Top soil $S_1 \geq 1$ m

Slags (Drainage layer) $S_2 \geq 0.5$ m

Marble sludge / CKD (waterproof layer)
$S_3 \geq 0.5$ m $k \leq 1 \cdot 10^{-8}$ m/s

Slags (gas-permeable layer) $S_4 \geq 0.5$ m

Scraps (Layer of regularization) S_5

Solid waste

Barrier soil laver

Natural or preparated subgrade
$S_6 \geq 1$ m $k \leq 1 \cdot 10^{-9}$ m/s

Figure 3. Example of alternative top cover

CONCLUSIONS

The experiment carried out proves that waste materials of different particle size distribution can be successfully used in the realization or in the completion of some layers of the cover systems for the different kinds of landfills.

The use of such materials, apart from determining an environmental double advantage - decreasing the consumption of natural materials and reducing the demand for volumetry of new landfills - also offers a technological advantage. In fact, the proposed materials supply comparable results and, in some cases, even better than the normally used raw materials.

REFERENCES

1. MILETIC, S. et al.: Building materials based on waste stone sludge, Fifth International Conference on the Environmental and Technical Implications of Construction with Alternative Materials "WASCON 2003", San Sebastián, Spain, 4-6 June, 2003.

2. LIND, B.B. et al.: Environmental impact of ferrochrome slag in road construction, The fourth International Conference on the Environmental and Technical Implications of Construction with Alternative Materials WASCON 2000, Leeds/Harrogate, United Kingdom, 31st May to 2nd June, 2000

3. ITALIAN D.LGS 13/01/2003 n° 36 (Waste landfill).

4. ITALIAN D.M. 13/03/2003 (Waste landfill).

5. TCHOBANOGLOUS, A. et al.: 'Integrated solid waste management', McGraw-Hill,Inc., 1993.

6. ITALIAN D.M. 05/02/1998.

7. DI MAIO, C., Onorati R.: Influence of pore liquid composition on the shear strength of an active clay. Proceedings VIII International Symposium on landslide, 26-30 June 2000 Cardiff, 1999.

8. DUCHESNE J., REARDON E.J.: Determining controls on element concentrations in cement kiln dust leachate, Waste Management No.18, 1998, pp339-350.

9. UNI 8520, Part 19^a – Determination of weight loss of coarse aggregates. 1984.

SESSION TWO:

GENERATION, PROCESSING, RECOVERY AND SUPPLY NETWORK

GENERAL CRITERIA TO ASSESS THE SUITABILITY OF DIFFERENT WASTES TO BE INCLUDED IN A CEMENTITIOUS MATRIX

L Fernández Luco

M Castellote

Institute of Construction Science Eduardo Torroja, CSIC, Madrid

Spain

ABSTRACT. Concrete can be considered as a multi-size composite material made of a cementitious matrix which binds inclusions of different sizes (aggregates). Depending on the shape, size and other characteristics, different recycling waste can also be included to conform the matrix itself or as a partial replacement for sand or coarse aggregate. The suitability of the immobilisation must be assessed from different points of view, considering both the hardened state and the fresh state. In the hardened state, the dimensional stability, elastic properties, compressive and tensile strength and durability, including environmental considerations, should be taken in consideration.

Should any unsuitable interaction occurs, it is always possible to apply a pre-treatment to the waste product to be recycled or to use chemical admixtures to overcome this situation. Sometimes, the design of a "special" concrete or mortar, which might exhibit special characteristics, would be a wise decision. This paper shows a general criterion for assessing the suitability of the inclusion of waste products in mortars and concretes and it will be illustrated by a case study.

Keywords: Concrete, Immobilisation, Residues, Electric-arc-dust, Strength, Durability, Leaching

Luis Fernandez-Luco is Civil Engineer and MICT. At present, he works at the Departmen of Structural Engineering and Mechanic of Composites, Institute of Construction Science Eduardo Torroja, CSIC, Madrid. His main interests deal with science and technology of concrete and mortars, with focus on performance-based specification and on-site testing.

Marta Castellote is PhD in Chemical Sciences by Zaragoza (Spain) University. At present she works as Researcher at the Department of Chemistry and Physic of Construction Materials, Institute of Construction Science Eduardo Torroja, CSIC, Madrid. Her main interest deals with the physico-chemical processes of transport phenomena in construction materials and concrete durability.

INTRODUCTION

The suitability of the partial substitution of a common concrete constituents by any by-product or recycled waste should be assessed from different points of view, considering that not only the hardened state is important but the fresh state and the transition period, regularly known as "setting period", as well. In the hardened state, the dimensional stability, elastic properties, compressive and tensile strength and durability should be taken in consideration. Environmental considerations are also very important and, in some cases, leaching tests and chemical characterisation should be included in the evaluation.

It is of outmost importance that the evaluation be complete, i.e., none of the basic considerations can be ignored to minimise the risk of failure in the use of a given residue. All these basic considerations have an equivalent importance in the evaluation process, and can be complemented by supplementary considerations which might give some help in optimising the overall process, although they are not essential to decide whether the residue can be included or not in a cementitious matrix.

FUNDAMENTAL CONSIDERATIONS

The preliminary assessment of the possibility for any given residue to be included in a cementitious matrix has to rely on the assumption that the long term behaviour of the composite material can be predicted from the results obtained from short-term testing. General knowledge related to the complex processed which govern the stability and durability of a cementitious matrix will certainly contribute to the assessment. Besides, a complete characterisation of the residue might show the possibility for it to react in the pore solution of a cementitious matrix.

As a consequence, deleterious or aggressive products might leach and/or expansive reactions might occur. From these considerations, the basic criteria should be based on the answers to the following questions:

1. Is there a possibility for the residue to degrade or expand? (based on the chemical and mineralogical composition of the residue)

2. Is the expansion capable of inducing some cracking of the concrete? (the kinetic of the expansive reaction is also relevant)

3. Is it possible for the concrete voids to accommodate the expansive products, without damage to the matrix?

4. Is it possible to give the residue a pre-treatment to reduce the deleterious effect of expansions? (Thus, the expansion is induced prior to introducing the material in the rigid matrix).

5. Is the composite material (concrete + residue) prone to leach some contaminants to the environment?

If the answers to these questions show that, regarding the long-term stability, the residue can be included in a cementitious matrix, new considerations arise.

GEOMETRICAL AND PHYSICAL CONSIDERATIONS

The first consideration involves geometrical and physical issues, such as size, shape, texture, density and absorption.

Size, shape and texture

The size of the particles being immobilised determine whether they can be consider as conforming the matrix itself or as a partial replacement for sand or coarse aggregate. The composition of the concrete has to be modified, on a volumetric basis, to prevent the particles of the residue to interfere with the mobility of the fresh concrete. Should the particles have a similar shape as the aggregates, the easiest criteria is to partially substitute aggregate by residue. On the other hand, if the residue has one or two dimensions very dominant (i.e. elongated or flaky), it would be required to "loosen" the aggregate structure to allow them to accommodate in between.

There are different practical procedures to loosen the aggregates structure; ACI -211 Mix Proportion Method defines a coefficient which ranges from about 0,44 to 0,87, to be applied to the dry-rod unit weight of the aggregate. The smaller the coefficient used, the more loosen the coarse aggregate structure.

An equivalent criterion is adopted by O. Okamura, one of the developers of the Self-compacting concrete concept. Okamura proposes this value to be limited to 0,5 to prevent the coarse aggregate from blocking.

The recommendations for steel-fibre reinforced concrete follow similar schema: to increase the amount of mortar to allow for some extra space for the fibres. From a practical point of view, depending on the size, texture and shape of the particles to be immobilised, some "loosening" effect would be advisable.

Density, Porosity, Absorption

Low-density particles might tend to float, while denser one would be prone to sink, leading to segregation and lack of homogeneity. Absorption might also affect the fresh state of the concrete. Lightweight aggregate concrete technology concept might be used to overcome these situations.

PHASE TO BE ANALYSED: CONCRETE, MORTAR OR PASTE

Once the geometrical and physical properties of the residue are analysed, it is advisable to decide which phase will be evaluated; paste, mortar or concrete itself. As a general rule, the maximum particle size of the phase under evaluation should be similar to the residue maximum size. This consideration might help in reducing time and labour cost, because it is easier to work with paste than mortar, while concrete would represent the largest effort.

If concrete is to be used as the immobilization matrix, the final stage of the evaluation should always include concrete evaluation, as the overall effect can not be assessed from results obtained from paste or mortar.

EFFECT ON THE FRESH STATE

The effect of the residue on the fresh state has to be assessed because a suitable workability has to be achieved. Should the geometry and size of the residue be inconvenient, it is always possible to treat the residue (by grinding, cutting, crushing, etc.) to modify its relative dimensions or to reduce its maximum size.

As a general rule, it can be stated that the smaller the size of the residue, the lesser the influence on the concrete properties, both in the fresh and hardened states. Nevertheless, as the maximum size of the particle is reduced, the specific surface increases, leading to a higher water demand. This inconvenient can easily be overcome by the use of suitable water reducing chemical admixture.

Exudation is usually reduced when fine particles are used and segregation is prevented by increasing the viscosity of the mix.

Setting time has to be monitored because the residue may induce some modifications on the behaviour of the concrete, either accelerating or retarding the mortar or concrete. The necessity for recovering a standard setting time has to be analysed, because in many cases, a retardation of setting would not be a great nuisance for the practical use, indeed.

EFFECT ON THE HARDENED STATE

To assess the effect of the residue on the hardened state, some considerations might help to focus and to reduce the testing program. Many different properties can be measured and/or defined for the composite material in the hardened state; In this case, only strength and dimensional stability have been taken into consideration, while durability will be considered in a separate section.

The effect of the residue on strength depends mainly on its intrinsic strength, its maximum size and proportioning (% of the residue in the composite material). Texture, absorption and shape may also influence strength, but their effect is less noticeable. Although elastic properties are a key feature to determine strain compatibility between the matrix and the inclusion, they are not usually measured. Besides, elastic modulus E can be associated to mechanical strength.

The m aximum s ize a nd the i ntrinsic s trength h ave t o b e c onsidered s imultaneously. If t he intrinsic strength of the residue is high (comparable to a common aggregate), it will not limit the strength of the concrete, and the major effect will be related to the quality of the interfacial transition zone (ITZ); the larger the aggregate, the more the effect on the ITZ.

If the residue to be included in the cementitious matrix has low intrinsic strength, the effect could be compared to the use of lightweight aggregate. The residue will limit the maximum strength of the concrete and will also determine the "optimum strength". Nevertheless, some lack of strength might be beneficial as a reduction in the elastic modulus will improve the mechanical compatibility with the matrix.

The reduction in strength will be associated with the amount of residues in the mix and its maximum size. To rapidly assess the "minimum strength level" to be expected, it can be assumed that the contribution of the residue is zero, i.e. a replacement of the residue by air.

If the residue is prone to expand while immersed in the matrix pore solution, the expansion can cause the matrix to crack, leading to irreversible damage. There are some standard tests to assess the dimensional stability (linear expansion of concrete prisms), but possibly the most sensitivity is achieved when the residue is finely ground (area exposed increases).

If the composite shows signs of deleterious expansions, a suitable pre-treatment should be designed and tested. This treatment should be relatively fast, cheap and easy to perform. Sometimes, an immersion in lime saturated water for a few days could be enough to develop the expansion while achieving a certain chemical stability.

On the contrary, some residues to be immobilised might show some pozzolanic or cementitious properties, thus contributing to strength. This is the case with fly ash, blast-furnace slag and silica fume, formerly considered as "residues" to be included in the cementitious matrix".

EFFECT ON DURABILITY

The effect the residue might have on concrete durability has always to be assessed. Different approaches can be combined: chemical and mineralogical characterisation, stability tests, and accelerated test for measuring any transport properties, and so on.

Nevertheless, it has to be kept in mind that it is somehow risky to extrapolate long term stability from accelerated tests. As it was said before, some judgement can be based on the composition and characteristics of the residue.

Apart from the intrinsic durability of the composite, the possibility of contamination of the environment has to be evaluated, by performing leaching test. Leaching itself might also alter the material and the eventual damage will depend on the composition of the fluid and the extent of leaching period. Usually, leaching progressively increases the matrix porosity.

If the residue is capable of diminishing the pH of the concrete, the potential corrosion of the steel reinforcement has to be carefully analysed. The identification of micro-structural and physicochemical changes induced by the residue is important to theoretically predict changes in the properties of the material.

PRACTICAL CONSIDERATIONS

It is difficult to give precise examples in a general paper, but the type of pre-treatments to be given to the residues to be immobilised are many and they widely depend on different circumstances.

There is also a need to be creative in the solution given to any immobilisation project. For instance, if the residue is lightweight, it induces a reduction in strength and there is some air-entraining effect, it would be wise to think about a lightweight concrete, with thermal isolating properties.

As an example, the use of domiciliary wastes was evaluated at the Portland Cement Institute of Argentina. As a non-negligible strength-loss occurred and the residue showed an important air-entraining effect, a special lightweight concrete was designed to build non-structural isolating panels.

As an example of a complete evaluation carried out at the Institute of Construction Sciences Eduardo Torroja, CSIC, Madrid, a case study dealing with the immobilization of electric-arc furnace dust is summarised in the following paragraphs.

CASE STUDY: IMMOBILIZATION OF ELECTRIC ARC FURNACE DUST (EAFD)

The steel industry generates a substantial quantity of Electric Arc Furnace Dust (EAFD). The chemical composition of this dust varies from one factory to another, and even for the same plant. The specific composition depends on the scrap material used in the process but in all cases this dust contains hazardous metals such as Pb, Cd, Cr, Cu or Zn. Therefore, the disposal of this waste has become a serious problem in the recent years [1,2] and several technologies have been developed in order to undertake it:

On the one hand, several processes for separation and recovery of selected metals have been developed [3-7]; on the other hand, EAFD are intended to be added in manufacturing cement clinkers [8-10], and finally, stabilization techniques are being applied to minimize the environmental risk associated to these wastes. Among them, one of the possibilities is to use this material as an addition to concrete [11-14]. In this case study, the results presented are summarised from [14], where an extensive study on the suitability of cementitious materials to immobilize EAFD is reported.

Characterization of the Waste

Provided that the composition of the EAFD can vary form one batch to another, depending on the scrap used in the process, it is necessary to determine the composition of the powder used, by its elemental analysis. In addition, the mineralogical composition, by X-ray diffraction (XRD) has also to be studied in order to detect any incompatibility with the cement. In this case, ZnO and PbO were detected, which imply a retarded setting, which was confirmed later.

In order to primarily assess the effect of this addition on a cementitious matrix, the results corresponding to two mixes of mortar, with Spanish cement type IV-B, are given. The reference dosage (without any addition), **R**, and the mix **M** in which a 24,2 % of cement + sand was replaced with EAFD.

Effect on the Fresh State

❖ Initial and final setting time

Providing the mineralogical study indicates the possibility of a delay in the setting time, the effect on the fresh state was studied. The results for the setting time, according to EN 196-3 are given in Table 1.

As expected, there is a delay in the setting time when EAFD is added, which is attributed to the fact that calcium ions, which are dissolved from the clinker minerals, are bonded by zinc ions to form calcium-hydroxozincate-hydrate [15]. Only after all the zinc is fixed, additional dissolved calcium ions can initiate the ordinary setting process.

Table 1: Initial and final setting times for the two mixes.

	R (without EAFD)	M (with EAFD)
Initial setting time (hr)	9.5	96
Final setting time (hr)	15.3	288

❖ Workability of the fresh mix

The effect of the addition on the workability of the fresh mix was assessed from two different approaches; although the mortar maintained enough workability to cast and compact standard moulds, some loss of workability was shown as compared to the Marsh Cone passing time for the reference mix M.

❖ Heat of hydration

The heat of hydration measurement has been carried out according to the UNE 80-118-86 standard. The results obtained are given in Table 2, where the averaged values are presented for 6, 12, 41,72 and 120 hours. The results in Table 2 imply that the heat of hydration diminishes when EADF is added, being the value obtained for mix M very low. This fact has been attributed to the retarded setting that allows the heat produced during hydration to dissipate.

Table 2: Results of the heat of hydration, for the reference and D3 mixes, according to the UNE 80-118-86 standard.

			TIME				
Mix	Calorimeter		6 hr	12 hr	41 hr	72 hr	120 hr
Reference	Average	KJ/Kg		89.3	227.98	243.23	244.91
M	Average	KJ/Kg	9.32	10.34	12.07	15.72	80.99

Effect on the Hardened State

❖ Mechanical strength

The average results from triplicate specimens on flexural and compressive strength at 7 and 28 days, for the different mixes, are shown in figure 1. The first data for mix M corresponds to the 12th day, as until that age the setting period was not completely finished. According to this retardation in the setting time, the 12-days strength is very low. However, after 28 days, mix M reached values for compressive strength that were only 12% smaller than those of the reference mix, R.

Table 3: Flexural and compressive strength at 7 and 28 days, for mixes R and M.

N/mm^2	FLEXURAL		COMPRESSIVE	
Mix	7 days*	28 days	7 days*	28 days
R	6.1	7.2	26	38.8
M	0.823	5.23	1.59	34.21

12 days for M

❖ **Dimensional stability**

Two different types of standardized tests have been carried out on the mixes: ASTM C-227-81, and UNE 146508-EX. According to the ASTM C-227-81 procedure, the specimens were stored for 7 days in a humid chamber at 96% HR. The data obtained at this age the reference length to calculate the relative expansion (12 days for M mix). After taking this measurement, t he s pecimens w ere k ept a t 3 8±2° C a nd 9 6% H R f or 9 0 d ays, t heir l ength being measured periodically during that period. According to the standard, the addition can be considered as potentially reactive if the expansion measured after 3 months is higher than 0.05%.

Following UNE 146508-EX standard, the specimens were stored for 7 days (12 days for M mix) in a humid chamber at 96% HR. After the curing regime, they were submerged in water at 80° C during 24 hr, being their lengths measurement the reference data to calculate the percentages of expansion. Then, the specimens were kept submerged in a 1 N NaOH solution at 80° C. According to the standard, after 14 days in the alkaline solution, if the expansion measured is lower than 0.1%, the aggregates can be considered as non potentially reactive. Just to be on the safe side, in this research, all the specimens were monitored up to 28 days.

The results from both procedures indicated that, although the expansion of the M mix was higher than the one corresponding to the reference mortar, the addition can not be considered as potentially reactive.

Effect on Durability

Total porosity and pore size distribution by Mercury Intrusion Porosimetry

The results obtained from Mercury Intrusion Porosimetry for the reference mortar and for the M mix are presented in Table 4, where the total porosity (percent in volume), the mean pore diameter and the density of the matrix, are given. In table 4, it can be noticed that the addition of EAFD induces an increase in the total porosity. However, it has to be noticed that the mean pore diameter is considerably lower in M than in R.

From this data it can be deduced that the addition of EAFD to the mix leads to a refinement of the microstructure that can be attributed to the packaging and filling of the voids by the powder, with a higher porosity in the range of gel pores.

Table 4: MIP microstructure parameters of the samples

MIX	TOTAL POROSITY (% vol)	MEAN PORE SIZE (μm)	DENSITY (gr/cm^3)
R	12.58	0.0539	2.129
M	16.25	0.0153	2.151

Determination of the effective chloride diffusion coefficient Dc.

The effective diffusion coefficient, D was calculated according to the procedure given in [16,17] on triplicate specimens. The average coefficient for mix R was 5.4 10^{-8}. cm^2/s and that of the mix M 1.88 10^{-8}.cm^2/s. Therefore, it can be said that the addition of EAFD reduces the diffusion of chloride ions through the resulting matrix, in agreement with the results of MIP.

Mineralogical changes (XRD results)

The mortar samples have been analyzed by X-Ray diffraction, using a CuKα Phillips PW1820 p owder d iffractometer, t o m ake a c omparative s tudy of t he c rystal p hases i n t he different mixes. From the results obtained it can be deduced that there has been an important change in relation to the amount of Portlandite. No Portlandite is present in mix M, which is attributed to the same mechanism that the delay in the setting time, as calcium ions are dissolved from the clinker minerals and are bonded with zinc ions present in the EAFD to form calcium-hydroxy-zincate-hydrate.

The development of this reaction was confirmed by the presence of calcium-hydroxy-zincate-hydrate in the diffraction pattern for the M mix. However, in spite of the reduction of the alkaline reserve of the resulting matrix, when applying phenolphthalein on to the specimens of the M mix, the indicator becomes pink, which indicates that the resulting matrix is alkaline enough, having a pH above 10.

Concerning the rest of species, as expected, there is a decrease in the amount of anhydrous cement as well as in the crystal phases of the calcium-silicate-hydrates, when EAFD is added.

CONCLUSIONS

From the results of this research it can be concluded that EAFD can be immobilised in cementitious matrixes as an addition to concrete. The addition tested in the present research, with the type of cement and mix proportion that is hereby studied does not seem to have a significant negative influence on the resulting matrix.

REFERENCES

1. United States Environmental Protection Agency, *Land disposal restrictions for electric arc furnace dust*, Fed. Regist, 56 (160) , **1991**, pp 41164-41178.

2. SUNDAL, J.; FUTORNICK, K, KRISHNAN, R., *Regulatory issues regarding the recycling of electric arc furnace dust and its use as a slag fluidizer*, Extr. Process. Tret. Minimization Wastes. Proc. Int. Symp. 1994. Ed by Hager, John P. Miner. Met. Mater. Soc. Warrendale, Pa, USA., **1994**, pp 637-644.

3. HEPWOTH, M.T.; TYLKO, J.K.; HAN, H., *Treatment of electric arc furnace dust with a sustained shockwave plasma reactor,* Waste Management and research, Vol 11, 5, **1993**, pp 415-427.

4. KEEGEL, J.F., *Methods for recycling electric arc furnace dust,* Journal of Cleaner Production, Vol 4, 3-4, **1996**, pp 260.

5. YOUCAI, Z.; STANFORTH, R., *Integrated hydrometallurgical process for production of zinc from electric arc furnace dust in alkaline medium,* Journal of Hazardous Materials, Vol 80, 1-3, **2000**, 223-240.

6. ABDEL-LATIF, M.A., *Hydrometallurgical recovery of zinc and lead from electric arc furnace dust using mononitrilotriacetate anion and hexahydrated ferric chloride,* Journal of Hazardous Materials, Vol 91, 1-3, **2002**, 257-270.

7. ANTREKOWITSCH, J. ANTREKOWITSCH, H., *Hydrometallurgically recovery of zinc from electric arc furnace dust,* JOM, Vol 53, 12, **2001**, 26-28.

8. HILTON, R.G., *Method for manufacturing cement clinkers, especially Portland cement clinkers, using stabilized electric arc funace dust as raw material,* USA patent n° 5853474 A, **1997**.

9. AOTA, J, *Electric arc furnace dust treatment*, Fuel and energy abstracts, Vol 38, Issue 1, January **1997**, pp 27.

10. XUEFENG, X.; YUHONG, T, *Application of electric arc furnace dust in cement production,* Gangtie, 33 (6), **1998**, 61-64.

11. FEMANDEZ   PEREIRA,   C.,   RODRIGUEZ-PIÑERO,   M.,   VALE,   J., *Solidification/Stabilization of electric arc furnace dust using coal fly ash. Analysis of the stabilization process.* J. of Hazardous Materials, B82, **2001**, pp 183-195.

12. aL-zAID, R.Z., AL-SUGAIR, F. H.; AI-NEGHEIMISH A.I., *Investigation of potential uses of electric-arc furnace dust (EAFD) in concrete.* Cement and Concrete research, Vol 27, **1997**, pp 267-278.

13. BALDERAS, A., NAVARRO, H., FLORES-VELEZ, L., DOMÍNGUEZ, O., *Properties of Portland cement pastes incorporating nanometer-sized franklinite particles obtained from electric arc furnace dust.* Journal of the American Ceramic Society, 84 (12), **2001**, pp 2909-2913.

14. CASTELLOTE, M., MENÉNDEZ, E., ANDRADE, C. ZULOAGA, P., Navarro, M and Ordoñez, M., *Radioactively contaminated electric arc furnace dust as an addition to the immobilization mortar in low and medium activity repositories*, Environmental Science and Technology (**2004**) in press.

15. LIEBER, W., *Influence of zinc oxide on the setting and hardening of Portland cements* (in German). Zement-Kalk-Gips, 20, **1967**, pp 91-95.

16. CASTELLOTE, M., ANDRADE, C., ALONSO, C., *Measurement of the steady and non-steady-state chloride diffusion coefficients in a migration test by means of monitoring the conductivity in the anolyte chamber. Comparison with natural diffusion tests*, Cement and Concrete Research, 31, **2001**, pp 1411-1420.

17. ANDRADE, C., *Calculation of chloride diffusion coefficients in concrete from ionic migration measurements. Cement and Concrete Research* 23 (3), **1993**, pp 724-742.

REUSING AND RECYCLING C & D WASTE IN EUROPE

V Corinaldesi

G Moriconi

Marche Polytechnical University of Ancona

Italy

ABSTRACT. Within the European Union, the construction and demolition wastes (C&DW) come to at least 180 million tons per year. Roughly 75% of the waste is disposed to landfill, despite its major recycling potential. However, the technical and economic feasibility of recycling has been proven, thus enabling some Member States (in particular Denmark, The Netherlands and Belgium) to achieve recycling rates of more than 80%. On the other hand, the South European countries recycle very little of their C&DW.

At present, recycled aggregates are mainly used untreated as obtained from demolition for excavation filling, roadbeds or floor foundation. However, several authors have studied the possibility of using recycled aggregates to manufacture structural concrete and, in 2002, the CEN/TC 154 Technical Committee drew the EN 12620 "Aggregates for concrete" in which artificial or recycled aggregates are considered beside natural aggregates for use in concrete.

Keywords: Construction and demolition waste, Recycled-aggregate concrete, Recycling.

V Corinaldesi is a Civil Engineer and has a Ph.D. in Materials Engineering; she is developing a research project on recycling of building materials at the Department of Materials and Environment Engineering and Physics at the Marche Polytechnical University of Ancona, Italy.

G Moriconi is Professor of Materials Science and Technology at the Marche Polytechnical University of Ancona, Italy and is the Head of the Department of Materials and Environment Engineering and Physics. He is the author and co-author of numerous papers in the field of cement and concrete technology, building materials performance and durability. He has been recently awarded by ACI for his contribution to the research development in the general field of concrete durability.

THE EUROPEAN MARKET OF C&DW

At m ore t han 4 50 m illion t ons p er year t he c onstruction a nd d emolition w aste form t he largest waste stream in quantitative terms within the European Union, apart from mining and farm wastes [1]. If one excludes earth and excavated road material the amount of construction and demolition waste generated is estimated to be 180 million tons per year.

Roughly 75% of the waste is disposed to landfill, despite its major recycling potential. This represents very large quantities (more than million tons per year) occupying existing landfills. H owever, t he t echnical a nd economic f easibility o f r ecycling h as b een p roven, thus enabling certain Member States (and in particular Denmark, The Netherlands and Belgium) to achieve recycling rates of more than 80% (see Table 1).

Table 1. C&DW practice in the European Union in 1996-97 [1]

MEMBER STATE	C&WD ARISING (m tonnes)	RE-USED OR RECYCLED (%)	INCINERATED OR LANDFILLED (%)
Germany	59	17	83
U.K.	30	45	55
France	24	15	85
Italy	20	9	91
Spain	13	<5	>95
The Netherlands	11	90	10
Belgium	7	87	13
Austria	5	41	59
Portugal	3	<5	>95
Denmark	3	81	19
Greece	2	<5	>95
Sweden	2	21	79
Finland	1	45	55
Ireland	1	<5	>95
Luxemburg	0	n/a	n/a
European Community	180	28	72

At present, the South European countries (Italy, Spain, Portugal, Greece) recycle very little of their C&DW. Their natural resources are of a sufficient quality and quantity to meet the demand for building materials at a moderate cost, thus implying a delay in the market for recycled materials to develop.

Nevertheless, mainly for environmental reasons, also in this part of Europe there is a growing interest in the possibility of recycling these materials. More recent data relative to the Italian market testified that almost 40% of C&DW were re-used or recycled in 1998 instead of 9% in 1996.

EUROPEAN RESEARCH ON RECYCLING C&DW

During the last twenty years the recycling of construction and demolition waste (C&DW) has emerged as a socio-economic priority within the European Union. In the late 70's European and Japanese members of RILEM (Réunion Internationale des Laboratoires d'Essais et de recherche sur les Matériaux et les construction), in particular Professors Yoshio Kasai and Torben C. Hansen, began a new field of research in recycling technology, concerning m aterials c oming f rom b uilding d emolition. T hey s tarted u p s everal r esearch projects in this field. In 1981 the RILEM designated a Technical Committee 37-DRC (Demolition and Reuse of Concrete), which worked up to 1988, when it was replaced by e new Technical Committee 121-DRG (Demolition and Reuse Guidelines) with the aim of drafting a guideline, published in 1993 [2]. Moreover, the results of this wide research programme have been published in the proceedings of three international symposia held by RILEM in Rotterdam (The Netherlands) 1985, in Tokyo (Japan), 1988 [3] and in Odense (Denmark), 1993 [4]. At present, a RILEM Technical Committee is operating since 2001 about the "Use of recycled materials in construction" (TC URM) with Prof. Charles Hendriks as chairman.

In the meantime, many other studies have been carried out on this topic by individual Universities and Research Centres spread out everywhere in Europe.

COMMUNITY POLICY ON C&DW MANAGEMENT

Since 70's, the European Community started up a program directed to the waste management by means of the 75/442 CEE directive and the following 78/319 CEE, 84/631 CEE, 91/156 CEE and 90/639 CEE.

Within t his g eneral p olicy, w hich h as a mong i ts m ain s copes t he i ncrease i n p roduction prevention and the waste reduction trough the new clean technologies' development, the problem o f C &DW assumes s ome r elevance s ince 1 992. In f act, i n t hat year, t he w aste coming from the building process were included among the priority waste flows and, in order to study them, a specific workgroup was established, the "Construction and Demolition Waste Project Group". Really, C&DW put management problems not so much for the presence of hazardous substances (such as asbestos, chromium, cadmium, zinc, lead, mercury), that are little or nothing present, but for the huge amounts produced.

The "Construction and Demolition Waste Project Group" was composed by agents of Member States and Organizations operating in the field (contractors, professional associations, disposers, ecologist movements, local authorities) and by experts. The main purpose of this group was the elaboration of a strategy for achieving a draft normative to be approved by the Council. The C&DW Project Group put its works (ended in June 1995) into practice by drafting two documents "Information" and "Recommendations".

The "Information" document gave a picture of the current situation concerning construction and demolition waste.

On the other hand, the "Recommendations" document suggested a series of measures and actions that, if undertaken by the various Member States, could widely develop the recycling of construction and demolition waste. The proposed strategy concerned the following points:

- prevention (besides to actions of education/information, it provided for a new way to design materials by keeping into account their disposal and recycling and their environmental impact);
- separation (it provided for a spreading of the selective demolition technology and for encouraging recycling and deterring disposal in landfill);
- treatment (it proposed the introduction of a system based on permissions and licences issued to those business involved in the production of construction and demolition waste. In this way, the qualified contractor, in order to get the licence, must indicate the produced waste amount, the adopted measure for treating it and its final destination);
- market (if P ublic A dministrations e xercise a m odel r ole a s a p urchaser o f P ublic Works the market for C&DW could develop).

The two documents had no legislative value, but they represented a valid technical-cognitive support useful for drafting a normative properly regulating the C&DW sector and strongly boosting their recycling. The European Community strategy, and the results of the researches carried out, sensitized the Member States which thought it right to acknowledge at national level the EU proposals. Therefore, each Member State in order to pursue the aims told by the European Community, adopted various political-economical tools that are giving encouraging results in terms of C&DW recycling.

In 1998 the Mandate M/125, concerning the execution of standardisation work for harmonized standards on aggregates, was issued by the Commission to CEN/CENELEC. This Mandate was subsequently correlated by several documents, dated 2002, in which artificial or recycled aggregates are considered together with natural aggregates:

- EN 12620 "Aggregates for concrete";
- EN 13043 "Aggregates for bituminous mixtures and surface treatments for roads, airfields and other trafficked areas";
- EN 13055-1 "Lightweight aggregates - Part 1: Lightweight aggregates for concrete, mortar and grout";
- prEN 13055-2 "Lightweight aggregates - Part 2: Lightweight aggregates for bituminous mixtures and surface treatments and for unbound and bound applications excluding concrete, mortar and grout";
- EN 13139 "Aggregates for mortar";
- EN 13242 "Aggregates for unbound a nd hydraulically b ound materials for use in civil engineering work and road construction";
- EN 13383-1 "Armourstone - Part 1: Specification";
- EN 13450 "Aggregates for railway ballast".

SOME EXAMPLES IN COMPARISON

As quite evident in Table 1, there is a strong difference in terms of recovery of construction and demolition waste within the various Member States.

The b est results w ere achieved i n t hose C ountries i n w hich s uitable c hoices o f p olitico-economical feature were carried out (such as waste management planning, targets of recycling arranging, new markets supporting, and so on).

In particular, a percentage of r ecovery of about 10% can be noticed for those Countries (Italy, France, Spain) in which specific interventions for promoting recycling are lacking, while this percentage becomes sharply higher for those Countries (The Netherlands and Great Britain) in which such interventions have been applied.

Afterwards, two meaningful examples are examined in detail:

- o The Netherlands i.e. the Member State with the best results in terms of recycling;
- o Germany i.e. the greatest producer of C&DW with fair results in recycling.

The m ain c ollection o f i nformation a bout t he q uantities o f C &DW a mong t he E uropean Union as well as the measures that each Member State has taken to improve the re-use and the r ecycling o f t his w aste s tream i s a r eport c alled "C onstruction a nd d emolition w aste management practices and their economic impacts" funded by the European Commission. The project was undertaken by Symonds Group (United Kingdom) in association with Argus (Germany), COWI Consulting Engineers and Planners (Denmark) and PRC Bouwcentrum (the Netherlands). The report describes the best practices in this field as well as the economics of the C&DW reuse and recycling [5].

Finally, also the Italian experience in terms of C&DW recycling is summarized.

The Netherlands

The Netherlands are the Country in which recycling and recovery of construction and demolition waste achieve the best results. These results and, in particular, the development of the C&DW recycling "culture" are strictly related to the nature of the Dutch territory, where unlike other European Countries, there is scarce availabiliss of both natural aggregates and areas for landfills. Therefore, the hurry to safeguard natural resources droved the Dutch Government to adopt political and economical measures steered into discouraging the disposal in landfill of demolition waste and into encouraging the recovery and recycling of building scraps.

Ever since 1996 a Dutch law compels the demolishers to part materials, such as asbestos, wood, masonry and concrete, during the storage. Moreover, many laws have been issued at both local and national level for promoting the selective demolition. In 1997, it was imposed the absolute prohibition from disposing in landfill of those C&DW that can be reused and only those companies having a specific licence can dispose in landfill of non-reusable C&DW. In addition, with the purpose of achieving the target of 90% C&DW recycling within the year 2000, the Dutch Government adopted several actions for discouraging the production of C&DW as well as for promoting their treatment and reuse.

The disposal in landfill of inert waste in The Netherlands is regulated by a system of taxes. The rates are variable depending on the region. Moreover, the Dutch Government offers incentives for those contractors which use recycled aggregate replacing natural aggregate for public works falling under the competence of Transport Ministry or Public Works Ministry. In the Dutch territory about 120 recycling plants are present with a total capacity of 16 million of tones and only the 15% of those plants are mobile while the others are fixed.

Table 2. C&DW production and recycling in The Netherlands in 1996 [5]

C&DW TYPE	PRODUCTION (m tons)	RE-USE (%)	RECYCLING (%)	INCINERATING (%)	DISPOSING (%)
Concrete, bricks…	10.48	0	93	1	6
Wood	0.26	0	50	10	40
Plastics	0.21	0	5	12	83
Metals	0.18	0	100	0	0
C&DW unselected	0.04	0	0	0	100
'core' C&DW	*11.17*	*0*	*90*	*1*	*9*
Excavation earths, stones…	6.20	0	40	0	60
Asphalt	2.72	72	28	0	0
Total	*20.09*	*10*	*66*	*1*	*23*

Since 1995, in the Netherlands are available Rules concerning the use of recycled aggregates in roadbeds and floor foundations draft by C.U.R. (Commissie voor Uitvoering van Research). In 1997, the old NEN 5905:1988 "Additives for concrete. Sand and gravel" was definitively replaced by the NEN 5905:1997 "Additives for concrete. Materials with a volumic mass of at least 2000 kg/m^3", including recycled aggregates.

Germany

Germany is the larger producer of C&DW in Europe. Nevertheless, as shown in Tables 1 and 3, only the 17% of such waste is recycled. Since 1996, in Germany a policy is carried out directed to favour the waste recycling instead of disposal in landfill. In particular, the recycling of C&DW is compulsory if technically possible and economically advantageous.

At local level, some measures in terms of waste management have been adopted for increasing waste recovery and decreasing their disposal in landfill. In some Länder (such as Hamburg and Mecklenburg-Vorpommern) the C&DW recycling achieved 80-90%.

Disposal in landfill of inert waste in Germany does not provide for taxes; however, for C&DW disposal a certain rate must be paid depending on nature and composition of the waste. In the German territory about 1600 recycling plants are present and about 700 landfills for excavation earth's disposal.

Since 1995, a Technical Rule is available (Anforderungen an die stoffliche Verwertung von mineralischen Restoffen/Abfällen – Technische Reglen), concerning the use of recycled aggregates in roadbeds and floor foundations and a Guideline was elaborated in 1998 by DafStb (German Committee for Reinforced Concrete), relative to their use in concrete [6].

Table 3. C&DW production and recycling in Germany in 1994-96 [5]

C&DW TYPE	PRODUCTION (m tons)	RE-USE (%)	RECYCLING (%)	INCINERATING (%)	DISPOSING (%)
Concrete, bricks…	45.00	0	18	0	82
Other 'core' C&DW	n.a.*	n.a.	n.a.	n.a.	n.a.
C&DW unselected	14.00	0	14	0	86
'core' C&DW	*59.00*	*0*	*17*	*0*	*83*
Excavation earths, stones…	215.00	n.a.	n.a.	n.a.	n.a.
Asphalt	26.00	n.a.	80	n.a.	20
Total	*300.00*	*n.a.*	*n.a.*	*n.a.*	*n.a.*

*) non available data.

The Italian situation
Also the Italian Normative, as the other Member States, acknowledged in the statutory law D.Lgs. 22/97 the European directive 91/156 CEE by introducing in the waste management the following priorities: prevention, then recovery and finally disposal. Nevertheless, the Italian situation reports a certain delay in C&DW recycling with respect to other European Countries (see Tables 1 and 4). The reasons can be:

-   the abundance of natural aggregates;
-   the low rates for disposal in landfills;
-   some hindrances to the use of recycled aggregates (such as the lack of up-to-date specifications and performance rules).

Table 4. C&DW production and recycling in Italy in 1995-97 [5]

C&DW TYPE	PRODUCTION (m tons)	RE-USE (%)	RECYCLING (%)	INCINERATING (%)	DISPOSING (%)
Concrete, bricks...	n.a.*	n.a.	n.a.	n.a.	n.a.
Other 'core' C&DW	n.a.	n.a.	n.a.	n.a.	n.a.
C&DW unselected	n.a.	n.a.	n.a.	n.a.	n.a.
'core' C&DW	20.00	6	3	0	91

*) non available data.

Many C&DW recycling plants are present in the Italian territory, both mobile and fixed. These plants, in particular the fixed, are characterized by a high technological level. In fact, not only the main objectives of crushing and separation are considered but also the quality control of input and output materials, the workers' safety, the dust reduction. Although the high quality of the recycled aggregates produced in Italy, the spreading of their use clashes with many obstacles and, in particular, the Technical Rules and the conservative attitude of designers and builders.

Fortunately, the new European Norms such as EN 12620, acknowledged by the UNI (Italian National Unification body) as UNI EN 12620 in October 2003, would open up the way for a wider recycling of C&DW, although it has been soon scaled down by the mandatory document for the Norm application.

CONCLUSIONS

The introduction of the new European Norms arisen from M/125 Mandate can give a useful common reference for all the European Countries and, in particular for the South European Countries which are behind, at the moment, in terms of C&DW reuse and recycling.

Another aspect which can strongly promote the C&DW reuse and recycling is the introduction of suitable taxes related to their disposal in landfill, following the positive experience of The Netherlands.

REFERENCES

1. EUROPEAN COMMISSION, Directorate-General Environment, DG ENV.E.3, "Management of Construction and Demolition Waste", Working Document N°1, 4 April 2000, pp. 1-26.

2.   RILEM TC 121, Specifications for concrete with recycled aggregates, 1993.

3.   KASAI, Y., Reuse of Demolition Waste, Proceedings of the Second International RILEM Symposium on "Demolition and Reuse of Concrete and Masonry", Volume Two, Chapman and Hall, London, U.K., 1988.

4.   HANSEN, T.C., Recycling of Demolished Concrete and Masonry, Report of Technical Committee 37-DRC on Demolition and Reuse of Concrete, RILEM Report 6, E & FN Spon, London, U.K., 1992.

5.   SYMONDS GROUP, ARGUS, COWI, BOUWCENTRUM, Construction and demolition waste management practices and their economic impacts, http://www.europa.eu.int/comm/environment/waste/studies/cdw/c&dw_report.htm, 1999.

6.   GRÜBL, P., RÜHL, M. German Committee for Reinforced Concrete (DafStb) Code: Concrete with Recycled Aggregates, Proceedings of the International Symposium on "Sustainable Construction: Use of Recycled Aggregate Concrete", Ed. R.K. Dhir, N.A. Henderson and M.K. Limbachiya, Thomas Telford Publishing, London, U.K., 1998, pp. 409-418.

CONSTRUCTION AND DEMOLITION WASTE IN SOUTH AFRICA

J J Bester

D Kruger

A Hinks

RAU University – Johannesburg

South Africa

ABSTRACT. The perception of recycled aggregates inferiority has lead to extensive research world wide on the viability of recycled aggregates in civil construction. The abundance of the primary aggregate in South Africa inhibits the real need for recycled aggregate hence it is not the initial choice when selecting an aggregate.

Although hosting the most recent World Summit on Sustainable Development in 2002, the South African government does not have any legislation or financial incentives promoting the use of secondary materials. Recycling Construction and Demolition Waste may not be the cheapest, quickest, or easiest method of forming an aggregate, but it is the choice that makes the most sense environmentally.

Keywords: Demolition waste, Construction waste, Crusher plant, Quality control.

J J Bester is a member of the Institute of Concrete Technologists and is a researcher in the Research Group for Polymers in Concrete at RAU University in Johannesburg, South Africa.

D Kruger is a professional Civil Engineer and head of the Research Group for Polymers in Concrete at RAU University in Johannesburg, South Africa.

A Hinks is a registered student in the Department Civil and Urban Engineering and is in his third year of study at RAU University in Johannesburg, South Africa.

INTRODUCTION

'Core' Construction and Demolition Waste (C&DW) can be defined as being those materials that are obtained from the demolition of an empty structure or the demolition of civil engineering infrastructure [1].

Reasons for recycling C&DW waste and using it as aggregates are as follows:

- There are increasing shortages of primary (virgin) aggregates, which lead to increased cost of primary aggregates.
- Landfills have decreased in capacities to cater for demolition waste.
- There are increasing negative effects of construction and demolition waste on the environment.
- It reduces illegal dumping.
- It may be economically viable.

DIFFERING PROPERTIES BETWEEN NATURAL AND RECYCLED AGGREGATE

Due to the adhesion of mortar to, and the recrushing of the aggregates, recycled aggregates have a rougher surface texture, lower specific gravity and higher water absorption than primary aggregates that are of similar size. The higher water absorption is due to recycled aggregates having increased porosity. The increased porosity also increases the effects of carbonation, as carbon dioxide can penetrate deeper into the concrete. This will necessitate better quality control on the production and use of recycled aggregate [2].

HOW CONSTRUCTION AND DEMOLITION WASTE IS CATEGORIZED

C&DW can be categorized as one of four main categories:

- Waste arising from total or partial demolition of buildings and/or civil infrastructure (Demolition Debris).
- Waste arising from the construction of building and/or civil infrastructure (Construction Debris).
- Soil, rocks and vegetation arising from land levelling, civil works and/or general foundations (Excavation Material).
- Road planning and associated materials arising from road maintenance activities (Scarifying) [1].

USES OF RECYCLED AGGREGATE

Once demolished concrete has been crushed, sieved and decontaminated it can be used as:
- General bulk fill.
- Fill in drainage projects.
- Sub-base material for road construction.
- Aggregates for the manufacture of concrete.

ENVIRONMENTAL PROBLEMS IN THE RECYCLING OF CONCRETE

The recycling of concrete presents both advantages and disadvantages. The advantage is that the substances are reused which would otherwise be classed as waste. The disadvantage include the intrusion of trucking into locations where this is undesirable, aesthetic concerns, and potential noise. Figure 1 shows the average percentage composition of demolition waste in developed countries.

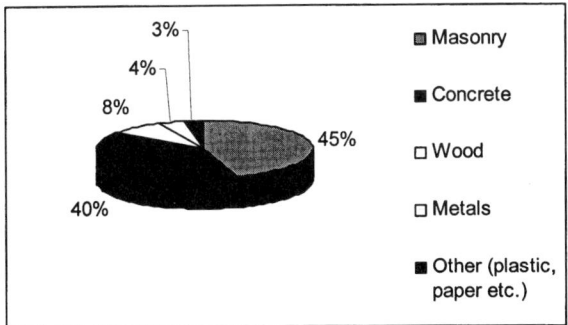

Figure 1: Average percentage Composition of Demolition Waste
in more Developed Countries [1]

SOME ADVANTAGES OF RECYCLING

Environmental

- By recycling the use of natural aggregates is reduced.
- Less waste is being landfilled.

Job Creation

- The employees of the recycling plant, including separators, plant operators etc.

Recycling of secondary materials

- The materials obtained from the C&DW can be recycled, e.g. steel, paper, wood etc.

TYPES OF CRUSHING PLANTS USED FOR RECYCLING

Mobile Processing Unit (usually on site)

- It consists of a mobile crusher, a pre-screen and an electromagnetic separator – see Figure 2.
- The end product obtained is dependent on the cleanliness and the homogeneity of the initial material.

- Difficulties may be experienced in obtaining a permit due to the dust created. This may be overcome by wetting the material to be crushed.
- Higher operation costs.

Figure 2: Malans Quarry Recycling Plant

Permanent Fixed Units (usually off site)

- Off-site crushers can be much larger and productive than on-site crushers.
- May consist of primary, secondary and possible tertiary crushers.
- Less control of input material.
- Lower maintenance cost.

SOUTH AFRICA'S SITUATION

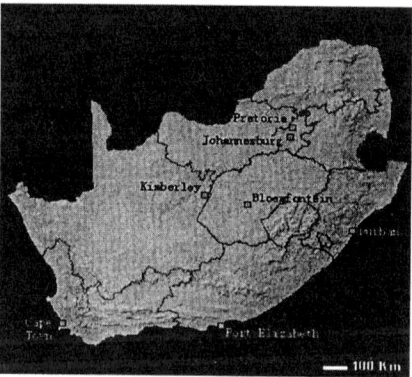

Figure 3: Map showing major Cities of South Africa

It is estimated that the South African Construction Industry generates about 5 to 8 million tonnes of C&DW per annum. The South African government, although hosting the most recent World Summit on Sustainable Development in 2002, do not have any legislation or financial incentives in place for promoting the use of secondary materials. The awareness levels of building construction practitioners are low in terms of research, knowledge of international trends and the use of such materials. It is for this reason that there is a lot of controversy about the perceived risk that secondary materials are inferior.

The record keeping and control of construction and demolition waste entering landfill sites throughout South Africa is, in general, not of a high degree of accuracy. Table 1 gives an indication of the waste practise in South Africa.

Table 1: Construction Waste Practice in South Africa [3]

PROVINCE	TOTAL WASTE Received in Landfills (tpa)*	REUSE	RECYCLING	ILLEGAL DUMPING
Gauteng	560 000	Extensive	None	Extensive
Kwa-Zulu-Natal	375 000	Extensive	Operational	Extensive
Western Cape	200 000	Extensive	Operational	Extensive
Eastern Cape	64 000	Extensive	None	Evident
Mpumalanga	Minimal	Extensive	None	Extensive
North West	Minimal	Extensive	None	Not Available
Free State	Minimal	Extensive	None	Not Available
Northern Cape	Minimal	Evident	None	Not Available
Northern	Minimal	Evident	None	Not Available

*tpa – Tonnes per annum

BARRIERS TO RECYCLING CONSTRUCTION AND DEMOLITION WASTE

Quick disposal of C&DW is preferred to a systematic processing and sorting for the production of recycled aggregates. This is due to the high initial capital cost to construct and maintain a recycling plant. There is a perception among consulting engineers that recycled aggregate is inferior in quality to primary aggregates although strict regulations are set by authorities for the quality of the recycled aggregate.

The abundance of the primary aggregate in South Africa, and the ease of mining it, inhibits the real need to use recycled aggregate, therefore recycled aggregate is not the first choice when choosing an aggregate. This however will become a problem when the primary aggregate is diminished and increasing charges are made for landfilling demolition wastes.

PROPOSED GOVERNMENT INCENTIVES

The following initiatives can be undertaken to promote the re-use and recycling of C&DW:

- Selective restriction or bans on the disposal of C&DW.
- The use of mono landfills for certain fractions of C&DW.
- The introduction of local, regional or national taxes on the disposal of recoverable C&DW.
- The introduction of subsidies for the use of C&DW.
- Financial support by government to research and develop projects.
- Tightening environmental and planning controls on disposal of C&DW.
- The provision of education and training support geared specifically towards C&DW.
- The availability of standards and norms applicable to recycled materials.

PROPERTIES OF RECYCLED AGGREGATE CONCRETE

Table 2 gives results of recycled aggregate concrete properties.

Table 2: Properties of Recycled Aggregate Concrete (1)

PROPERTIES	NATURAL AGGREGATE	RECYCLED AGGREGATE	
Density (kg/m^3)	2500 - 2610	2340 – 2490	
Water Absorption (%)	0.1 – 1.0	8-12% Higher	
Modulus of Elasticity (E)		14 – 28% lower	
		% Comp.	CBR
California Baring Ratio (CBR)		92.8	42
		97.8	66
		100.8	142
Plasticity Index (P.I)		N.P. – 4	
Drying Shrinkage		50% Higher	

QUALITY CONTROL OF RECYCLED AGGREGATES

In order to create an aggregate of uniform consistency, not only needs source control of the raw materials to be done, but the method of stockpiling at the crushing plant needs to be controlled. Stockpiling needs to be according to size, composition and quality. Table 2 also compares natural aggregate properties to those of recycled aggregates, and it can be seen that there is a marked decrease in the quality of product.

Noise, vibrations and dust always accompany the operation of a crushing and screening plant. Some good locations to construct a crushing plant are next to a ready-mix plant, on a landfill site or in an area where there is a lot of on-going demolition.

PROJECTS RECENTLY COMPLETED MAKING USE OF RECYCLED AGGREGATE

South African suppliers, such as Malans Quarries, Bradis Demolition and Ross & Sons Demolition, all predominantly operational in the Western Cape, supplied material for the following high profile projects in South Africa:

In the late 1990's and early 2000's, Malans Quarries supplied construction material for the use in road and parking lot sub-bases in the following projects in the Western Cape area:

- Westlake Office Development
- The Cape Convention Centre
- Century City and Ratanga Junction
- Grand West Casino

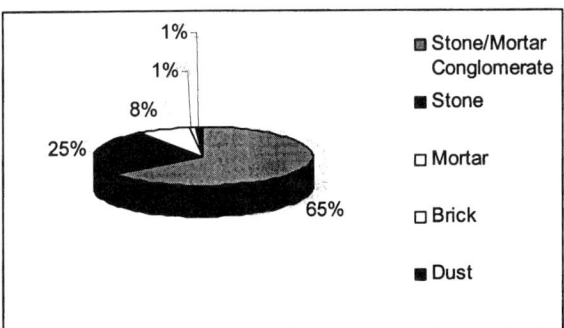

Figure 4: Graph showing percentage Composition of Recycled Aggregate produced by Malans Quarries (1)

Ross & Sons Demolition

In the late 1990's and early 2000's Ross & Sons Demolition supplied:

- The crushed material from the seawall in Cape Town was used as a base for Franki-Pile to work from.
- The same as above applied for the Arabella Hotel.
- Use of sub-base on the corner of Pollsmoor Road.

IN-HOUSE RECYCLING

Due to rising costs of primary aggregates companies found it more economical to purchase a crusher and crush products not meeting specifications. The product is then reincorporated into the manufacture of their product. Some companies also found it economically viable to use C&DW as partial replacement for virgin material.

Typical in-house Operation Process:

- The C&DW is fed onto a vibrating screen to separate the fines
- The contaminants are removed simultaneously (by hand, sieve or magnet)
- The C&DW is then fed by means of a conveyor belt to the primary crusher
- Prior to crushing, the stone passes through a stationary electromagnet
- The conveyor then leads to a secondary crusher where the material is broken into 10mm aggregate.
- The aggregate now passes through a vibrating screen where stones smaller than 10mm are removed

South African companies that apply this method are Columbia DBL, Cape Brick, Steco Concrete Works, Inca Bricks and Crammix Bricks, all in the Western Cape.

CONCLUSIONS

With the growing production rate of Construction and Demolition Waste, and the increasing difficulty in obtaining licenses for mining natural aggregate, recycling C&DW should be the step taken forward. Research is being done to analyse the properties of recycled aggregate and to find the suitability of recycled aggregate for use in structural concrete as well as for other applications such as general bulk fill, fill in drainage projects and sub-base material in road construction.

The authors believe that some of the reasons why aggregate recycling is more active in the Western Cape region of South Africa is that the region has limited natural aggregate resources remaining and the region is environmentally sensitive. It is also arguably the most popular tourist destination in South Africa.

REFERENCES

1.   AYERS M.J, Recycled Aggregate, Usage and Barriers within Western Cape, University of Cape Town, South Africa, 2002.

2.   HANSEN T.C., Recycling Demolished Concrete and Masonry, Report 6, Cement and Concrete Institute, 1992.

3.   MACOZOMA D.S., A. Benting, Construction Waste Management, CSIR, South Africa, 1999.

A BIOMECHANICAL PROCESS FOR THE SPECIFIC MATERIAL-FLOW TREATMENT OF RESIDUAL WASTE – PERSPECTIVES FOR CONCEPTS OF DECENTRALISED WASTE DISPOSAL IN GERMANY

M Wittmaier

University of Applied Sciences Bremen

Germany

ABSTRACT. The deposition of untreated waste in municipal landfills is a burden on the environment. Following a transitional period that ends in 2005, only waste that has been rendered largely inert will be permitted in German landfills. Thus, thermal or biomechanical treatment will be necessary for domestic and domestic-like waste from trade and industry. Biomechanical treatment can be designed to follow one of two different courses: a) as large a proportion of the waste as possible is rendered inert in preparation for deposition in a landfill, or b) as large a proportion of the waste as possible is prepared for utilisation as a secondary raw material or secondary fuel. In this paper both kinds of biomechanical treatment are discussed, before studies of a material-flow oriented process are presented as an illustration of the results that can be achieved. The material flow balance of the plant in question shows that the proportion of waste requiring disposal in a landfill can be significantly reduced, in this case to about 12% of the original amount. Study results are presented for the quality of the secondary fuel produced and the importance of pollution reduction in secondary fuel is discussed. A material flow oriented process must be capable of reacting flexibly to changing market conditions. This is illustrated taking the biomechanical waste treatment plant in Rügen as an example of the changes that were necessary to it and concepts for its further development. Modular design of the biological treatment stage facilitates the treatment of smaller amounts of waste (10.000 Mg/a) and shows the way towards concepts of decentralised waste treatment.

Keywords: Domestic waste, Domestic-like waste from trade and industry, Biomechanical waste treatment, Secondary fuel, Material flow oriented treatment of residual waste, Decentralised waste treatment.

Dr M Wittmaier is the director of the Institute for Recycling and Environmental Protection (Institut für Kreislaufwirtschaft GmbH), University of applied sciences Bremen, Germany. His research interests include the biological waste treatment and various aspects of sustainable waste management.

INTRODUCTION

It has long been known that depositing untreated municipal waste in landfills and the associated degradation of organic substances by microorganisms results in the production of landfill gases and highly polluted seepage water [1], both of which pose considerable environmental problems. Although much progress has been made in the construction and operation of landfill sites (landfill gases and seepage water, for example, can be dealt with by technical means [2]), many problems relating to the deposition of municipal waste remain unsolved. An example: in modern landfills, in which untreated municipal waste is deposited, the drainage system often becomes impaired by incrustations formed through microbial conversion of organic and inorganic substances contained in the highly polluted seepage water. While those parts of the drainage system accessible to flushing and clearing measures can be kept operational, the function of inaccessible sections of an incrusted drainage system cannot currently be restored [3]. A large proportion of municipal landfills have incrustation-related drainage problems.

In response to the problems that arise from the landfill of municipal waste with a high organic content, in Germany in 1993 the Technical Regulations for the disposal of Municipal Waste in landfill sites [4] and in 2001 the Waste Landfill disposal Act [5]) were enacted, along with legislation regulating the construction and operation of landfill sites, which specified stricter requirements also in respect to the waste to be deposited. In particular, the proportion of organic material, i.e. of readily and moderately degradable organic substances, and the resulting microbial activity (gas production potential over 21 days and respiratory activity over 4 days) were limited by setting maximum values for ignition loss. The consequence of this legislation is that waste with a high organic content, such as domestic and domestic-like waste from trade and industry, must be treated prior to landfill. Such treatment can be thermal or biomechanical.

In the area of biomechanical waste treatment a number of different processes have been developed and introduced in recent years. The different processes are differentiated into those which aim to treat as large a proportion of the waste as possible for landfill, and those which aim to separate as large a proportion as possible for utilisation (e.g. as secondary fuel or raw materials), while minimising the proportion requiring disposal via landfill.

This paper deals generally with biomechanical methods of residual waste treatment. In addition, using the example of the biomechanical waste treatment facility, Rügen, on the Baltic Sea, a concept for decentralised waste treatment will be presented. The facility is characterised by it's modular design, which allows its biological treatment stage to be operated economically at a capacity of just 10,000 – 15,000 Mg/a. This is significant when one considers that most methods of biomechanical residual waste treatment require a throughput of 50,000 Mg/a or more for their economic operation. The process is based on a material flow oriented concept designed to utilise as large a proportion of the waste as possible and to leave just a small amount requiring landfill.

METHODS

A material flow analysis was conducted by evaluating all material input and output flows. Weight loss from the biological treatment stage was calculated by deducting all output flows from total waste input. Since the dried waste, following biological treatment, is subjected to further mechanical treatment, which itself has a slight drying effect, the weight loss from biological treatment is in fact a little less than the figure given.

Water content was determined according to DIN 18123. The determinations of chlorine and sulphur were based on DIN 51577, Part 1/DIN EN ISO 10304-1. Pentachlorophenol was determined following extraction after E DIN 38407-F 10. PCB determination was after DIN 38414-S 20. Calorific value was determined after DIN 51900, heavy metal content following digestion in aqua regia after DIN 38414-S 7. Arsenic, lead, cadmium, chrome, copper and nickel were determined after DIN EN ISO 11885 (E 22), mercury after DIN EN 1483 (E 12).

RESULTS

Method of biomechanical treatment for residual waste

The biological treatment of residual waste can be carried out aerobically or anaerobically, whereby the latter method offers the possibility of producing biogas, a regenerative source of energy, from the waste's organic content. However, Germany's statutory requirements for landfill waste [4, 5] cannot be met by anaerobic treatment alone. Thus, once the transitional period has expired in 2005, further treatment will be necessary, including waste treated in existence anaerobic treatment plants, either through further, aerobic biological treatment or through thermal treatment. Anaerobic methods of treatment are used only to a limited extent, and even then usually in combination with an aerobic stage of biological treatment. Thus, what follows deals mainly with aerobic methods of biological treatment.

Biomechanical methods of waste treatment for residual waste can be divided into two principally different kinds. Classic biomechanical waste treatment is designed to prepare as large a proportion of the waste as possible for landfill, while modern, material-flow oriented methods of biomechanical waste treatment are designed to make material or energetic use of as much of the waste as possible.

Since, as a rule, even simple methods of biomechanical waste treatment produce at least a ferric metal fraction and possibly also a calorifically valuable fraction, the basic units of most biomechanical treatment concepts are similar (see Figure 1). In a mechanical pre-treatment stage the waste is coarsely reduced in particle size, broken up in preparation for biological degradation, and thereby partially homogenised. Biological treatment follows, followed by further mechanical treatment, by means of which all the waste, or a particular fraction, is conditioned in preparation for subsequent utilisation or disposal (incineration or landfill).

Material flow oriented biomechanical waste treatment, illustrated by the Rügen plant

In the material-flow oriented treatment of residual waste, efforts are focused on producing waste fractions which can be utilised for their energetic or material value. In the particular form of biomechanical stabilisation used at the facility on Rügen since 1999, biological treatment is shortened to just one week and serves merely to dry the waste. The dried waste is thus rendered more resistant to storage and has better separating properties, which are important for subsequent mechanical processing to specific material fractions.

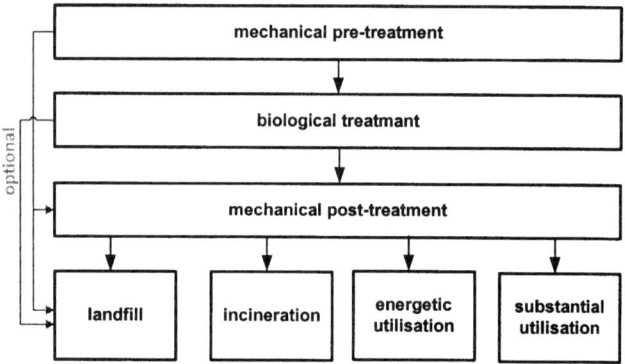

Figure 1. Areas of a mechanical-biological treatment of residual waste treatment.

The heart of the system consists of a number of special containers (see Figure 2) for the biological treatment of the waste, which because of their modular nature can be expanded or reduced as needs require. The process is illustrated in the form of flow diagrams in Figures 3 and 4. Initially the incoming waste is received in a covered area, where it is roughly sorted, and large unwanted or interring components removed. Subsequently the waste is reduced in particle size (< 300 mm) by a machine which tears large aggregates of material apart, opens bags and homogenises the waste in preparation for further processing.

Following comminution, the waste is filled into the treatment containers, which are then hermetically sealed to contain the gases produced, which are collected and subsequently cleaned. Currently, the waste air is cleaned using a biofilter. When the transitional period allowed by the Federal Emissions Act [6] for existing plant expires this will be replaced by a thermal-regenerative afterburning plant for offgas.

What occurs during waste stabilisation is similar to what occurs during composting, but with different flow volumes and process control. An air flow carries oxygen (atmospheric air) into the system, where it is used by microorganisms to oxidise organic waste contents and results in heat production. The waste is heated up, just as a pile of compost is, and this causes large amounts of water to enter the vapour phase, which is conducted away in the air flow. In this manner the waste is dried.

Following biological treatment (drying) the stabilised waste is transferred to a bunker, where any remaining interfering components are removed. The waste is then transported via conveyer belt to a drum screen which separates it into a > 50 mm fraction and a < 50 mm fraction. Ferrous and non-ferrous metals are subsequently removed from the < 50 mm fraction. The fine grain can then be optimally disposed of as waste or reunited with the high-calorific > 50 mm fraction. Such mixing however reduces calorific value and quality of this fraction as a secondary fuel.

Figure 2. Modular, and thus expandable, system of containers for the mechanical-biological treatment of waste (System provider: Nehlsen AG, Bremen, Germany)

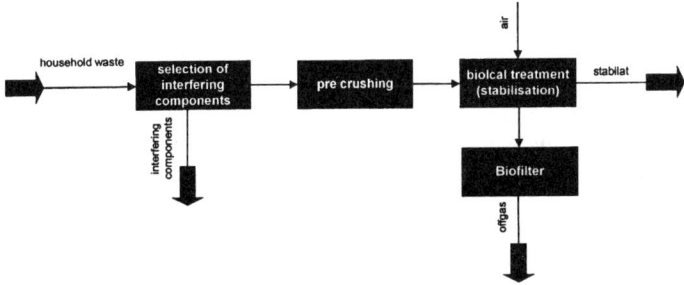

Figure 3. Basic flow diagram of preliminary mechanical treatment and subsequent biological treatment of waste

The > 50 mm fraction is sieved again using a ballistic separator (< 50 mm) and also separated into a heavy and light fraction. Components that cannot be reduced in particle size and those containing harmful or interfering substances are removed in the heavy fraction. Ferrous and non-ferrous metals are removed from the light fraction, which is subsequently reduced to < 50 mm. In order to comply with the demand that secondary fuel contain as low a metal content as possible, following this comminution stage, ferrous metals are again removed.

The secondary fuel fraction is subsequently pressed and loaded into transport containers, which take it about 150 km for use at a cement works. The ferrous metals separated from the waste are also used. The non-ferrous metals are so contaminated with other materials, as composites etc., that the aluminium content is too low to allow its economic recovery. This would require an additional mechanical processing stage in order to increase aluminium content. For the time being this fraction is disposed of by landfill.

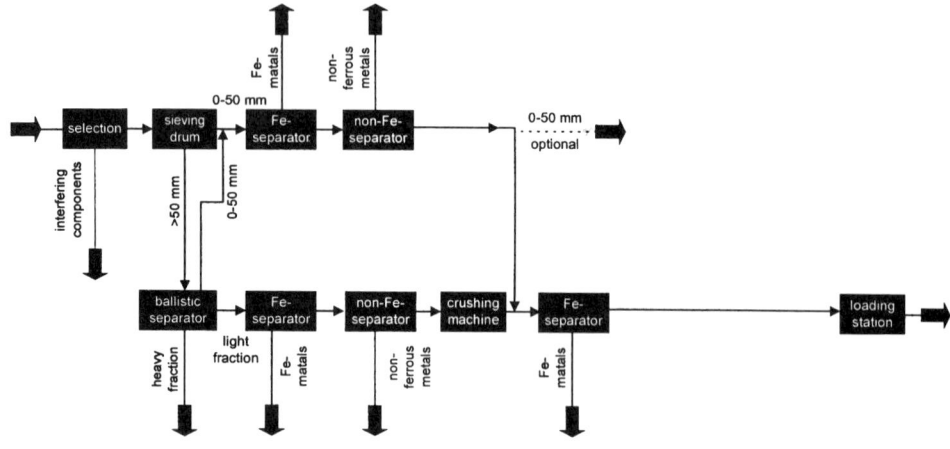

Figure 4. Basic flow diagram for the preliminary mechanical treatment and the biological treatment stage of residual waste – situation in 1999

The original concept for biomechanical waste stabilisation on Rügen had the main purpose of drying the waste in order to increase its calorific value and improve its processing properties. As large a proportion as possible was to be used as a secondary fuel. In the contracts concluded with Rügen's municipal authorities it was agreed that no more than 30% of the incoming domestic and domestic-like waste from trade and industry would be disposed of by landfill or incineration.

Plants for material-flow oriented waste treatment should be designed to be as flexible as possible and readily capable of expansion. This is because the demands placed on output flows can change, and for the sake of optimising the economics of plant operation by opening up possible new markets for the fractions produced.

In the case of Rügen, within just two years, there have been major changes to the market situation, which required rapid responses. The requirements for the secondary fuel changed to the effect that it had to have a higher calorific value and density. In addition, an opportunity arose to make use of the fine grain fraction (< 10 mm) as an intermediate covering layer for a landfill. The plant had to be correspondingly adapted to cope with these new demands.

Figure 5 shows the original basic flow diagram with the necessary extensions added in light blue. The first alteration, which, however, became necessary because of technical problems, was in the drum screen section. Insufficient comminution of the waste prior to stabilisation had resulted in the drum screen becoming clogged, with subsequent operational problems here and with machinery and conveyers further down the line. In consequence, an extra separator place was installed for the > 200 mm fraction, followed by another comminution machine which reduced particle size to < 80 mm. Also, the second screening gauge was reduced from 50 mm to 40 mm.

In order to produce a < 10 mm fraction for use (until 2005) as an intermediate landfill covering layer, a gyratory riddle was installed into the processing line for undersize material (now 0 – 40 mm). The 10 – 40 mm fraction is reunited with the processed oversize material (40 – 200 mm). In the process line for oversize material (40 – 200 mm), an air separator was added behind the last ferrous metal removal point. Here a heavy fraction is separated out for landfill disposal in order to increase the quality of the secondary fuel fraction. A pelletizing press was also installed, not to produce pellets, but merely to increase the density of the secondary fuel to approximately 0.3 kg/l.

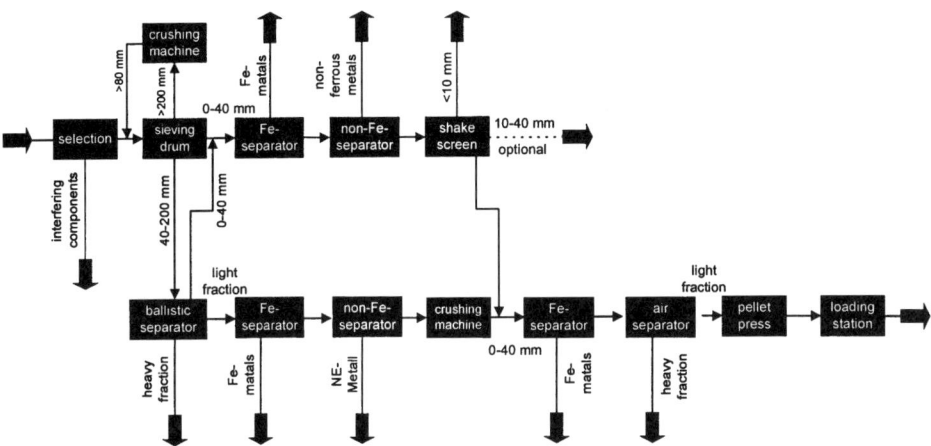

Figure 5. Basic flow diagram for preliminary mechanical treatment and the biological treatment stage – situation in 2002.

Material flow balance

Figure 6 shows the material flow balance for the Rügen plant. From the domestic and domestic-like waste from trade and industry delivered to the plant a sorting clamshell removes 2.2% as gross interfering components (carpets, mattresses, large batteries, large pieces of metal etc.). At 6.2% the proportion of ferrous metal recovered was relatively high, which, however, is largely explained by the structure of the district of Rügen, which, in addition to its c. 75,000 permanent inhabitants, during the Summer months counts approximately 5 million overnight stopovers by tourists, which increases the island's effective Summer population by up to 50%. The high ferrous metal content of the municipal waste is a consequence of changed consumer behaviour on behalf of the tourists.

The < 10 mm fine grain fraction is also relatively high at 21%. This is explained in part by the waste's increased ash content during the Winter months from coal and wood fired stoves that are still in common use. This material, however, gives an earthy impression and is used as an intermediate covering layer at a landfill.

The biggest fraction comprising the material flow balance is the secondary fuel fraction, at 34.2%, which is used in a gasification plant together with coal to produce synthesis gas, which in turn is used in a cement works to fire the calcination furnaces. Biological treatment results in the removal of water and some CO_2, through biological mineralisation of organic waste components, and a 27 percent reduction in weight. Besides the interfering components and the unusable waste associated with non-ferrous metals that are removed, there is also the heavy material fraction from the air separators, which must be disposed of by landfill. The amount of waste requiring landfill in 2002 could thus be reduced to just 11.6 percent of the waste delivered to the plant. This residual waste was deposited in a landfill for municipal waste

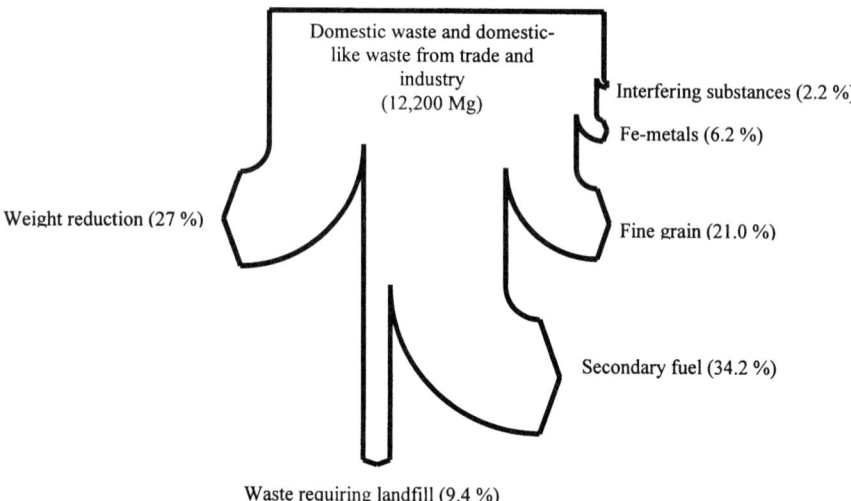

Figure 6. Material flow balance for biomechanical waste treatment, Rügen, Germany.

Quality of secondary fuel

Quality and marketability of the output streams produced are decisive for the successful operation of the plant. Because of the amounts involved and the quality demands placed on it, the secondary fuel fraction is of particular importance.

Figure 7 shows the calorific value of the secondary fuel fraction as determined in the last 13 declaration analyses. The values can be seen to fluctuate considerably, a consequence of the inhomogeneity of the incoming waste. However, all the analyses show a calorific value well above 11,000 kJ/kg, which is the minimum value for secondary fuel laid down by German law [7]. In 2002, mean calorific value of the secondary fuel fraction was approximately 17,000 kJ/kg.

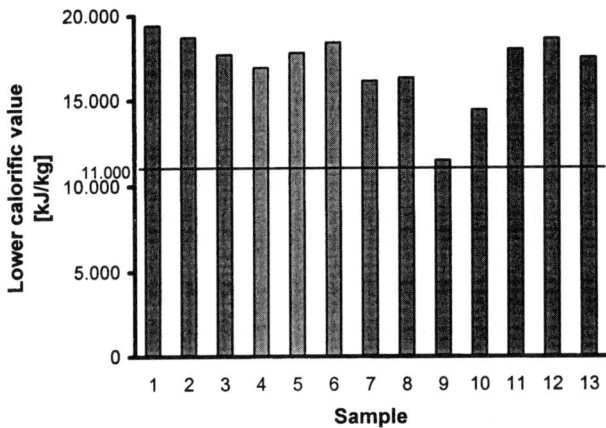

Figure 7. Lower calorific value of secondary fuel fraction
(declaration analyses, 1999 – 2002, per 1000 Mg).

Of decisive importance for its use as a secondary fuel, apart from its calorific value, is the fraction content of harmful substances. Thus, a further goal of biomechanical waste treatment is to reduce its content of pollutants. Figure 8 shows by way of example the fluctuation in copper content in the secondary fuel fraction. Although the reduction in copper was generally good, i ts c oncentration c an f luctuate widely d ue t o i nhomogeneity o f t he i ncoming w aste. The copper concentrations reported in some of the declaration analyses are close to the permitted limit, meaning that further efforts must be made to reduce copper content of the fraction.

For other pollutants, such as chlorine and sulphur (see Figure 9) the picture is more uniform. Most importantly, the concentrations are well below the limits laid down for the particular plant was the secondary fuel fraction is used. Since, for reasons of cost and because the plant making use of it may not always be available, thus making it desirable to be in a position to supply other plants, which may well have stricter limits for pollutants, it is necessary to develop reduction strategies for these pollutants too. These may involve both technical measures and/or measures relating to the whole field of waste management.

Figure 10 shows the degree to which the limits are approached for a number of pollutants at the plant making use of the secondary fuel fraction. It can be seen that for organic pollutants such as PCB's and PCP's, but also for chlorine, sulphur and a number of heavy metals, the limits are so far from being reached that there is no need for any action to be taken. However, for a number of heavy metals, in particular for mercury, chrome, lead and copper, the mean concentrations are between 40 and 60 percent of the specified limits. Because of fluctuations resulting from inhomgeneity of the incoming waste, as shown for the example of copper, it is possible t hat i solated a nalyses m ay w ell e xceed t he l imit. T hus, a f urther r eduction i n t he concentrations of these substances is necessary.

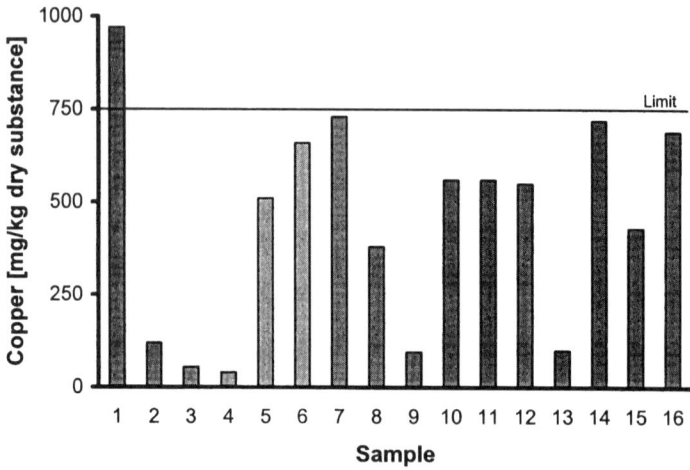

Figure 8. Copper content of secondary fuel fraction (declaration analyses 1999 – 2002).

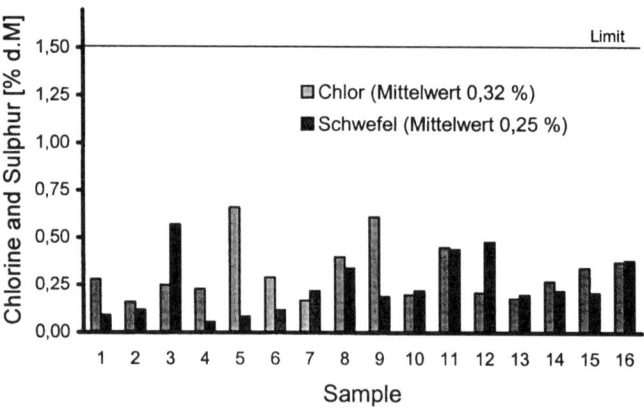

Figure 9. Chlorine and sulphur content of secondary fuel fraction (declaration analyses 1999 – 2002).

DISCUSSION

Utilisation before disposal – perspectives for material-flow oriented mechanical-biological treatment of municipal waste

Material-flow oriented mechanical-biological waste stabilisation serves as a form of preliminarily waste treatment prior to its utilisation or disposal. Hereby the raw waste is shredded and stabilised biologically (dried), before having interfering materials and metals, including pollutants (e.g. heavy metals) removed and being shredded further so as to make it suitable for use as a substitute fuel. Depending on the selected sieve gauge, calorific values of between 13,000 and 18,000 kJ/kg can be produced. The secondary fuel fraction can be further processed i f d esired, f or e xample, m ade i nto p ellets. In p rinciple, additional m aterial f low fractions can be produced, such as inert material, wood, glass, plastics, fine grain, etc., provided there is a market for them and economic conditions permit.

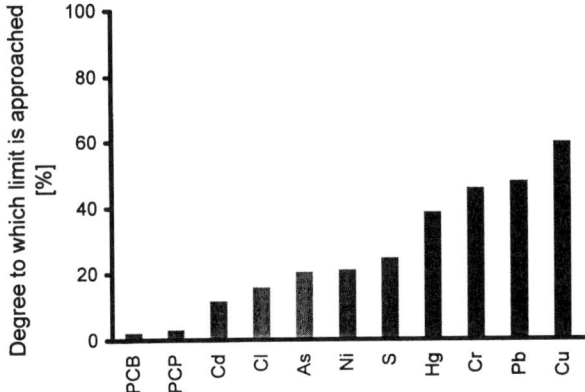

Figure 10. Degree to which limits are approached for particular pollutants at the plant using the secondary fuel fraction

Plants offering material-flow oriented, mechanical-biological waste treatment thus offer development potential and can be adapted to changing market conditions. In the concept current being pursued at the plant on Rügen, just metals, a secondary fuel fraction and a fine grain fraction a re p roduced for u tilisation. U sing the energy content of substitute fuel c an serve to conserve fossil fuels and reduce net emissions of greenhouse gases. Waste transport can be reduced through decentralised waste treatment.

The example of mechanical-biological waste treatment on Rügen shows that its application to waste stabilisation reduces waste disposal costs compared with earlier practice, while at the same time reducing the amount of waste requiring landfill to about 12% of what it would otherwise be.

Mechanical-biological waste treatment as an element of decentralised waste management concepts
Where municipal waste is disposed of in landfills, waste management is traditionally decentralised in structure. Towns and rural districts, as the authorities responsible for waste management, own landfill sites, in which waste, usually from the surrounding area, is deposited.

It is only since the changes made possible to the structure of waste management through implementation of the Technical Regulations for Municipal Waste [4] that Germany's traditionally decentralised waste management structure, at least in rural areas, has been questioned. This had to do largely with the assumption that in future thermal treatment alone could provide both an ecological and an economical means of waste disposal.

At present, waste incineration plants can only be operated economically on a large scale, with capacities of 100,000 Mg/a or more. Due to the technical demands placed on such waste incineration plants, plans for a change to centralised waste management strategies have also been widely discussed for rural areas. Many traditionally decentralised and self administered municipal and district authorities responsible for waste management in rural areas were left feeling rather uncomfortable at the prospect.

Centralised treatment plants (thermal and mechanical-biological), because of the economies of scale, initially offer the lowest specific costs for waste treatment. A disadvantage of large plants, however, has proved to be the difficulties involved in operating them at full, and thus at their most economical, capacity. There are a number of recent examples of such plants operating well below capacity, thus creating specific treatment costs that are much higher than those originally planned.

Another disadvantage of centralised waste management, particularly in rural areas, is the long distances that the waste has to be transported to the treatment plant. This has negative consequences for the environment, for the roads and for costs. Exactly how high the transport costs are can only be determined for each individual case, taking into account waste volume, transport distances and other parameters. A simple transhipment of waste, however, will incur costs of at least 5 to 10 €/Mg. Then transport from the point of transhipment to the treatment plant will add another 5 to 10 €/Mg to costs, producing total transport costs of at least 10 to 20 €/Mg. The costs of transport and handling can considerably reduce or even nullify the potential cost savings of a large-scale, centralised treatment plant.

Decentralised waste management concepts are more in tune with traditional, established structures, particularly in rural areas. Compared with centralised concepts, they offer a number of advantages (Local wealth creation from the disposal of waste locally, under the jurisdiction of the district authorities, where it is collected; positive effects on local employment; possible use of existing waste management structures (personnel, plant...); greater freedom for local self-administration; better acceptance by the population; local conditions can be better taken into account for small plants than for large ones; lower investment costs for the individual plants and more clarity regarding risks; Avoidance of waste transport).

The method of material-flow oriented mechanical-biological waste stabilisation presented here can be operated economically on a scale of 50,000 – 60,000 Mg/a upwards. The modular system of biological stabilisation that is presented, depending on local conditions, can be operated economically in plants working at capacities beginning at about 10,000 Mg/a. This is shown by the example of mechanical-biological waste treatment on Rügen, which was designed for a capacity of 10,000 – 15,000 Mg/a and in 2002 processed just over 12,000 Mg.

REFERENCES

1.   SENIOR, Erik: Microbiology of landfill sites. CRC PressBoca Raton, Florida, USA, 1990.

2.   BILITEWSKI, B., HÄRDTLE, G. and K. MAREK, K.: Waste Management, Springer Verlag New York, 1995.

3. BRUNE, M., RAMKE, H. G., COLLINS, H. J. and HANERT, H. H.: Incrustation Processes in Drainage Systems of Sanitary Landfills. Third International Symposium, Sardinia, Italy, 1991.

4. Technical Regulations for Municipal Waste (TA-Siedlungsabfall): Third General Administrative Regulation relating to the Waste Management Act (Dritte Allgemeine Verwaltungsvorschrift zum Abfallgesetz). 1st edition, situation on 1 June 1993. Cologne: Bundesanzeiger Verlags-Ges. mbH, 1993.

5. Regulation on the environmentally friendly landfill of municipal waste (Verordnung über die umweltverträgliche Ablagerung von Siedlungsabfällen (Abfallablagerungsverordnung - AbfAblV), BGBl. I p. 305, 2001.

6. Thirtieth regulation on the implementation of the Federal Emissions Protection Act relating t o p lants f or t he b iological t reatment o f waste (Dreißigste V erordnung z ur Durchführung des Bundes-Immissionsschutzgesetzes - 30. BImSchV - Verordnung über Anlagen zur biologischen Behandlung von Abfällen), BGBl. I p. 305, 2001.

7. Law to promote waste recycling and ensure the environmentally friendly disposal of residual waste (Gesetz zur Förderung der Kreislaufwirtschaft und Sicherung der umweltverträglichen Beseitigung von Abfällen - Kreislaufwirtschafts- und Abfallgesetz - KrW-/AbfG), BGBl. III 2129-27-2, 1994.

HAZARDOUS MATERIALS IN DEMOLITION WASTE- HOW TO IDENTIFY THE MATERIALS AND PLAN AND CONDUCT THE DEMOLITION WORK

N Strufe

RAMBOLL A/S

Denmark

ABSTRACT. Optimum environmental and feasible recycling of Demolition Waste materials requires "clean" materials. This requires that during the demolition process a planned and concise sorting of the different materials be carried out, thereby preventing any mix of the waste with hazardous materials. Therefore, an initial identification of the presence of these materials is required. To estimate possibly quantities of hazardous materials being present a process based on the examination and establishment of facts can be followed, including the period in time of erecting the building, historical data on materials used, sampling and testing. The identification of hazardous materials affects the economic implications of the planning and sorting of the materials, and the technological possibilities (limitations) in carrying out the work. In this article examples are given on the hazardous substance PCBs in Open Uses found in buildings.

Keywords. Demolition waste, Hazardous materials, Sustainable recycling, Investigation, Historical data, Planning and conducting, Selective demolition, Sampling and testing, Sorting.

Mr Niels Strufe has a background as a Civil Engineer and holds a M.A. in Area Studies on Africa from the University of Copenhagen. He is specialised in Project Development, Assessment, Analyses and Studies, as well as Conducting of Selective Demolition, and Recycling of Construction and Demolition Waste projects. Mr. Strufe has also extensive global experience from Management in Post-disaster Reconstruction and Recovery programmes, Emergency Protection of Building Structures and Cultural Heritage, Damage Assessment, as well as experience with Management of Hazardous Waste in general.

INTRODUCTION

The only way to enable optimum environmental and feasible recycling of Demolition Waste materials is by ensuring "clean" materials. Mixed demolition waste requires special treatment and costly disposal of unnecessarily large quantities of materials. A pre-condition for "clean" demolition waste is careful sorting of the waste at source as an integrated process of the demolition work.

This requires that during the demolition process a planned and concise sorting of the different materials be carried out, thereby preventing any mix of the waste with hazardous materials, including materials such as asbestos, lead, mercury, PVCs, solvents and adhesives containing PCBs, etc. Therefore, an initial identification of the presence of these materials is required.

Experiences show that the only way to identify the presence of hazardous materials is through thorough investigation of the historical data on the building or structure, followed by sampling and testing of the material in question and laboratory analyses. To estimate possibly quantities of hazardous materials being present it is recommended to follow a process based on the examination and establishment of facts, including the period in time of erecting the building, historical data on materials used as well as materials sampling and testing.

CONSTRUCTION AND DEMOLITION WASTE (C&DW)

Most C&DW had traditionally been landfilled, frequently in the same landfills as used to dispose of municipal solid waste (MSW). Furthermore, the volume of C&DW, most of which is inert, is roughly equal to that of MSW. Given the increasing scarcity of landfill space, and the increasing costs of improved environmental protection involved in modern landfill engineering and management, as well as the landfill directive of EU (Directive on landfill of waste, 99/31/EEC) has resulted in a significant change in the volume and nature of waste accepted at Europe's landfill sites, as well as it has led to improvements in design and operating standards, also in the aftercare of new and existing landfills. It is obvious that action to re-use or recycle C&DW would reduce the proportion going to landfill, thereby relieving the pressures on MSW disposal as well as respecting the hierarchy of waste management practices set out in the Framework Directive on waste (75/442/EEC as amended by 91/156/EEC) and the Fifth Environmental Action Programme.

A study from the European Union (Symonds, 1999) shows that, in the EU as a whole, arising of so called 'core' C&DW alone amount to around 180 million tonnes each year. This is over 480kg per person per year. Adding construction waste, road planings and excavated soil and rock to this figure more than doubles the total weight and volume of C&D material to be managed.

Despite the factors mentioned in this article, which might affect the composition of C&DW in future, it is clear that the inert (or decontaminated) fraction which is suitable for crushing and recycling as aggregate will continue to be the largest component, typically concrete, bricks/tiles/clay materials, and asphalt. Today recycling of C&DW is a well-established and profitable business in many countries within the EU and abroad. However, recent findings during the last years have discovered that C&DW might contain harmful substances that are unwanted in the recycling process. Also, specific waste directives and legislation strictly regulate the handling and disposal requirements for many of these substances.

HAZARDOUS SUBSTANCES IN C&DW

Hazardous materials and substances used in buildings and constructions can be selected from a long list of undisireable substances made by e.g. the European Commission, the United Nations, various national protection agencies and so forth. The harmful materials and substances to be found in buildings and structures shown in table 1 is a good example.

Table 1. Example of identified building components and materials
with harmful substances (the list is not complete).

SUBSTANCE	BUILDING COMPONENTS AND/OR MATERIALS
Asbestos	Roofs and tiles Glue Sound deadening sealing Fire resistant sealing Wall plaster
PVC	Gutters and pipes
Lead	Roofs and tiles Electrical cables
Cadmium	Plastic (e.g. cable, pipes and plates etc.) Occurring with zinc Occurring with concrete
Mercury	Fluorescent tube Switches and relays (Electrical installations) Others (e.g. concrete)
Nickel	Stainless steel Surface treatment
Chromium	Stainless steel Others (e.g. painted surfaces)
Copper	Cables and wires: - Permanent installations - Temporary installations Roofs, pipes etc. Screws, locks etc. Pigments and dyes
Zinc	Gutters/pipes and galvanised products Plastic (especially gutters and pipes)
PCB	Small capacitors and electric installations Double-glazed windows (glue) Sealant (softener) Paint (pigments) Fire resistant additive (paint, glue/binder)
Chlorinated paraffins	Plastic (in general) Sealant (softener) Others (e.g. glue)
CFCs	Thermal insulation – PUR foam Other insulation materials
HCFCs and HCFs	Thermal insulation – PUR foam Foam for joints
Sulphur hexafloride	Soundproof windows

Common for the above listed is their toxicity either to the environment and/or on the human health. It should be noticed that combinations and mixtures of above materials might also be found. In the EU Council Directive 91/689/EEC establishes a list of criteria (in its Annex III) to be used when the hazardousness of wastes is being determined. A list of hazardous wastes has been published as Council Decision 94/904/EEC (the hazardous waste list) including later amendments.

Only a few materials which may be classified as C&DW are invariably hazardous as defined in Directive 91/689/EEC or Decision 94/904/EEC. One of the most obvious examples of this small group, and certainly the one which is most frequently cited, is asbestos-based insulation. Other materials may be hazardous because they display one or more of the characteristics used in the Directive's Annex III to define hazardousness (such as toxicity or flammability). These characteristics may only be revealed under specific circumstances, and it may be possible to avoid those circumstances. Also, some other waste materials, which are typically found in relatively small amounts in C&DW (such as paint and plastics), although not necessarily hazardous, are not inert either. For the sake of the much larger inert fraction, such materials should be kept separate from the inert fraction if at all possible. If they are not, it may not be possible to treat the main bulk of the materials as inert, or in the worst cases, the whole of the bulk has to be treated as hazardous waste, with the derived cost implications.

One of the keys to maximising the yield of C&DW-derived aggregates and recycling is separation of materials at source, predominantly through selective demolition and good management of construction sites. The goal must be to minimise the waste quantities requiring special treatment and disposal, leading to optimum recycling and environmentally friendly disposal of building waste materials.

It is therefore necessary to plan and conduct demolition work as selective demolition. This is in principal carried out as the reverse of the construction process and follows the following principal procedures as shown in table 2:

Table 2. Principle procedures (phases) in Selective Demolition

PRINCIPAL PHASES IN SELECTIVE DEMOLITION
Removal of remains and non-fixtures (e.g. furniture, machines, etc.).
Stripping, comprising internal clearing, removal of doors, windows, roof components, installations, water, heating and electric installations and so forth. This leaves only the bearing and main structures of the building in place.
Demolition of the bearing and main structures.

All three phases shown in table 2 can involve handling of hazardous materials. At the demolition site there are always likely to be some materials (such as asbestos and mercury in electric equipment), which are hazardous in their own right. Residues of hazardous substances manufactured, used or stored at the (e.g. industrial) site may also remain. Where possible these should be removed from the site prior to demolition activities commencing or separated out during the selective demolition.

If they have become impregnated into the fabric of the building it may be possible to neutralise or treat them in situ prior to demolition. Table 3 gives the (minimum) components of a demolition plan in relation to hazardous materials.

Table 3. Components of a Demolition Plan

A DEMOLITION PLAN SHOULD AS MINIMUM CONTAIN DETAILS ON
A   The amount of material(s) containing hazardous substances,
B   The amount of material(s) containing hazardous substances per building (if more than one!),
C   The use of the building (e.g. private housing, industry etc.),
D   Time plan for the work,
E   Cost analysis of the work including the waste management.

However, first step is to define whether hazardous materials exist in the building, where, and what kind of substances as this directly affects the handling techniques and disposal options available.

IDENTIFICATION OF HAZARDOUS MATERIALS, EXAMPLE ON PCB

In the following Table, an example on identifying the presence of a particular hazardous substance, PCBs in Open Use (also called 'Open-ended' applications), found in buildings and the processes of demolition and handling of the hazardous materials.

Table 4. Definition of PCBs in Open uses

DEFINITION OF PCBS IN OPEN USES
Open systems (open uses) are applications in which the PCBs are in direct contact with their surroundings and thereby may be easily transferred to the environment. Direct PCB contact with the environment is of greater concern for open uses than it is for closed applications. Generally, closed and partially closed systems contain PCB oils or fluids. The PCBs in open systems typically take on the form (type of media) of the product in which they have been used as an ingredient. Therefore, PCBs in open uses may be found as constituents in products ranging from paint, plastic, and rubber, impregnated wood, concrete, plaster, and so on.
The typical classification system for PCB is based on *classification by concentration*, 0.005 % (50 ppm) of PCBs, is the ultimate threshold specified in the Stockholm Convention.

PCBs have been used for a great variety of applications in open use, for example: varnishes, waxes, synthetic resins, epoxy and marine paints, coatings, flame retardant, etc. A rule of thumb that could be to take into consideration is the technical qualities in relation to wearing capabilities and weather quality. PCBs in open uses have often been used in wet rooms (walls and flooring), where high wearing quality was required (e.g. stairs), and on surfaces exposed to harsh weather conditions. Concerning the identification of materials containing PCBs in open uses, one major problem is the missing data and information relating to the material. Electrical equipment (e.g. capacitors) containing PCBs in Closed Uses is typically marked or labelled w ith d ata a nd i nformation (serial a nd p roduction l ine n umbers, d ates a nd s pecific contain etc.) allowing determination of whether or not the component contains PCBs.

PCB-containing materials and components present in buildings are typically not marked with production d ata. E xamples i nclude p aint a nd p laster a s well a s e lastic f illers, i mpregnated wood, flooring materials and adhesives.

Here the only way to identify the *possible presence* of PCBs is through investigation of the historical data on the building or structure, followed by sampling and testing of the product of suspicion by technical analyses. The only way to accurately identify the presence and content of PCBs is to take samples and analyse them. The use of field kit sets is an option but can only provide confirmation that a material does *not* contain PCBs. To *verify* the presence of PCBs and accurately identify the specific content (measured in ppm) only analytical laboratory testing in an accredited laboratory can be recommended.

The PCB concentration levels are also pertinent in determining the necessary disposal method. Most countries regulating PCBs require that products containing high concentrations be disposed of by specific methods, for example by incineration. PCBs have knowingly been used in building components since 1950 and until the end of 1980. The open use of PCBs was stopped in many countries during the early 1970s. Production of PCBs declined rapidly during t he 1 970s as u ses b ecame m ore a nd m ore r estricted, v irtually ceasing b y the early 1980s, and is believed to have finally ceased in the mid/end of the 1980s.

Buildings from the period 1956-1973 have a relatively high possibility of containing PCBs. Buildings constructed before this period possibly contain PCBs in building components added to those buildings during repair and renewal work during or before the period mentioned. Following the mid 1970'ties the open uses of PCBs declined significantly. However, in buildings electrical devices containing PCBs (capacitors), such as neon lamps etc., might have been installed until the end of 1980.

To enable an estimation of possible quantities of PCB-containing materials being present it is recommended to follow a process based on the examination and establishment of facts. This includes the following steps and identification of:

    I.   The time period of the building being erected and/or repaired/renewed
    II. The possibility of PCBs being present in the building
   III. Historical data available on the building and the materials used
   IV. Are PCBs present in the material of suspicion? (by use of sampling and testing)
    V. Verification of the presence or non-presence of PCBs, hereby enabling decision-making on disposal methods for the material in question.

The following Figure1 illustrate the identification process, and the principal process of demolition and monitoring and control mechanism by a public authority is shown in Figure 2.

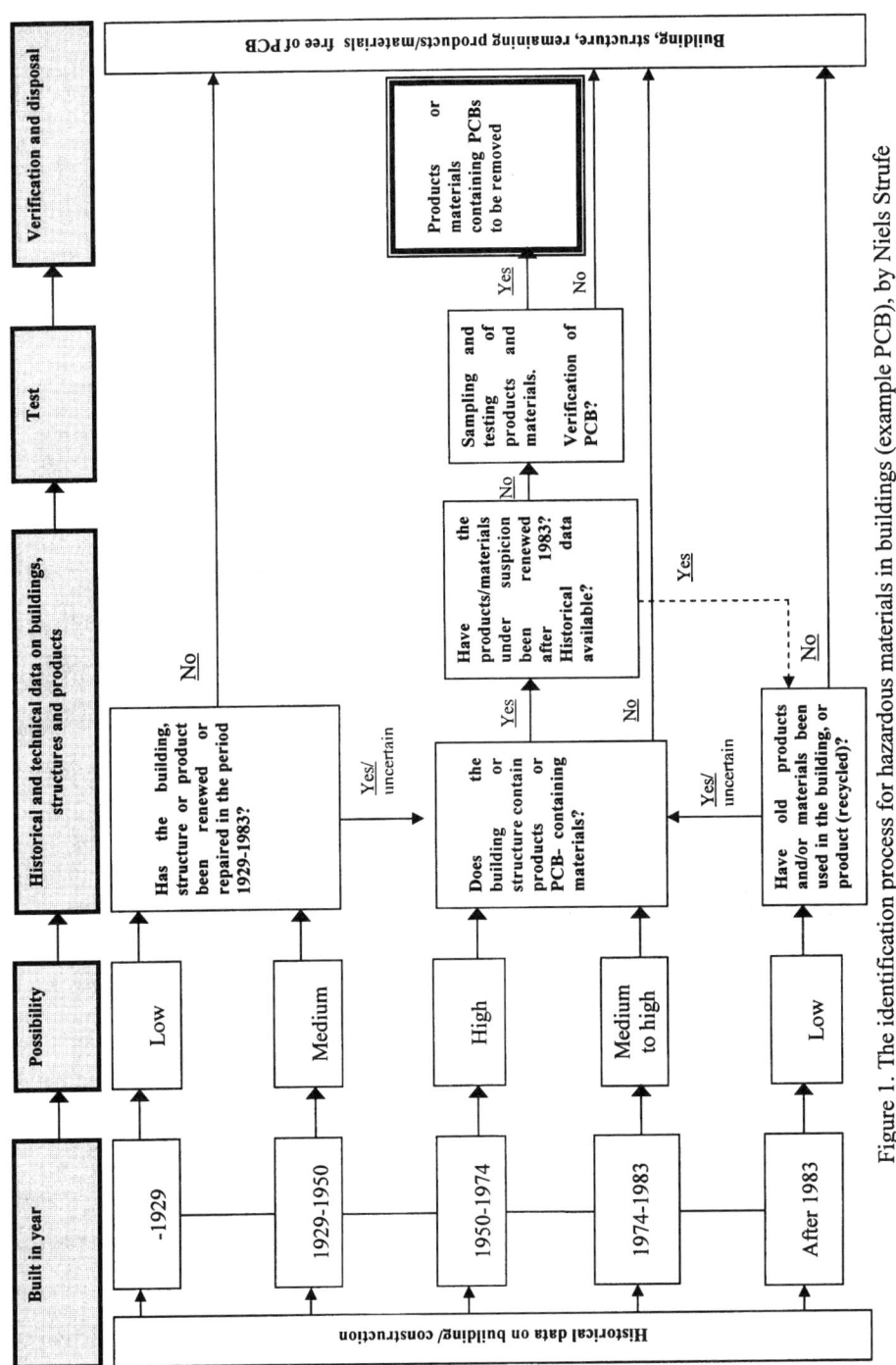

Figure 1. The identification process for hazardous materials in buildings (example PCB), by Niels Strufe

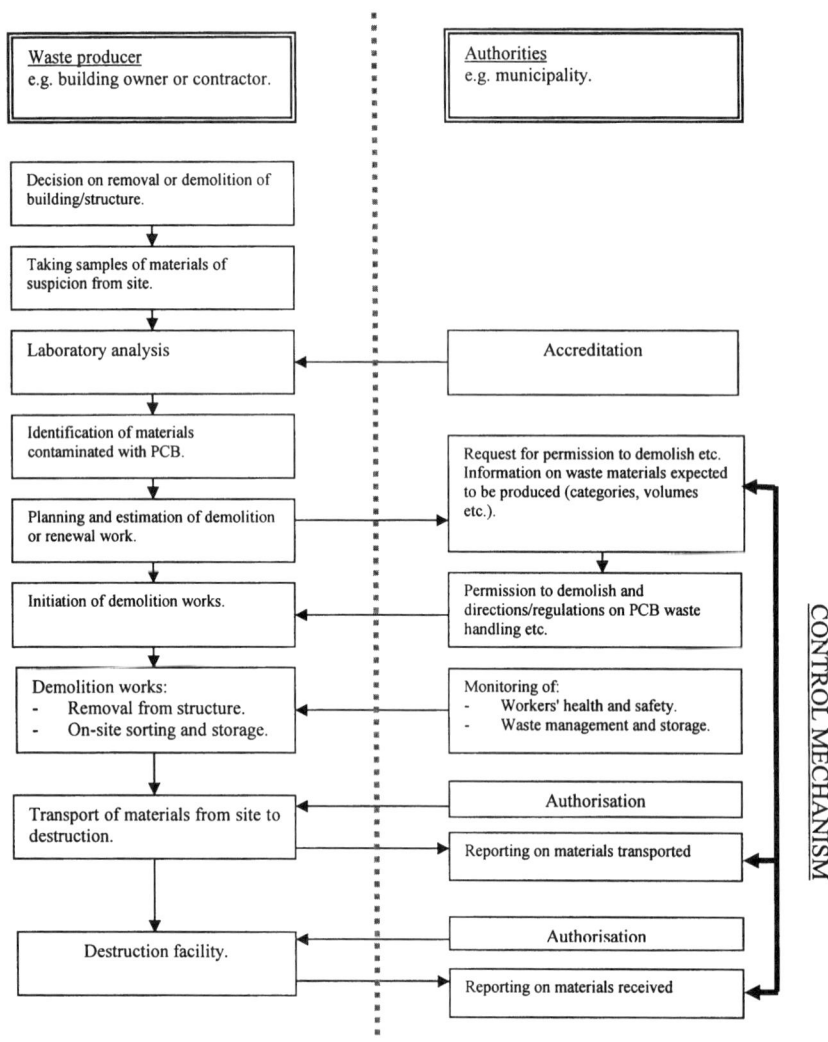

Figure 2. The process of demolition and monitoring and control mechanism, by Niels Strufe

ECONOMIC IMPLICATIONS

Control of hazardous wastes and others which should be collected separately ought to be easier on construction sites than on demolition sites (because the management's knowledge of the materials should be better, being under their control and of their choice). Economies of scale mean that organising and controlling hazardous materials should be easier on large construction sites than on small ones. On demolition sites, matters can be greatly helped if a proper pre-demolition survey is organised.

Owing to the additional work required in relation to Selective Demolition and sorting of the different categories of materials, the processes of demolition will necessarily become more timely and expensive. On the other hand "clean" material, for example for recycling, will have a high degree of purity and thereby lead to cost savings (income). These savings are obtained through a better quality of the recycled materials, no re-sorting (in the recycling process) and so forth. Savings can also be obtained through saved transport (e.g. by reuse/recycling on site) and reloading of vehicles, as well as by reduced disposal fees.

The decisions, which determine the style and detail of the demolition process, may be taken by the original owner of the site, or by a new owner (planning to erect a new structure), or by a contractor or consultant working directly or indirectly for either of these two parties.

As a whole this method results in an economic and environmental optimum, and maybe even profit, to the actual project. The selective demolition method therefore contains an economic incitement that, used in the right way, can work as a catalyst for implementing the method into the demolition industry.

Referring to the C&DW report from European Commission DG ENV (Symonds Group Ltd., 1999) the following general model can be used as a tool to determine selective demolition and separate handling whenever:

$$Vm(Tm + Dm) > V1(T1+R/D1- SV1)+V2(T2 + R/D2 - SV2)...Vn(Tn+R/Dn-SVn)+Es$$

Where: Vm = Volume of unsorted C&DW
Tm = Cost of transporting unsorted waste to disposal site
Dm = Cost of disposing of unsorted waste
$V1$ = Volume of inert waste
$T1$ = Cost of transporting inert waste to recycling or disposal site
$R/D1$ = Cost of recycling / disposing of inert waste (including quality control)
$SV1$ = Sale value of recycled product (if relevant)
$2 ...n$ = Other sorted waste streams
Es = Additional costs of separate demolition and materials sorting

The additional costs associated with separate demolition and materials sorting will include any additional costs created by time and/or space constraints, as well as costs for laboratory analysis of hazardous materials as required. On all sites the cost of labour, but more particularly the costs of machinery, will affect them. In order to try to compare the decision making process on different site types and in different Member States, the labour cost component needs to be expressed as EURO wage rate and time and productivity factors. The same thing needs to be done for fixed and variable machinery costs to facilitate the asking of a series of different 'what if?' questions.

HEALTH AND SAFETY ASPECTS

To assure that the selective demolition work in general can be performed with proper quality and minimal spreading of hazardous substances to the surroundings, and of course workers health and safety protection, following requirements should be taken in to consideration.

- The qualifications and experience by the contractor to be sufficient
- The equipment and technical know how available to be sufficient
- That the site and working platform(s) are designed in a proper manner
- That sufficient protective measures are taken to prevent spreading to surrounding environment and for workers health. This could include sealing of the area/room the work shall be done (quite similar to present methods of Asbestos removal).
- Quality assurance plans are made and include identification of critical periods and moments in the work as well as rolling quality control system and documentation.

The removal of materials containing hazardous materials beforehand of the demolition of the bearing structures of a building is considered not to be a problem of significance.

Static stability and safety during work might affect the method and use of tools as well as the line of processes during selective demolition. The importance is in the first place of course safety during work, but always having in mind the separation of different materials and avoiding mixing and "pollution" of materials and the surrounding environment. Generally, the methods and techniques in selective demolition ensure that this is not to be considered a problem.

This applies also for the bearing structures and the static stability of the building or structure in question. As long as the hazardous substances and contaminated materials have been identified prior to commence of work, and taken in to consideration during the planning of the work, this matter is considered not to be a particular problem.

Dealing with hazardous and health risk substances, it is recommended to always initially establish contact to the relevant local authorities before work commences. One reason is that local regulations and/or conditions might exist.

CONCLUSIONS AND RECOMMENDATIONS

In the industrial world we have – and continue - to use large quantities of chemical substances and additives in all types of products, including substances that threatens our health. The use of these substances is typically grounded in order to enhance technical properties, e.g. extend a product's life, weather capabilities and so forth. Some of these substances have already been shown to have an impact on the environment and/or human health. Therefore the authorities have made regulations or bans to counter these impacts. A large number of hazardous materials and substances are known to be present in buildings and structures and that eventually they end up in C&D waste. With the aim of sustainable Recycling of C&D waste, the problem is to find and identify these substances, and then remove and dispose of the materials in a proper and safely manner, as well as avoiding that unwanted materials and substances enters in to the process and market of reuse and recycling of building materials.

The general conclusions are that the quantities of harmful substances in C&D waste is an issue already attaining increased focus and attention. It is also clear that the issue and the problems of sorting and handling hazardous materials in C&DW can be dealt with if a number of essential steps are taken.

A number of recommendations can be given as follows:

➢ Prioritisation and detailed assessment of the identified harmful substances.
It is recommended that effort be put into the development of detailed information about the quantities and streams of the hazardous substances in C&D waste in order to plan and prioritise future initiatives.

➢ Technical investigations.
It is recommended that technical investigations be performed to reveal the possibilities for (clean) removal and handling methods of building materials containing harmful substances to assure optimum health and environmental safe disposal and destruction.

➢ Information and future initiatives.
It is recommended to develop and provide information and education to the construction and demolition sector in relation to the handling of harmful substances in C&D waste. This should include the development and implementation of effective methods for the identification, removal, and handling of harmful substances.

Finally, when the technical foundation for a more comprehensive handling of C&D waste is established, it is recommended that guidelines and rules are formulated, e.g. as an addendum to local waste regulations, and thereby ensuring the effective and safe implementation of new or improved methods of C&D waste handling, and hereby assuring the Sustainable Recycling of C&D waste.

REFERENCES

1. BAYERISCHES LANDESAMT FÜR UMWELTSCHUTZ, *Kontaminierte Bausubstanz, Erkendung, Bewertung, Entsorgung,* BayLFU, 2003.

2. DEMEX CONSULTING ENGINEERS A/S, *Andre problematiske stoffer i bygge- og anlægsaffald - Kortlægning og prognose for affaldsstrømme samt undersøgelse af bortskaffelsesmuligheder, Fase 1,* DEMEX Consulting Engineers A/S, March 2002.

3. HELSINKI COMMISSION, *PCBs: A compilation of information, derived from HELCOM Recommendations, EU-Directives, UN-ECE-LRTAP, UNEP and OSPAR, and analysis of appropriate measures aiming at safe handling and reduction of releases of PCB from PCB-containing equipment in use,* Helsinki Commission, Baltic Marine Environment Protection Commission, May 2001.

4. BASEL CONVENTION, *Technical Guidelines on Wastes Comprising or Containing PCBs, PCTs and PBBs (Y10),* Basel Convention on the Control of Transboundary

Movements of Hazardous Wastes and Their Disposal Adopted by the Conference of the Plenipotentiaries, 22 March 1989, 1998.

5. SYMONDS, Report to DGXI, European Commission, *Construction and Demolition Waste Management Practices, and Their Economic Impacts*, Symonds, in association with ARGUS, COWI and PRC Bouwcentrum, February 1999.

6. ØKOBYGG, *Identifisering af PCB i norske bygg*, PCB-Guide, Økobygg, Norway, 2002.

7. UNEP, *Compilation of EU Dioxin Exposure and Health Data*, Summary Report, UNEP,Basel, October 1999.

8. SFT, *PCB i Bygg', SFT Fakta, nr. 1730/2000*, Statens Forurensningstilsyn, Oslo, 2000.

9. Web: U.S. Environmental Protection Agency PCB Home Page, www.epa.gov/opptintr/pcb

10. Web: PCB Web site, Sweden, www.sanerapcb.nu/

11. Web: PCB Web site, Norway, www.pcb.no

CHEMICAL MINERALOGICAL INVESTIGATION OF CONSTRUCTION AND DEMOLITION WASTE MATERIALS

E Marrocchino   **G Bianchini**

R Tassinari   **C Vaccaro**

University of Ferrara

Italy

ABSTRACT. EU Directives and Italian Legislation are encouraging recycling of construction and demolition waste materials, provided by continuous urban redevelopment to face the pressing landfill shortage (i.e., limited area for storing, with related increase of the storing costs), and to limit the quarrying activity of alluvial sediments that often induce environmental problems. Attention has been focused on chemical-mineralogical and technical (physical-mechanical) features of these materials, to provide recycling strategies, ecologically correct and economically advantageous. In this light, we have studied construction and demolition waste materials from a landfill of Ferrara, performing: a) careful grain-size sorting b) chemical (XRF) and mineralogical (XRD) analyses of the different grain-size fractions. This approach has allowed us to recognise particular grain-size fractions that can be profitably re-utilised by the building industry.

Keywords: Construction-demolition waste, Chemical-mineralogical characterisation, Recycling.

Dr E Marrocchino is a researcher fellow and Reader in the Earth Sciences Department of Ferrara University. Her research interests include petrographical–mineralogical and geochemical characterisation of natural sediments and their industrial applications.

Dr G Bianchini is a researcher fellow and Reader of "Rocks and Industrial Minerals" in the Earth Sciences Department of Ferrara University. His research interests include both petrology and geochemistry.

Dr R Tassinari is the lab technician responsible for the XRF and ICP-MS labs of the Earth Sciences Department of Ferrara University. His research interest is the analytical geochemistry of a wide range of geological and environmental materials.

Professor C Vaccaro is an Associate Professor in the Earth Sciences Department of Ferrara University, where she is involved in research activities concerning applied petrography and geochemistry.

INTRODUCTION

Research and development of new building materials re-utilising, construction and demolition waste debris (C&D [1]) is an important and complex issue requiring a multidisciplinary approach that involves analytical investigation of the chemical mineralogical characteristics and identification of the physical-mechanical properties as well as marketing evaluation of the C&D waste materials and their re-cycled compounds.

With regards to the chemical-mineralogical characterisation the coupled use of X-ray fluorescence (XRF) and X-ray diffraction (XRD) could provide a powerful tool to identify any predisposition of C&D waste materials to be employed as first-order material in the building activities.

More specifically, in this contribution we have studied samples from a landfill located in Ferrara (NE Italy), performing laboratory sieving and the subsequent characterisation of the different grain-size fractions.

SAMPLING AND ANALYTICAL METHODS

Sampling was performed in two different periods (February and September) to evaluate possible temporal variations in the composition of the deposited materials. In the investigated C&D plant the inert material was divided in two different masses:

- the first consisted of the original untreated raw material arriving at the landfill (called TQ), and characterised by extremely variable grain-size and composition (bricks and terracotta, concrete, asphalt) processed through a preliminary separation of undesired material (plastic, metal, wood, paper);
- the second consisted of material crushed and screened in three different grain size classes: 8-4 cm, 4-1 cm (hereafter named MD), < 1 cm (hereafter named FN).

In particular, during February two samples were collected from the first cumulus of material (TQ1-Feb, TQ2-Feb), and two from the masses that had been crushed and sorted (MD1-Feb, FN1-Feb); during September one sample was collected from the first cumulus (TQ3-Sept) and some others from the sorted fractions (MD2-Sept, MD3-Sept, FN2-Sept).

Each sample was subsequently grinded and sieved in laboratory, obtaining the following grain-size classes:
- >4 mm (coarse gravel),
- 4-2 mm (fine gravel),
- 2-0.6 mm (coarse sand),
- 0.6-0.125 mm (medium sand),
- 0.125-0.075 mm (fine sand),
- <0.075 mm.

The different grain-size fractions were investigated through X-Ray fluorescence (XRF) and X-Ray diffractometry (XRD) analyses to evaluate their chemical and mineralogical composition, providing data that should be considered in order to develop correct recycling strategies.

XRF chemical analysis was carried out (utilising a Philips PW1400 spectrometer) on the different grain-size fraction of each sample, to investigate their chemical composition in terms of major (SiO_2, TiO_2, Al_2O_3, Fe_2O_3, MnO, MgO, CaO, Na_2O, K_2O, P_2O_5 expressed in wt%) and trace element (Pb, Zn, Ni, Co, Cr, V, Th, expressed in parts per million) components. Accuracy and precision have been estimated (with the repeated analyses of certified standards) to be better than 10%.

XRD mineral investigation was performed with a Philips PW1860/00 diffractometer, using graphite-filtered Cu-Kα radiation (1.54 Å) to recognise the constituent mineralogical phases. Diffraction patterns were collected in the 2 θ angular range 5-50°, with a 5sec/step (0.02° 2θ̃)

DISCUSSION

In order to assess the best way of re-using the considered recycled inert materials, the different grain-size fractions of each sample have been analysed by both XRD and XRF methods. XRD analysis (Figure 1) shows that the separated C&D fractions are constituted by different amounts of quartz, calcite, dolomite, feldspar, muscovite/illite and chlorite, as well as calcium-aluminium-iron hydroxides and hydrous silicates (typical phases of cement materials) and minor amounts of gehlenite and wollastonite (typical phases of bricks and terracotta). Hazardous mineral phases such as amianthus (asbestos) have not been recorded. Even if this investigation does not permit a quantitative determination of the mineral abundance, the relative heights of the different peaks suggest that quartz is more abundant in the sand fractions, and phyllosilicates such as illite/muscovite and chlorite tend to increase in the fraction characterised by finer grain-size.

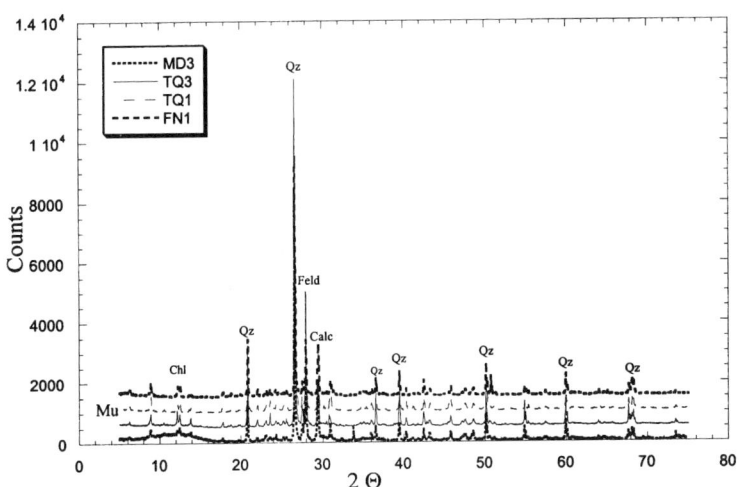

Figure 1. Selected XRD patterns of medium sand fractions (0,6-0,125 mm). Abbreviations: Qz = quartz; Feld = feldspar; Calc = calcite; Chl= Chlorite; Mu = muscovite/illite.

The separated fractions obtained by sieving were analysed by XRF, and the relative chemical data have been preliminarily treated by a multivariate cluster analysis method that allow to elaborate simultaneously a wide number of variables. This approach is useful to reduce the multi-dimensionality of the data to an imaginable and "plottable" two dimensional diagram, and to highlight similarities between different samples, thus providing a preliminary sample grouping. In particular, Euclidean distance-cluster analysis demonstrates that sorting and sieving leads to a progressive homogenisation (Figure 2).

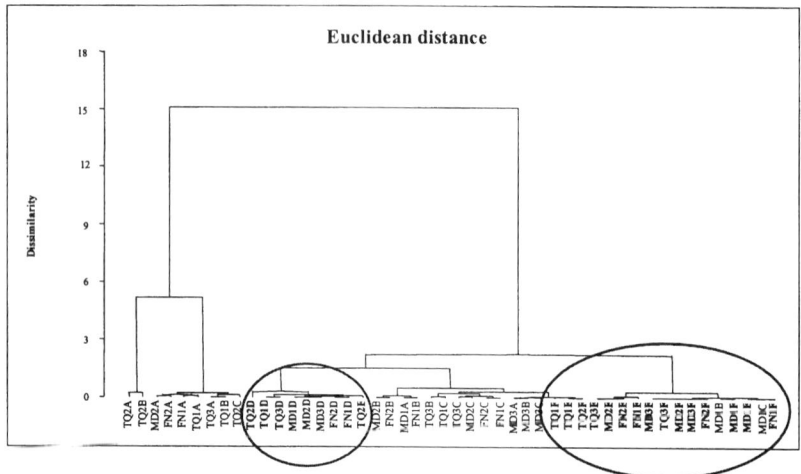

Figure 2. Dendrogram resulting from model based on hierarchical cluster analysis, using Euclidean agglomerative algorithm.

From Figure 2, it can be observed that "medium sands" (labels ending with D) form quite an homogeneous group. "Fine sands" and fraction <0.075mm (labels ending with E and F respectively) also seem to be homogeneous groups. On the other hand, bivariate diagrams, with SiO_2 concentration as variation index, are necessary to pinpoint chemical differences (in terms of major and trace components) among the different samples and the different grain size fractions. It can be observed that, in all the investigated samples, a peak of SiO_2 concentration characterises the medium sands (0.6 mm-0.125mm) fraction; on the contrary, this grain size fraction displays lower content in CaO, Fe_2O_3 and Al_2O_3 compared with the finer fractions (Figure 3). Abundance of transition metals such as Ni, Co, Cr, V, Zn, Pb tends to increase in the finer fractions, as they are plausibly trapped by clay minerals.

Considering the different grain-size fractions, it can be envisaged that the medium sands (0.6 mm-0.125mm) represent a homogeneous group, showing analogies with natural sediments of comparable grain-size. In particular, if compared with natural sands from the Ferrara area (authors data), these recycled inert materials are richer in CaO and poorer in Al_2O_3-K_2O. This simply means that the 0.6 mm-0.125mm fraction of our recycled inert samples contains more carbonate and/or more calcium-bearing aluminium-iron hydroxides and hydrous silicates, and less clay minerals than the natural sands of the area. The content of hazardous elements, such as the transition metals of these recycled inert samples, is comparable to or even lower than that recorded in the natural sands. These elements are not associated with meta-stable phases, as they are plausibly concentrated within the clay fraction. This means that these harmful elements cannot be easily leached and released in solution (this statement should be verified with proper leaching tests as proposed by [2, 3]). In this light, these recycled sands could be directly used in the preparation of new mortar and concrete.

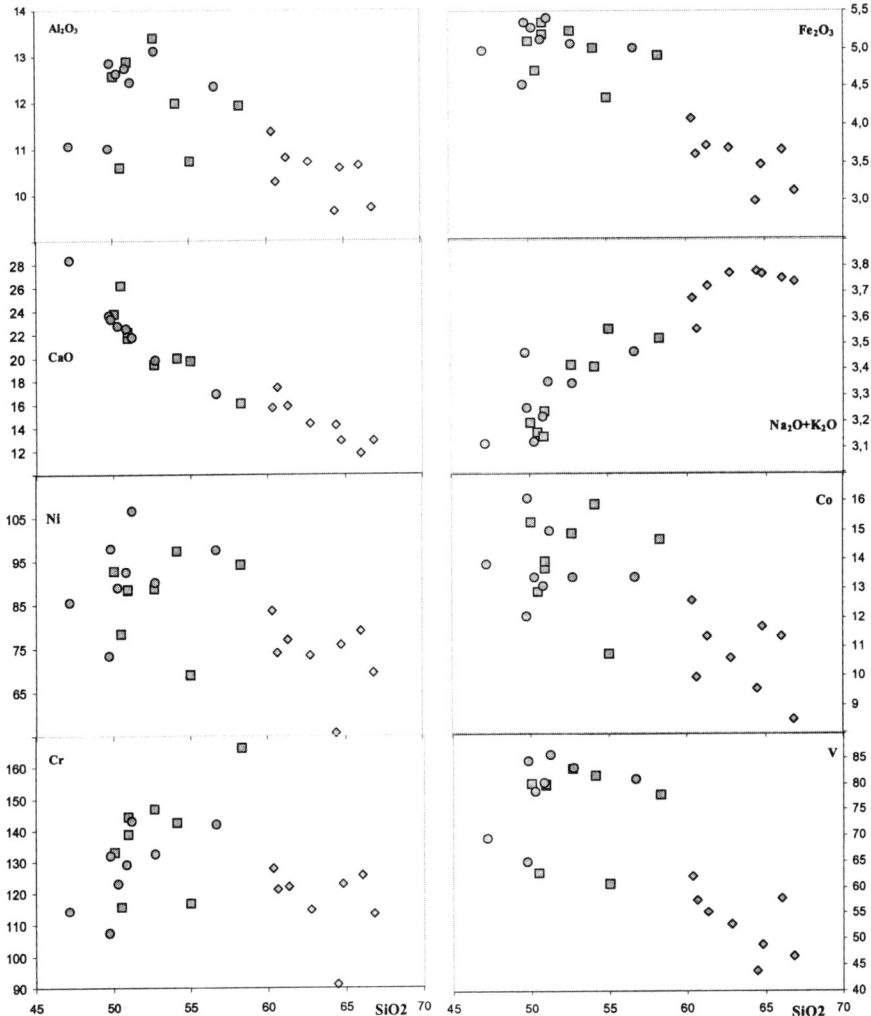

Figure 3. Binary diagrams of CaO, Al₂O₃, Fe₂O₃, (Na₂O+ K₂O), Ni, Cr, Co, V vs. SiO₂ .
Symbols: ◆ = recycled inert material with grain-size between 0,6-0,125 mm; ■ = recycled inert material with grain-size between 0,125-0,075 mm; ● = recycled inert material with grain-size < 0.075mm.

Chemical analyses of finer fractions (0.125mm-0.075mm; <0.075mm) are also quite homogeneous and, if calculated on anhydrous basis, reveal a comparative enrichment in Al₂O₃ with respect to the composition of the other grain-size classes.

Compositions, plotted in the Fe₂O₃-(Na₂O+K₂O)-(CaO+MgO) triangular diagram (Figure 4), do not overlap the terracotta compositional fields (Maiolica, Cottoforte, Gres), thus precluding their possible employment as starting materials for bricks and tiles preparation. However, we suggest their possible re-cycling as a raw material in cement preparation.

To test this hypothesis, chemical compositions have been reported in the phase diagram CaO-Al₂O₃-SiO₂, in which components that make up cements are easily represented (Figure 5).

It has to be noted that suitable composition for the cement preparation, enclosed in the sub-triangle C₃S, C₂S, C₃A [4], could be obtained by blending~ 40% of the considered recycled inert materials with~ 60% of lime. This cement prepared with high temperature calcinations (~ 1400 °C) of the fine C&D fraction could be employed to bind the coarser fractions in order to prepare moulded bricks and pavement blocks [5].

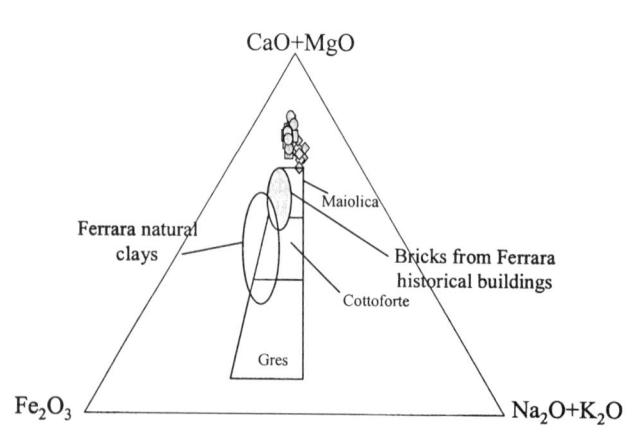

Figure 4. Fe2O3-(Na2O+K2O)-(CaO+MgO) triangular diagram.

Figure 4 reports compositions of the analysed recycled C&D materials as well as compositional fields of typical terracotta Symbols: ◆ = recycled inert material with grain-size between 0,6-0,125 mm; ■ = recycled inert material with grain-size between 0,125-0,075 mm; ● = recycled inert material with grain-size < 0.075mm.

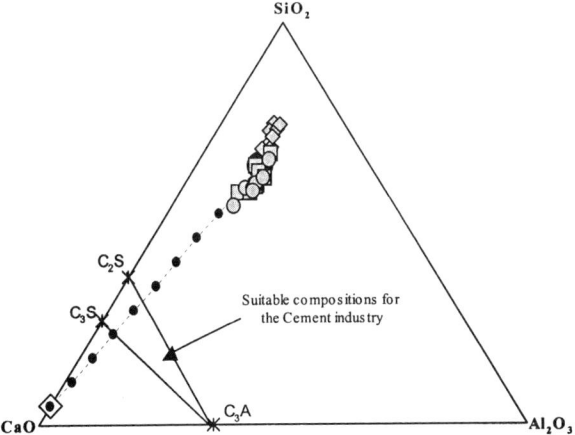

Figure 5. Ternary diagrams describing the system CaO-Al2O3-SiO2.

Figure 5 reports the composition of mineral phases typically recognised in cements: tricalcium silicate (alite; C3S), dicalcium silicate (belite; C2S) and tricalcium aluminate (C3A). In accordance with Manning (1995), raw material compositions required for cement preparation a re i ncluded i n t he s ub-triangle C 3S-C2S-C3A. S imilar c ompositions c an b e achieved by blending lime (~ 60%) with compositions comparable to those of the recycled inert fractions 0.125mm-0.075mm and <0.075mm. Symbols: ◆ = recycled inert material with grain-size between 0,6-0,125 mm; ■= recycled inert material with grain-size between 0,125-0,075 mm; ● = recycled inert material with grain-size < 0.075mm; asterisks=phases classically recognised in cements; white diamond = lime composition; dotted line= mixing calculation.

CONCLUSIONS

This study indicates that, through an accurate crushing and sorting of C&D material, it is possible to obtain grain-size fractions with roughly homogenous chemical and mineralogical composition.

In particular, chemical-mineralogical characterisation of different grain-size classes, obtained through laboratory sieving of C&D waste material, has allowed us to recognise a particular grain-size fraction (0.6mm-0.125mm: medium sands) that can be directly re-utilised in the preparations of new mortars and concrete items. We also propose a possible reuse of the finer fractions in cement preparation, after suitable mixing with lime and subsequent calcination process.

Therefore, similar investigations would provide background knowledge useful for further applications of these recycled materials similar to those proposed by [5-9], thus contributing t o m inimize t he exploitation o f n atural a ggregates, a nd t o face t he p ressing landfill shortage problem.

REFERENCES

1. POON, C.S., YU, A. T.W., NG, L.H. 2001 On-site sorting of construction and demolition waste in Hong Kong. Resources, Conservation and Recycling 32, 157-172.

2. TRÄNKLER J.O.V., WALKER I., DOHMANN, M. 1996 Environmental impact of demolition waste –an overview on 10 years of research and experience. Waste Management 16, 21-26.

3. WAHLSTRÖM, M., LAINE-YLIJOKI, J., MÄÄTTÄNEN, A., LUOTOJÄRVI T., L., KIVECÄS, L. 2000 Environmental quality assurance system for use of crushed mineral demolition wastes in road constructions. Waste Management 20, 225-232.

4. MANNING D.A.C., 1995 Cement and Plasters. In "Industrial Minerals" Editor Chapman & Hall, 141-155.

5. POON, C.S.; KOU, S.C.; LAM, L. 2002 Use of recycled aggregates in moulded concrete bricks and blocks. Construction and Building Materials 16 (5), 281-289.

6. LIMBACHIYA M.C. LEELAWAT T., DHIR, R.K., 2000 Use of recycled concrete aggregate in high-strength concrete. Materials and Structures 33 (233), 574-580.

7. SAGOE-CRENTSIL K.K, BROWN T., TAYLOR A.H., 2001 Performance of concrete made with commercially produced coarse recycled concrete aggregate. Cement and Concrete Research 31 (5), 707-712.

8. AJDUKIEWICZ A .. K LISZCZEWICZ A ., 2 002 Influence o f r ecycled a ggregates o n mechanical characteristics of HS/HPC. Cement and Concrete Composites 24 (2), 296-279.

9. OLORUNSOGO F.T., PADAYACHEE N., 2002 Performance of recycled aggregate concrete monitored by durability indexes. Cement and Concrete Research 32 (2), 179-185.

SESSION THREE:

RECYCLING FOR REUSE AS AGGREGATES

IMPLICATIONS OF USING RECYCLED CONSTRUCTION DEMOLITION WASTE AS AGGREGATE IN CONCRETE

S Chandra

Gothenburg University

Sweden

ABSTRACT. There is critical shortage of natural aggregates. This is aggravated with the increase in the new construction. At the same time, there is enormous amount of demolished concrete from deteriorated and obsolete structures generating waste material. It creates disposal problem and is an ecological and environmental problem. Recycling concrete is a viable alternative to using natural aggregate in concrete construction. But care is to be taken in designing the concrete as the quality of concrete vary very much depending upon the status of the parent concrete from where they have been produced.

Keywords: Cohesion, Deterioration, Demolished concrete, Fly ash, Pore structure, Recycling, Recycled concrete aggregate

Dr Satish Chandra is working as associate professor at the Institute of Environmental Science and Conservation, Department of Conservation, Gothenburg University, Sweden. He has 30 years of experience in the field of concrete technology. His research interest includes the recycling and reuse of waste materials, industrial by-products, chemical admixtures, polymers and the ancient building materials.

INTRODUCTION

According to the earth summit in Rio de Janerio in 1992 sustainable waste management is defined as " an economic activity that is in harmony with the earth's ecosystem" . The concept of sustainable waste management includes, first and foremost, the judicious use of natural resources, which are rapidly decreasing. Secondly, it is necessary to reduce the energy consumption that is associated with carbon dioxide gas emission, which is the primary cause for t he w ell k nown " green h ouse e ffect". A t t he s ame t ime, t here i s a c ritical s hortage o f natural aggregates for production of concrete, which is increasing with new constructions. Besides, there is enormous amount of demolished concrete from deteriorated or obsolete structures generating waste material. It creates ecological and environmental problem. One way to solve this problem is to use them as aggregates. The green house effect can be reduced by partially replacing Portland cement, which consumes energy in production and releases CO_2 by the supplementary cementing materials; such as fly-ash (PFA), blast furnace slag (BFS), condensed silica fume (CSF).

Recycling concrete is a viable alternative to using natural aggregate in concrete construction in many instances, particularly those in which the natural aggregate would have to be transported some distance and there is a problem disposing of old and rejected new concrete.

RECYCLED CONCRETE AGGREGATE

Recycled aggregates reviewed in this paper are from two sources; i) produced from rejected precast elements and cubes after testing strength, and ii) from demolished concrete buildings. In the first case, the aggregates are covered with clean cement paste and are mechanically sound. In the other case, depending upon the status of deterioration, the aggregates have cement paste contaminated with salts produced during the process of deterioration. The recycled aggregates produced by crushing the concrete have irregular form and are covered with old cement paste (Figure.1).

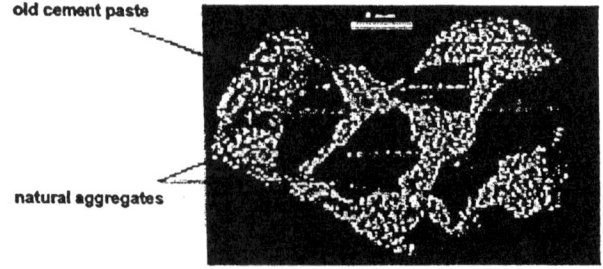

Figure 1. Illustration of the old cement paste adhering to the natural aggregates [1]

In the first case the aggregates are produced by crushing the concrete and grading them to different sizes. Whereas in the second case pretreatment is to be done, and the quality of the aggregates is to be very carefully assessed. The relevant tests are bulk specific gravity, water absorption, abrasion loss and soluble salts.

CRUSHED CONCRETE AS COARSE AGGREGATE

Hanssen and Narud [2] have tested coarse recycled aggregate made from three classes of concrete high, medium and low strength. The properties of the aggregate are shown in Table 1 and the mix proportion in Table 2. It gives an impression how the properties of the aggregates are influenced by the quality of the parent concrete.

Table 1. Properties of natural and recycled coarse aggregates

TYPE OF AGGREGATE	SIZE, mm	SPECIFIC GRAVITY.	WATER ABSORPTION %	LA ABRASION loss %	BS AGGREGATE CR. VALUE %	VOLUME % mortar*
Natural Gravel	4.8	2.50	3.7	25.9	21.8	0
	8.16	2.62	1.8	22.7	18.5	0
	16.32	2.61	0.8	18.8	14.5	0
Recycled (H)	4.8	2.34	8.5	30.1	25.6	58
High	8.16	2.45	5.0	26.7	23.6	38
	16.32	2.49	3.8	22.4	20.4	35
Recycled (M)	4.8	2.35	8.7	32.6	27.3	64
Medium	8.16	2.44	5.4	29.2	25.6	39
	16.32	2.48	4.0	25.4	23.2	28
Recycled (L)	4.8	2.34	8.7	41.4	28.2	61
Low	8.16	2.42	5.7	37.0	29.6	39
	16.32	2.49	3.7	31.5	27.4	25

* V- volume percent of mortar attached to natural gravel particles

Table 2. Mix proportion of concrete

MIX COMPONENTS	kg/m^3 (lb/ft^3)		
	H	M	L
Cement	110 (691)	215(362)	138(233)
Water	165(278)	151(255)	165(278)
Fine aggregate	545(919)	696(1174)	951(1603)
Coarse aggregate	1266(2134)	1290(2174)	1117(1883)
Effective W/C	0.40	0.70	1.20

All recycled concrete were made with basically the same mix proportion as original concrete except that the small adjustment were made in order to compensate for differences in density between natural gravel and recycled aggregate. All recycled concretes required an increase of approx. 10 litre of mixing water/m^3 of concrete, after correcting for water absorption of recycled a ggregate. Cement content was increased in order to maintain the same effective water- cement ratio for recycled concrete as for original concrete.

The recycled aggregates have lower density, higher water absorption, higher abrasion loss and higher B.S crushing value. These are inferior to natural gravel because considerable amounts of old mortar always remain attached to gravel particle in recycled aggregate. The mortar can be 30% for 16-32 mm aggregates to more than 60% for 4-8 mm fraction. This high amount can have deleterious effect on drying shrinkage, creep and other durability properties. However, this is not always the case [3].

High Strength Concrete with Recycled Coarse Aggregate (RCA)

Limbachiya et al. [4] have studied the high strength concrete made by the addition of recycled aggregates, produced by crushing the rejected precast elements. It is reported that RCA had 7 to 9% lower relative density and 2 times higher water absorption than the natural aggregates (NA), reflecting the porosity of the hydrated cement paste attached to RCA. The mechanical properties were somewhat lower than NA. The general trends observed indicated that up to 30% coarse RCA had no effect on the ceiling strength and thereafter a gradual reduction occurred with increasing RCA content. At 28 days (Figure 2a), the maximum attainable strength for concrete with 100% RCA was less than 75 N/mm^2 compared to over 80 N/mm^2 for concrete containing up to 30% coarse RCA. In Addition, for concrete made with 100% coarse RCA and natural sand, the start of strength ceiling effect were observed at a slightly higher water to cement ratio (0.28) than that containing up to 30% coarse RCA (0.25). Similar strengths were noted at 60 and 90 days (Figure 2b).

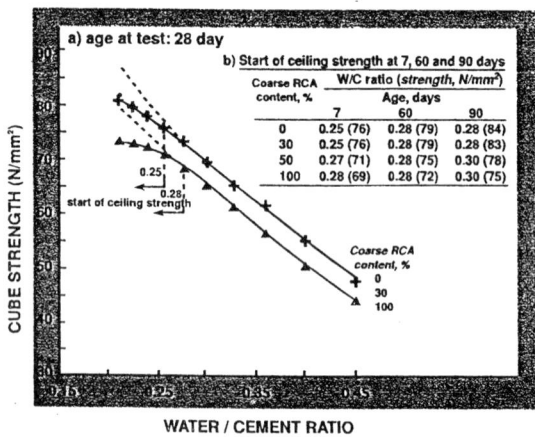

Figure 2. Ceiling strength at a) 28 days and b) 7, 60 and 90 days of RCA and NA [4]

Chloride Ion Diffusion and Freeze-thaw Resistance

The coefficient of chloride diffusion has shown that the use of up to 100% coarse RCA has no negative effect on the chloride diffusion of the resulting concrete. These results suggest that the high strength RCA concrete mixes perform as well as those containing natural gravel, with the differences in the D values between these concretes less than 1.0 x 10 -11 m^2/s (Figure 3).

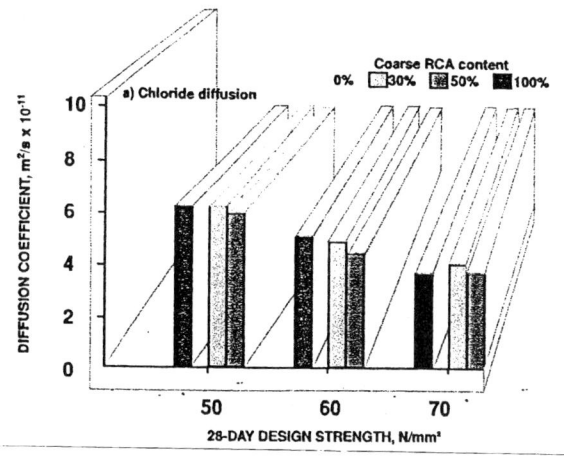

Figure 3. Effect of RCA on chloride diffusion [4].

It is shown that the concrete produced using up to 100% coarse RCA had durability factor in excess of 95%, showing good freeze-thaw durability. Uchikawa and Hanehara [5] have mainly attributed the shortcomings in performance of the recycled aggregate concrete to the pore structure of the previously hardened cement paste, which is partially cohering to surface of the recycled aggregate. It is porous, and absorbs water during concrete mixing. Owing to which W/C ratio around the paste is increased,. The newly produced cement paste is not sufficiently incorporated into the recycled aggregate and the previously hardened cement paste. This can be solved by using high range water reducer or by prewetting the stockpiles.

Recycled coarse aggregates from rejected precast element concrete can be used for producing concrete without much variation in the mix proportions. There is some decrease in strength which can be compensated by extra addition of cement and use of high range water reducers these are even possible to be used in high performance concrete. However, the suitability of RCA derived from different sources will have to be assessed individually.

CRUSHED CONCRETE AS FINE AGGREGATE

Ravindrarajah and Tam [6] have studied concrete made by the addition of recycled concrete fine aggregates in place of sand. Aggregates were produced by crushing the concrete made using natural sand and crushed granite having a nominal maximum size of 20 mm. The course aggregates used were also crushed granite with a max size of 20 mm. The physical properties of the aggregates are given in the Table 3. The crushed concrete fines (CCF) consisted of mortars and crushed stones of different sizes below 5 mm. The particles were irregular and angular in shape and the surface texture was partially rough and porous. In the two mixes about 10% by weight of the fine aggregate was replaced with a PFA. The water to cement ratio was 0.55 for all mixes, but for the mixes incorporating PFA, the water to binder (cement+PFA) ratio became 0.45 (Table 4).

Table 3. Physical properties of aggregates

PROPERTIES	CRUSHED GRANITE	NATURAL SAND	CCF
Fineness modulus	6.98	3.59	3.30
Specific gravity, SSD basis	267	261	232
Apparent density	2.68	2.64	2.54
Water absorption	0.30	0.63	6.20

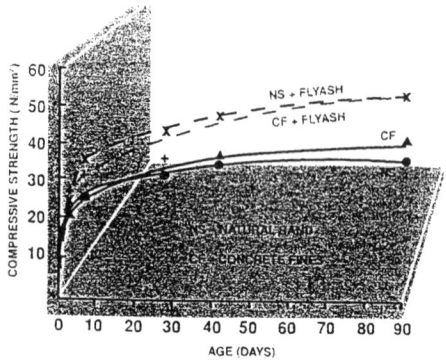

Figure 4. Development of compressive strength with age for concretes with natural sand and concrete fines [6].

Table 4. Mix proportions of concrete

MIX PROPORTIONS (kg/m³)					TYPE OF FINE AGGREGATE	W/C+PFA
Cement	PFA	Water	Aggregates			
C	PFA	W	Fine	Coarse		
386	-	212	875	875	Natural sand	0.55
382	-	210	860	860	Concrete fines	0.55
386	87	212	788	875	Natural sand	0.45
382	83	210	777	860	Concrete fines	0.45

The compressive strength developed up to the age of 90 days for water cured concrete (Figure 4). At later ages, the compressive strength was not affected by the type of fine aggregates used. It is seen that at the age of 7 and 90 days the strength of concrete with crushed concrete fines was about 0.85 & 1.12 % than for the control concrete made with natural sand. When 10% of the fine aggregate was replaced by PFA, the corresponding value changed to 0.88 and 1.01 respectively. The increase in strength is attributed to the decrease in W/B ratio and to the pozzolanic reaction of PFA, apart from its filler effect which densifies the pore structure [7]. Other researchers [3, 8, 9] have reported up to about 15% drop in the 28 days compressive strength with the use of crushed concrete fines. It is further reported that the ratio of tensile strength or flexural strength to compressive strength was not affected. Modulus of elasticity of concrete at the age of 28 days was reduced by 15% to 20%. Compressive strength-pulse velocity relationship was affected considerably and for the same strength the pulse velocity was lower.

Drying Shrinkage and Freeze-thaw Resistance
Drying shrinkage of concrete was increased by about 40%. These deleterious effects were significantly reduced by the use of pulverized fly ash (10%) (Figure 5). Freeze-thaw resistance of these concrete is reported to be somewhat superior, and the mixtures with 10-20 percent substitution by fly ash had a greatly reduced D-cracking potential.

Compressive and flexural strengths of recycled concretes tend to be slightly less than those of comparable mixes with natural aggregates, however, strengths above the minimum normally required are still easily obtained with proper mix design.

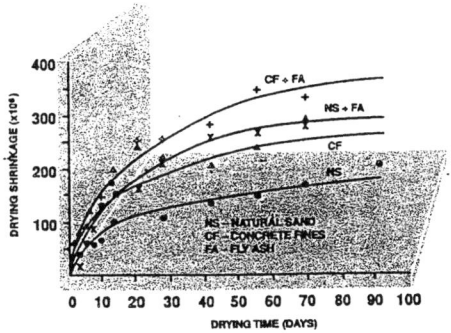

Figure 5. Drying shrinkage of concretes with natural sand and concrete crushed fine aggregate after 28 days moist curing [6].

WHY PHYSICAL PROPERTIES OF CONCRETE PREPARED USING THE MATERIALS RECOVERED FROM CONCRETE WASTE DETERIORATES?

Aggregates Surface Characterisation
Recycled aggregates produced from the hardened waste concrete have significantly different shapes and cohesion of hardened cement paste from those of original aggregates. Since the aggregates are worn out and rounded during the treatment and handling, the particle size distribution shifts to the small diameter side. It is shown that a part of the hardened cement paste cohering to the surface of aggregate is composed of Ca-rich porous structure which might have previously formed a transition zone in concrete and the other part of it has average composition.

The increase in the porosity of the interface, interfacial transition zone is due to higher water absorption of the coherent cement paste, which accumulates on the ITZ and therefore, more water is needed for equalizing slump. Cracks are observed, which are considered to be caused during the crushing process.

Fly Ash Refines the Pore-structure

It is well known that the fly ash refines the pore-structure of normal concrete by densification. Some recent studies have shown that it even refines the pore-structure of recycled aggregate concrete by reducing the macro-pore volume. [10,11]. This has beneficial effects on the mechanical properties such as compressive, tensile, and bond strength so much so that a concrete with performance similar to an ordinary concrete mixture prepared with natural aggregate is possible to achieve except for somewhat lower stiffness of the recycled aggregate concrete [12-14].

The recycled aggregate have higher content of fine materials, and water absorption. Consequently they require much lower water to cement ratio (W/C) than natural aggregates in order to achieve comparable concrete strength class. However, less W/C reduction is required when fly ash is added to concrete mixture, because of its pozzolanic activity. Anyway in the presence of recycled aggregate, a superplasticizer has to be used. It is observed that coarse RCA can be used in a range of high strength concrete mixes with satisfactory engineering properties, namely compressive strength, flexural strength etc. However, shrinkage and creep strains were found to increase with RCA content in the concrete, which are compensated to some extent by the use of fly ash.

AGGREGATES FROM DEMOLISHED CONCRETE BUILDINGS

The coherent cement paste in the aggregate from the demolished concrete is contaminated with salts. However, it is a matter of discussion whether the salt contamination in the aggregates will hamper the durability properties of the concrete. One view point is that if the chlorides are present from the beginning and are bound, then these are not detrimental. The other argument is that these have deleterious effect anyway. In a study on the concrete from the highway, where high amount of rock salt is used, recycled material was found to contain less than 1.2 kg/m^3, compared to the critical sodium chloride (NaCl) level of 2.4 kg/m^3 used for bridge decks. It is emphasized that while assessing the salts, it is not only the salt content of the recycled aggregate, but the total amount of NaCl in the concrete is to be calculated.

Ii is necessary to check the sodium chloride content of recycled material, which may have excessive salt and calculate the salt content for the new mix. It should also be borne in mind that the addition of fly ash and blast furnace slag will significantly decrease the deleterious effect, and hence their use is recommended.

Recycling and Alkali-Aggregate Reactivity
The consequences of using RCA, which has suffered from alkali-aggregate reaction (AAR) as an aggregate in new concrete, have not been well defined. In this regard several questions are to be answered. How severe is the reaction? Has the reaction gone to completion, i.e., have the reactive constituents been consumed? It is not so simple problem. Use of low alkali cement in the new concrete may not prevent further alkali-aggregate reaction with the recycled material because the reaction may continue within the recycled material between the old mortar and aggregates. If the reaction is taking place between the recycled materials, it may be that no level of alkali in cement will be low enough to prevent the reaction. There are certain speculations; use of limestone aggregates may reduce the AAR, mineral admixtures like fly ash, blast furnace slag and silica fume may reduce the AAR. Reduction in the recycled aggregate size may also be helpful in controlling the reaction problem.

Risk for Corrosion of Steel in Recycled Aggregate Concrete

Contrary to ordinary concrete, where penetration of chlorides is caused by external sources, outside the concrete, recycled aggregate concrete has an additional source of chlorides inside, when chloride contaminated concrete is used as aggregate inside the new concrete. This inside c hloride t hat c an p enetrate i nto t he s urrounding c oncrete m atrix m ay i nduce p itting around the steel reinforcement. It depends upon the porosity of the recycled aggregates. Higher is the porosity of the recycled aggregate, the more chlorides can penetrate into the surrounding cement matrix. A cyclic washing and drying procedure can reduce the chloride content of recycled aggregate [15]

FIELD EXPERIENCE

Eerland Recycling Services [16], The Netherlands has the experience of making aggregates from demolished concrete. They have finished a project at Shiphol Airport and are working on another project in the United Arab Emirates. Aggregates from the demolished concrete have been tested. It was found that the salt contaminations are under the threshold values. Thus these can be used for producing the concrete. Nevertheless, use of blast furnace slag, and fly ash is recommended.

CONCLUDING REMARKS

Recycled concrete coarse aggregates from rejected precast elements and cast concrete cubes can be used in making concrete with no significant effect on mixture proportions or workability compare to the control mix. Recycled concrete requires more water, thus to produce a concrete of the same strength as with the use of fresh aggregate, about 5% more cement is expected to be added. Coarse RCA can be used in a range of high strength concrete mixes with satisfactory engineering properties, namely compressive strength, flexural strength etc. However, shrinkage and creep strains were found to increase with RCA content in the concrete, which are compensated to some extent by the use of fly ash. It is important to recognize that there is a need to introduce standards for recycled aggregates, and demonstrate that these materials can be used successfully in practice.

REFERENCES

1. MESBAH, H.A. AND BUYLE-BODIN, F, Efficiency of polypropylene and metallic fibre on control of shrinkage and cracking of recycled aggregate mortars, Construction and Building Materials, 1999, pp 439-447.

2. HANSEN, T.C., A ND N ARUD, H., Strength of recycled concrete made from crushed concrete aggregate, Concrete International, January, 1983, pp 79-83.

3. BUCK, A. D., Recycled concrete as a source of aggregate, ACI Journal, Proceedings, v. 74, no. 5, May 1977, pp 212-219.

4. LIMBACHIYA, M. C., LEELAWAT, T., AND DHIR, R. K., Use of recycled aggregate in high strength concrete, Materials and Structures vol. 33, November 2000 pp 474-580.

5. UCHIKAWA, H, AND HANEHARA, 1996, Recycling of Concrete Waste, Proceedings of International Conferemce- Concrete in the Service of Mankind, 24-28 June. 1996, Dundee, Scotland, U.K.

6. RAVINDRARAJAH, R.S AND TAM, C.T., Recycling concrete as fine aggregate in concrete, Int. Journal of Cement Composites and Lightweight Concrete, Volume 9, Number 4 November, 1987,pp 235-241.

7. SOROKA, I., SEETER, N., The effect of fillers on the strength of cement mortars, Cement and Concrete Research, vol.7, no.4, July 1977, pp 449-56.

8. MALHOTRA, V.M., Use of recycled concrete as aggregate, CANMET Report 76-18 1976, Canada, Energy Mines and Resources, Canada, Ottawa.

9. RASHEEDUZZAFAR, and KHAN, A., Recycled concrete, A source of new aggregate, ASTM Journal, Cement Concrete & aggregate, Volume. 6, No. 1,1984, pp17-26.

10. CORINALDESI, V, ISOLANI, L, AND MARCONI, G., Use of rubbles from building demolition as aggregate for structural concretes" Proc. 2nd National congress on "Valorisation and r ecycling of industrial waste, M . Pelino, ed. Edizioni Graphic Press, L'Aquila, Italy, 1999, pp 143-153.

11. CORINALDESI, V, AND MARCONI, G., "Role of chemical and mineral admixtures on performance and economics of Recycled Aggregate Concrete" Fly ash, silica Fume, Slag and Natural pozzalona in concrete, Madras, India, July 22-27, 2001, Ed V.M.malhotra, ACI Int. SP-199, Farmington hills, USA, SP 199-50, vol, 2, 869-884, 2001.

12. CORINALDESI, V., TITTARELLI, F., COPPOLA, l., AND MARCONI, G., 2Feasibility and performnance of recycled-aggregate concrete containing fly ash for sustainable building Development", "Sustainable development of cement and concrete", Proc. 3 days symp. CANMET/ACI, San Francisco, USA Sept. 16-19 2001, ed. V.M.malhotra, SP 202-11, 161-180, 2001.

13. CORINALDESI, V., AND MARCONI, G., Rubbel processing to manufacture structural recycled aggregate concrete, proc. 5th Int sump. Cement and Concrete, Shanghai, China, Oct. 28-Nov. 1 2002, ed, Wang Pei Ming, Tongjo University Press, Shanghai, China, vol. 2, 796-802, 2002.

14. CORINALDESI, V., AND MARCONI, G., Durability of Recycled aggregate concrete incorporating high volume fly ash, proc. 9th Int. Conf. on " Durability of Building materials and components, Brisbane, Queens land, Australia, March 17-20 2002, Paper 71, 2002.

15. FRIEDL, L-. VOLKVEIN, A., AND SCHIESSL, P., The risk of corrosion of steel in recycled aggregate concrete, in proc. Sixth CANMET/ACI International Conference, Durability of Concrete, ed. V.M.Malhotra, Thessaloniki, June 2003. pp 1055-1073.

16. EERLAND, D.W, Eerland Recycling Operations, Geldermalssen, The Netherlands, 2000, Personal communication).

17. Eerland Recycling Operations Precast Concrete Paving Blocks, Advisory Note, 006, October 1990.

'WASTE NOT WANT NOT' –
A UK SHOWCASE FOR AGGREGATE RECOVERY

A Holmes

F M Conway Limited

United Kingdom

ABSTRACT. F. M. Conway Limited, based in Dartford, Kent, is investing in a 14 strong fleet of unconventional mobile concrete batch mixer lorries in order to take full advantage of it's annual 250,000 tonne Recycled aggregate flow. A special high-pressure water washing plant, the only one of its type in the UK, for cleaning and screening up to 80t/hour of dirty raw feed material into reusable sand and aggregate has been installed.

This huge investment follows on from the purchase in 2001 of a conventional 350t/hr crushing and screening plant, and in 2002 of a mobile Foamix plant, which has the ability to produce either cement or bitumen stabilised road base materials. This specialised equipment was also joined in the spring of 2004 by a unique purpose built plant which will recycle up to 95% of drainage waste from it's Cleansing Division activities and transfer the recovered grit and sand to the washing plant for cleansing and subsequent use in concrete production. The 14-acre Dartford Recycling facility is believed to be one of the largest and most advanced in Europe.

Keywords: Aggregate washing, Crushing and screening, Drainage waste, Foamix cold-lay macadam, Mobile concrete batching, Recycled aggregate for use in concrete or asphalt production.

Andrew Holmes MIHT is Business Development Manager for F. M. Conway Limited. He has over 30 years experience in road recycling and has been responsible for carrying out some of the first insitu road recycling projects on trunk road schemes in England. His expertise has taken him to the Falkland Islands where in 1988 he was largely instrumental in securing a multi-million pound pavement reconstruction project on the basis of an alternative recycling bid. He was the author of a technical paper on road recycling that was published in the Municipal Engineers Journal in 1988.

INTRODUCTION

In excess of 200 million tonnes of virgin aggregate is quarried for use in the UK construction industry each year. It takes 60,000 tonnes of aggregate to build just 1 mile of motorway. Just two of many statistics that many will find surprising but the impact that this massive industry has on the environment can be devastating. A scarred landscape and a choked road network are just two of the consequences. The construction industry is reliant on aggregate for its very survival but few can dispute that the pressures are mounting for it to change. Planning consent to extend or create new quarries is virtually impossible to gain. Sustainability must be the key to its future.

So why hasn't recycling become more popular sooner?

F. M. Conway Limited embarked upon its first Recycling initiative back in the early 1990's. By crushing concrete and asphalt arisings from it's Contracts in London it found that sub-base materials could be produced to a pretty good standard and used as a direct replacement for conventional virgin aggregate products. This development seemed to have potential at the time but to persuade Local Authority Engineers to accept it as an alternative to the tried and tested equivalents proved difficult and the company was forced to abandon it's plans. The hardware was disposed of but the concept remained very much at the forefront of the company's long-term plans.

Instruments of change began to appear a few years later with the introduction of the Landfill Tax and then in 2002 with the Aggregate Levy. Agenda 21, National indicators highlighting the proportions of recycled aggregates used in construction, Government Procurement Targets and the Landfill Directive have all contributed to the momentum of change. More recently spiralling costs of landfill disposal, greatly reduced numbers of 'special waste' sites and legislation prohibiting the co-disposal of 'special waste' with inert is without doubt the main driving force behind the innovation and development of the company's unique Drainage Waste Treatment Plant.

A wealth of Government initiatives to reduce waste and increase recycling are also helping to change minds. Not for profit organisations such as Remade and WRAP are essential persuaders and enablers. So too is the introduction of Specifications for recycled products such as BRE Digest 433, the amendments to the Specification for Highway Works in 2001 permitting the use of recycled aggregates in bound and unbound applications and a number of TRL Reports. New testing methods and regimes together with manufacturing protocols for Quality are essential tools in the process of confidence building. So can we safely say that the 'dig and dump' years have finally gone forever? The company firmly believes that recycling is finally here to stay.

It can be seen from the company's recycling portfolio that the level of investment and innovation that it has committed to recycling is second to none. It's vision is to become self sufficient in aggregate requirement by producing high quality materials that are equal or better in performance to their virgin counterparts, materials that it's Clients can specify and use with confidence. It is the company's belief that it is close to achieving this goal and that it's commitment and dedication to the 'art of recycling' will enable F. M. Conway Limited to remain at the forefront of recycling technology in the UK for many years to come.

CONSTRUCTION AND DEMOLITION WASTE - AGGREGATE RECOVERY

Feedstock control is paramount in the business of recycling. Each load is carefully inspected on the weighbridge by a series of remote cameras enabling weighbridge personnel to scan the vehicles content before acceptance. Contaminated material is rejected. The vast majority of arisings are generated through it's own Term Maintenance Contracts where strict guidelines are established as to the suitability and consistency of feedstock. The remaining feedstock balance is sourced from a select number of reputable suppliers who are not only bound by the above guidelines but with whom the company have established reciprocal trading arrangements in order to safeguard supplies. Approved loads are stockpiled in a designated area to await processing.

Feedstock is loaded into the combined crusher and screener by mechanical loading shovel. The material is fed on to a vibrating feeder deck that then passes it over a 35mm to 50mm grizzly grid. All material passing the grid falls on to a primary finger screen that splits the raw feedstock at 14mm. Material passing the 14mm grid is extracted on to a side discharge conveyor whilst material retained on the grid falls into the crusher where it is reduced in size to less than 75mm. Immediately after crushing the resultant aggregate joins the plus 14mm on to a feeder plate which carries it to the main feed conveyor. On its way to the double deck screening plant the aggregate passes under a powerful, hydraulically adjustable magnet that removes all items of ferrous metal. The aggregate then passes over two further screen deck's that classifies the material in to 65mm and 45mm sizes. Oversized aggregate retained on the double deck screening plant is automatically returned to the crusher on a return conveyor.

A slewing conveyor automatically stockpiles up to 500 tonne via its unique photosensitive slewing mechanism. The plant is capable of crushing concrete or rock lumps in excess of 1.0 m3 in size. With its unique screening deck the plant can process up to 350 tonne per hour. The plant is programmed to enable samples to be taken at intervals for quality control purposes regardless of the grade of material being produced.

The fully computerised plant has an inbuilt management system that allows daily interrogation to determine throughput and also to identify any faults that may develop during the working shift. Two modems link the system to the manufacturers monitoring centre where technicians are often alerted to potential problems in advance thus avoiding lengthy and costly periods of downtime. Both mechanical and electrical faults are pinpointed as they occur thereby eliminating the need for labour intensive and expensive fault finding activity.

Materials produced as standard are Type 1 granular sub-base and 20mm crushed asphalt.

AGGREGATE WASHING

The installation of an aggregate washing plant was the final link in the aggregate processing chain and has enabled the company to produce high quality Recycled aggregates that are thoroughly clean and free from contaminants therefore rendering them ideally suitable for use in concrete production or other high value applications. The plant has also enabled the company to reduce the volume of imported primary aggregate by as much as 150,000 tonne per annum.

The plant is designed to wash and size aggregates at breakneck speed. Up to 100 tonne per hour of dirty aggregate can be washed and sized into 4 standard grades that comply fully with new European Standards. The plant is capable of washing raw feedstock or crushed aggregate and is intrinsically linked to the adjacent Drainage Treatment Plant on which it is almost entirely dependant for its supply of clean water.

The unique hydro wash system uses high-pressure water jets to loosen the fine particles from the course aggregate. Further cleansing takes place as the aggregate is conveyed to the screen decks prior to sizing.

Used wash-water containing solids less than 63 micron is transferred to the Drainage Treatment Plant, via a series of underground pipes, to be thoroughly cleansed and then returned to the washing plant for the process to be repeated. The amount of fresh water taken from the mains for the washing process is minimal and is generally only necessary during hot periods in the summer months. A continuous requirement of 55 m3/hr of clean water for washing has been achieved through a process of intricate mathematical balancing calculations to e nsure t hat a n adequate c lean w ater s upply i s a lways available. All c ollected rainwater from the facility is fed into the Drainage Treatment Plant to assist in maintaining this balance. It is estimated that the unique design of these two plants saves the use of 24 million gallons of mains water per annum.

The plants designers have given serious consideration to operating noise levels through every stage of its conception making the plant one of the quietest of its kind available and therefore making it eminently suitable for operation in the semi-urban environment.

CONCRETE PRODUCTION FROM CLEAN RECYCLED AGGREGATE

Mobile concrete batching vehicles have been specifically designed to manufacture any grade of concrete or mortar at the point of delivery using recycled aggregate. This rapidly expanding fleet of volumetric trucks and state of the art aggregate washing plant means that the company can now satisfy the demand for concrete made from recycled aggregate. With highly trained operators and an effective calibration regime the company has put quality at the top of its priorities. An independent testing house carries out frequent testing and monitoring of the concrete produced.

The vehicles are capable of carrying the constituents required for 8 cubic metres of concrete. Unlike conventional ready-mix trucks the sand and gravel and cement are carried in a dry format ready to be mixed when needed. There is a separate silo for storing the cement and a built in tank to carry the water. A facility exists for carrying additives on board so that the customer's requirements may be met in full. The system is fully automated.

The controls that regulate the mix and rate of delivery are situated at the rear of the vehicle and via a remote keypad the operator is able to stand alongside the machine in full control and ensuring that the concrete is placed exactly where the customer requires it. Mix proportions are set to deliver the desired grade of concrete and slump but can be varied to suit the customer's needs at any time allowing multi-drops of varying concrete grades at different locations within any construction site. The components are mixed into fresh concrete by the action of an auger and conveyor and the concrete is placed by an adjustable shoot mechanism.

The main benefits of mobile concrete batching are:

- Facilitates the use of washed Recycled aggregate
- Cost effective and flexible delivery
- Multiple drops of varying grades of concrete are possible
- The customer is only charged for the quantity and grades taken
- Deliveries as small as 1 m3 are permitted
- No 'part load' charges
- No wastage
- Category of mix can or slump can be adjusted at any time
- Manufactured in accordance with BS EN 206

DRAINAGE TREATMENT AND CONSTITUENT RECOVERY

The company's unique Drainage Treatment Plant is designed to process and treat drainage waste and to render its constituents safe and suitable for reuse in a variety of applications. Previously all drainage waste was disposed of in landfill but now less than 5%, mainly litter, necessitates this kind of disposal.

The plant is capable of treating 30,000 tonne of drainage waste per year, thus diverting it from a rapidly reducing number of landfill sites that are licensed to accept it. This innovative solution to the treatment of drainage waste not only reduces the need for long distance transportation to the nearest special waste site but also enables the recovery of valuable aggregates, organics, hydrocarbons and even scrap metal.

The plant also performs another valuable function. Intrinsically linked to the Aggregate Washing Plant (AWP), water required for washing is continually recycled and cleansed as it flows through the Drainage Treatment Plant (DTP) and is then pumped back to the AWP for repeat use. Drainage waste is often up to 60% water and this together with a unique system of rainwater collection from the entire site helps to maintain the volume of water needed to satisfy the demand of the AWP, particularly through periods of hot weather.

So what is drainage waste? Its constituents will vary from district to district but generally analysis has identified the following constituents that are typical of that recovered from a city environment: Sand, organics and silt, emulsified oil – hydrocarbons – heavy metals, scrap metal, hypodermic needles, litter and water.

The DTP automatically separates all of these constituents. Sand is recycled for use in concrete production. Organics and silt are pressed into a cake format via a system of belt-presses for use in topsoil. Hydrocarbons are recovered and can be used to enhance the calorific value of incinerator fuel. Scrap metal is transferred for scrap. Hypodermic needles are identified and removed for disposal as medical waste. Water is cleansed beyond current legislative levels and recycled within the DTP and AWP system. Surplus water can be utilised in drain jetting activities.

So how does the DTP work? Believed to be the first of its kind in the UK the DTP works via a complex system of washing, grit classification, screening, dissolved air flotation and dewatering. The wash drum separates any material larger than 10mm and also washes clear any oil and organics. Particle sizes less than 10mm are conveyed to the grit classifier where any remaining organics or volatiles are removed, rendering the resultant grit clean and suitable for use as an aggregate A fine screen is used to remove any floating solids such as plastic or other litter. Having removed the grit from the body of the waste the remaining material is subjected to a process of dissolved air floatation and oil separation that removes the hydrocarbon content and very fine silt. The remaining sludge is drawn off to the Sludge Dewaterer w here i t i s s creened a nd s queezed t o r emove m oisture a nd p roduce a s emi d ry material.

The company believes that the system on which it's DTP is based is the only environmentally acceptable method of drainage waste treatment available today. It is particularly suited for location in the urban environment where other known settlement techniques are precluded. The main benefits of this system are:

- ❏ Environmentally preferred solution to the problem of drainage waste
- ❏ Thoroughly cost effective by greatly reducing transportation costs
- ❏ 95% of the waste deposited in the DTP is reused facilitating 'Closed Loop Recycling'
- ❏ Designed to be virtually silent in operation its location is not site sensitive
- ❏ Fully bunded and enclosed the plant is safe and poses no risk of leakage into ground water or water courses
- ❏ Improved utilisation and efficiency of cleansing fleet allowing the disposal of drainage waste at any time throughout the working day
- ❏ Special on site storage and collection service available for remote locations

FOAMWAY COLD-LAY BITUMINOUS MACADAM

Foamway is a cold-lay bituminous macadam manufactured from 100% recycled aggregate, bound together by a revolutionary foamed bitumen technique that boasts many environmental benefits including greatly reduced CO_2 emissions and energy requirement during the course of its manufacture, greater versatility in its use and most importantly sustainable credentials that contribute greatly to the nations desire for waste minimisation and sustainability in construction.

Proven over many years around the world this technology has been refined and improved to meet the demands of today's Highway Engineer in the UK and has gained confidence with many of the company's Clients as a cost effective, durable, high strength bituminous bound road base or binder course capable of withstanding the most punishing of environments and meeting the most demanding structural design requirements.

The company has invested heavily in the very latest technology that enables it to produce quality, consistent and competitively priced cold-lay macadam for use in all highway applications including heavily trafficked trunk road reconstruction with design traffic of up to 30 MSA. Foamway is equally effective in lesser demanding circumstances such as in binder course or road base renewal applications on both classified and unclassified roads, car parks, cycleways, in trench reinstatements or even as a temporary ramping material on regular resurfacing schemes.

The German designed plant forms an integral part of the overall Recycling Plant configuration within the facility but having mobile capability means that special projects can be undertaken anywhere in the UK if required. The primary constituents for a successful Foamway mix is a well-graded aggregate (Type 1 sub-base is ideal), 100 penetration grade bitumen, Ordinary Portland Cement (OPC) and water. Deficiencies in grading characteristics, for example, can be overcome by the controlled addition of fines in the form of Pulverised Fuel Ash (PFA) or similar.

The Foamed Bitumen technology works in a similar way to that of asphalt production. Hot bitumen is injected simultaneously with a small percentage of water into the mixing chamber. As the two liquids interact the water c auses the hot bitumen to foam and consequently to expand in volume to approximately 20 times its original volume. As soon as mixing commences the bituminous foam begins to collapse and deposits a residue of bitumen around the fine particles (nominally 75 Micron). As soon as laying commences and at the time of compaction the bituminous coated fine particles combine to encapsulate the larger aggregate to form a bituminous mortar. Further kneading and compaction assists the bitumen migration still further.

During production the required constituents are all physically linked to the Foamway plant via pipes and conveyors and a computer programme accurately controls the percentage additions to ensure that a consistent product is made and one that meets the requirements of the specification.

It is essential that all physical characteristics of the material to be treated are known and fully taken into account. Trial mixes, produced under controlled laboratory conditions, are often used to verify the suitability of the material for treatment with foamed bitumen and to predict dynamic stiffness values in relation to varying percentage of binder addition. A strict curing regime is employed to ensure that results are credible and to aid correlation. The company has developed two specific grades of Foamway for varying applications.

Structural grade can be designed to produce stiffness values ranging from 2,500 Mpa to 6,000 Mpa depending on the design traffic requirements. Structural grade Foamway is designed to be laid within hours of manufacture to prevent any premature cementing of the product prior to compaction. The product is capable of withstanding high traffic loading within 1 hour of laying making it an ideal alternative to conventional bituminous macadam or asphalt and despite its early high strength characteristics remains flexible in terms of its long term performance and moisture susceptibility.

Storage grade, as its name suggests, can be kept for up to 1 month under sheeted conditions and is ideal for footway and lightly trafficked applications.

As part of a continual product improvement programme the company's Foamway cold-lay macadam is tested at intervals in accordance with Department of Transport recommendations by an accredited independent testing house.

Rhinopatch Insitu Pavement Repair System
Rhinopatch is an insitu pavement repair system that is fast, effective, long lasting and cost effective. This means that disruption to the road user is kept to an absolute minimum whilst enabling the Engineer to carryout more repairs with less money. Developed by Asphalt

Systems International Limited, F. M. Conway Limited is an Authorised Contractor and is licensed to operate this system throughout the South-East of England.

The system utilises proven technology to repair potholes and crazed areas on both the highway and footway if required. The process involves heating, reworking, rejuvenating where appropriate and the addition of new asphalt if necessary to make good the repair. The system is quiet; no hydraulic breakers are needed so the process has considerable benefits in urban areas or when night-work is essential. Heating the area of repair is carried out by a bank of adjustable propane fuelled heaters specifically designed to transfer heat effectively and safely to the road surface without fear of danger to the operator or the general public. After reworking the effected area new hot asphalt is used to make up any deficiencies and to reinstate the repair to the correct level. Compaction is carried out by dead weight rolling.

Rhinopatch has many distinct advantages over conventional repair methods.

- A passive repair system that exhibits no joints
- Quiet and unobtrusive – ideal for inner city repairs at night
- Long lasting due to welded joints at base and sides of repair
- Rejuvenates and Recycles the existing asphalt without harmful oxidisation
- Fast and efficient
- Improved resistance to rutting and deformation
- Water tight repair

CONCLUSIONS

The company believes that through innovation it has achieved its goal of 'Closing the Loop on Recycling' and is confident that it's achievements are commercially viable and sustainable. 'Waste not want not' is a phrase that was commonly used in by gone days during periods of hardship and deprivation but surprisingly it is still appropriate to us today. Whilst we now talk in terms of 'Waste minimisation and recycling' the message is still the same. With legislation gathering pace at home and abroad the construction industry is left in no doubt that the days of 'dig and dump' are gone for good.

RECYCLING AND RECONSTITUTION OF CONSTRUCTION AND DEMOLITION WASTE

R K Dhir

K A Paine

T D Dyer

University of Dundee

United Kingdom

ABSTRACT. The annual production of construction demolition (C&D) waste in the UK amounts to 109 million tonnes. This accounts for over 60% of the UK's total waste. Rather than send C&D waste to landfill, increased environmental awareness has led to pressure to use these materials as a resource. There are two ways in which C&D waste can be developed for use in concrete; (i) as recycled aggregates (a commonly perceived approach) or (ii) as a cement component (a novel approach). This paper reports on recent research undertaken by the Concrete Technology Unit (CTU) at the University of Dundee to promote these methods for using C&D waste. Research, carried out since 1996, to show the suitability of coarse recycled aggregates in concrete is described, including a full-scale demonstration that used recycled concrete aggregate as the sole coarse aggregate in a concrete pavement designed for a XF4 exposure class. In addition, the paper reports on research to render C&D waste suitable for use as a cement constituent. The processing options examined are ultrafine grinding and thermal processing. Preliminary results of an ongoing study are presented, along with a discussion of their significance and of the environmental impact implications.

Keywords: Construction and demolition waste, Recycling, Reconstitution, Recycled aggregate, Recycled concrete aggregate, Thermal processing, Reactive powders

Professor Ravindra Dhir is Director of the Concrete Technology Unit, University of Dundee. He specialises in binder technology, permeation, durability and protection of concrete. His interests also include the use of construction and industrial wastes in concrete to meet the challenges of sustainable construction.

Dr Kevin Paine is a Lecturer in the Concrete Technology Unit, University of Dundee. His research interests are concerned with the use of recycled materials and industrial by-products as value-added cement constituents, additions and aggregates in concrete.

Dr Tom Dyer is a Lecturer in the Concrete Technology Unit, University of Dundee. His research largely involves recycling of waste materials in concrete, the study of chemical reactions of cement and related materials, and environmental life-cycle analysis.

INTRODUCTION

Currently at 109 million tonnes per annum, construction and demolition (C&D) wastes represent the largest waste generated from any single industrial sector. It accounts for over 60% of the UK total waste produced and is equal to 66% of the 165 million tonnes of primary aggregates used by the construction industry. As such it is a major sustainability concern nationally and actions are urgently required to develop a thoroughly sound technical knowledge base for enabling appropriate use of the material, rather than disposing of it in landfill or recycling as fill and unbound material

This paper describes the research undertaken since 1996 by the Concrete Technology Unit, University of Dundee. The main thrust of this work has been to create a sound technical base that is dependable and assists the construction industry in utilising C&D waste as a valuable resource in concrete and concrete products.

Of several attributes, one that makes concrete environmentally friendly is its potential to absorb many industrial wastes as part of its cement or aggregate component. Indeed, C&D waste is ideally suited for such applications, provided, of course, the use can be deemed to comply with appropriate existing or newly developed product specifications.

RECYCLING OF CONSTRUCTION AND DEMOLITION WASTES

BS EN 12620:2002 and BS 8500-2:2002
Since 2002, the use of C&D wastes has been permitted as recycled aggregate in BS EN 12620 and specification for their use in concrete is given in BS 8500-2. They are classified into two groups, RCA (comprising >95% crushed concrete) and RA (< 95% crushed concrete), based on composition (Table 1).

Table 1 Compositional requirements for coarse RCA and RA in BS 8500-2

REQUIREMENT	RCA	RA
Grading – conform to	Same as natural aggregate to BS EN 12620	
Maximum content masonry, % m/m	5	100
Maximum fines, % m/m	5	3
Maximum lightweight material, % m/m[§]	0.5	1.0
Maximum asphalt, % m/m	5.0	10.0
Maximum other foreign materials, % m/m (e.g. glass, plastics, metals)	1.0	1.0
Maximum acid-soluble sulfates SO_3, m/m	1.0	1.0[‡]

[§] density < 1000 kg/m^3

[‡] an amendment to BS 8500-2 made in October 2003 has replaced the limit of 1.0% on acid-soluble sulfate content with a requirement that the suitability of RA with respect to maximum acid-soluble sulfate content be determined on a case-by-case basis.

RCA in BS 5328 Designated Mixes
The initial work of the CTU during 1996-1998 [1] investigated RCA. Research was undertaken to examine and develop methods for the use of RCA in BS 5328 designated

concrete mixes, GEN1, GEN3, PAV1, PAV2 and RC30, i.e. concrete of strength classes ranging from C8/10 to C35/45. This represents 60% of the total UK concrete market.

RCA from six sources of concrete, ranging from airport pavements and rejected precast structural elements to lower quality kerbs and pavier concretes, were compared and it was shown that the initial source of concrete had little affect on the quality of RCA. For all classes of concrete, 30% replacement of natural aggregate (NA) with coarse RCA had only a modest influence on performance in terms of fresh properties, engineering performance and durability (permeation, carbonation, abrasion and freeze/thaw) [5]. At higher replacement levels, equal performance to natural aggregate concrete could be achieved at equivalent 28-day strength (Table 2), but this required a small reduction in w/c ratio. A mix proportioning method for this was developed [1].

Table 2 Comparison of engineering and durability performance
of natural aggregate (NA) and RCA concretes of equal 28-day cube strength

PROPERTY	DESIGN STRENGTH N/mm^2	CONCRETE TYPE			
		NA	RCA, % Coarse		
			30	50	100
Engineering Property					
(a) Elastic modulus, kN/mm^2	30	25.5	26.0	25.5	25.0
	35	27.0	26.5	26.0	26.0
(b) Flexural strength, N/mm^2	30	4.4	4.3	4.3	4.5
	35	4.7	4.6	4.6	4.7
(c) Shrinkage, μm	30	596	600	625	673
Durability Property					
(d) Carbonation depth, mm	30	21.0	21.0	20.0	18.5
	35	18.0	18.0	17.5	16.5
(e) Freeze/thaw durability factor, %	35*	99	102	100	97
	45*	96	96	97	99
(f) Abrasion depth, mm	35*	0.69	0.83	0.69	0.78
	45*	0.49	0.51	0.48	0.55
(g) Sulfate expansion, % x 10^{-5}	10	47	48	47	63
	20	38	35	54	56

(a) and (b) tested at 28 days, (c) 20°C, 55%RH, 90 days (d) 4% enriched CO_2 at 20°C, 50% RH (20 weeks), (e) ASTM C666-A, (f) Modified BCA method, (g) 0.3g/l Na_2O_4 (180 days exposure), *air-entrained concrete

It was also shown that fine RCA could be used in very low quantities but that it caused difficulties with the stability of the mix and the strength of the resulting concrete. For these reasons BS 8500-2 does not permit use of fine RCA.

RCA in High Strength Concrete
To further promote the use of RCA, its use in concrete of class C40/50 or greater was considered [1] and a range of engineering and durability properties investigated. Coarse

RCA from rejected structural precast elements was used, and it was shown that use of 30% RCA had no effect on the ceiling strength of concrete (Table 3). Furthermore, high strength RCA concrete gave equivalent performance to natural aggregate concrete of the same strength. This should give engineers the additional confidence required to use RCA in structural concrete.

Table 3   Water/cement ratio at which ceiling cube strength occurs for NA and RCA concrete

CONCRETE TYPE	W/C RATIO AT WHICH CEILING CUBE STRENGTH OCCURS (Ceiling cube strength, N/mm^2)			
	7 days	28 days	56 days	90 days
NA	-	0.26 (76)	0.28 (79)	0.28 (84)
RCA				
30% coarse RCA	-	0.26 (76)	0.28 (79)	0.28 (83)
100% coarse RCA	0.24 (64)	0.28 (69)	0.30 (72)	0.30 (75)

Resolving Application Issues for the Use of RCA

Although the above research had shown that appropriate RCA could be used in normal and high-strength concretes, there were concerns expressed by industry with regard to: (i) methods for measuring chlorides, (ii) the risk of alkali-silica reaction (ASR), (iii) risks with using RCA from high alumina cement (HAC) concrete, and (iv) hydrated gypsum powder. Research was carried out to investigate and resolve these application issues [2].

It was shown that the chloride contribution of RCA is most appropriately measured by an acid-soluble test; water-soluble techniques, as used for natural aggregates, were found to underestimate the availability of chlorides in the cement paste attached to RCA. Subsequently, BS 8500-2 has adopted an acid-soluble test (BS 1881-124) for determining chloride contents.

In the first edition of BS 8500-2, RCA was treated as a high-reactivity aggregate. However, research has shown that sodium oxide equivalent value of the vast majority of RCA falls below 3.5 kg/m^3; the limit for normal reactivity aggregates. As a result, in an amendment to BS 8500-2 in 2003, RCA was deemed to be a normal reactivity aggregates and guidance on its use given. The research on RCA from HAC concretes (a type of concrete used particularly extensively in the 1960s) has shown that it gives similar performance to Portland cement RCA. There are, therefore, o concerns with use of HAC RCA.

Finally, it was shown that as long as RCA passed the requirement for a maximum acid-soluble content of 1% in BS 8500-2, there were no concerns with hydrated gypsum powder.

Demonstrating the Use of RCA

To promote and further encourage the use of RCA in concrete, full-scale site demonstrations have been carried out to establish case-studies suitable for long-term monitoring of in-situ performance [3]. One example, is the second phase of development work at a biotechnology park, which comprised of a new access road, treatment of boundary walls and ground preparation. This required require removal of the existing factory floors and the underground foundations to what was formerly the largest Jute Mill in Dundee. This 19th century concrete was crushed to produce RCA for use in a concrete pavement.

The properties and characteristics of the RCA are given in Table 4. Demolition was carried out in a controlled manner to avoid contamination of the RCA with masonry from brick piers and other potential sources of foreign material. Screening was also carried out. The acid-soluble sulfate content of 0.6% was lower than the limit in BS 8500: Part 2, however the acid-soluble chloride content of 0.12% was higher than that recommended (BS 882) for use in r einforced concrete. For t his r eason a d ecision w as m ade t o construct a n u nreinforced pavement. The only steel in the pavement, therefore, were 400mm long stainless steel dowel bars at the expansion and contraction joints.

Table 4 Mean properties of RCA used in pavement construction

PROPERTY	TEST	20-10 mm	10-5 mm
Particle density, kg/m³	BS EN 1097-3	2445	2440
Loose bulk density, kg/m³	BS EN 1097-3	1200	1260
Water absorption, %	BS EN 1097-3	5.9	5.5
Acid-soluble sulfates, %	BS EN 1744-1	0.60	0.62
Acid-soluble chlorides, %	BS 1881-124	0.12	0.12
Water-soluble chlorides, %	BS 812-117	0.00	0.00
TFV, soaked †, kn	BS 812-111	151	-
AIV, soaked †, %	BS 812-112	22	-
Concrete setting time*, min	ASTM C403M-99		
Initial		340	
Final		460	

† *14-10 mm only*   * *20-5 mm (50% NA)*

Because of the high water absorption of the RCA, it was suggested following laboratory tests that to ensure the required concrete performance, it would be necessary to use the RCA in a saturated condition. Therefore, during the pavement construction phase, the stockpiles of RCA were maintained in a wet condition by frequent spraying.

The main pavement section was approximately 95m in length, and consisted of 19 panels (5m x 6.1m), whilst the turning circle at the end of the pavement (radius of 13m) consisted of 6 larger panels. For sustainability reasons the aim was to construct the majority of the pavement using RCA, As a control on the performance, natural aggregate (either (i) gravel aggregate or (ii) whinstone aggregate) was used for a small number of panels.

The pavement consisted of a 150mm deep RCA concrete with an imprinted surface, laid on top of a 300mm deep recycled aggregate (RA) sub-base (Figure 1). Construction joints were formed between each 5m panel and expansion joints at every third panel (i.e. every 15m). Joints w ere f ormed b y use o f 4 00mm l ong s tainless s teel b ars (in a n a ttempt t o m inimise corrosion problems from the relatively high chloride content of the RCA). No other reinforcement was used in the pavement.

The pavement was competed in December 2002, and the finished pavement is shown in Figure 2. Results of tests on cubes and other specimens taken from selected mixes to check for cube strength, flexural strength, drying shrinkage and accelerated freeze/thaw resistance showed good performance.. The mean cube strength of 35N/mm^2, and mean flexural strength of 3.9 N/mm^2 were consistent with performance of the laboratory mixes. Virtually no freeze/thaw scaling was detected after 56 cycles, thus demonstrating the excellent performance of air-entraining agent and polypropylene fibres working in tandem.

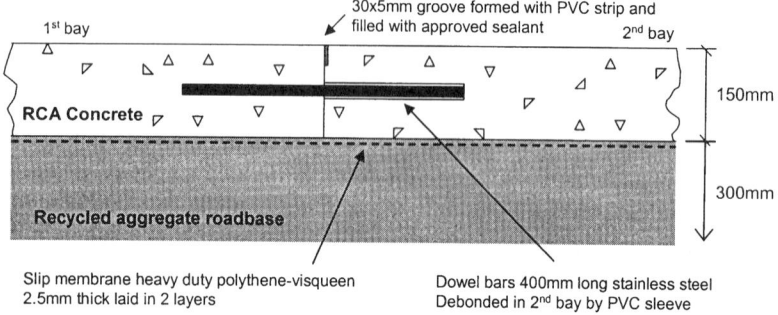

Figure 1 Cross-section of pavement panels and detailing of formed contraction joints

To date the pavement has been subject to natural conditions for eighteen months and from visual inspection all sections appear to be performing equally well. Note, however, that to date (April 2004) the pavement has not been open to traffic. The client was happy with the surface finish achieved, particularly on the RCA sections and the colour of the pavement has weathered to match the initial phase of development constructed using block paving. Indeed, due to the success of this demonstration the client has used RCA and patterned concrete in a number of other recent projects in the city of Dundee.

Figure 2 Completed RCA concrete pavement

Use of C&D Wastes as Recycled Aggregate (RA)

The distinction between RCA and RA (material containing less than 95% crushed concrete) in BS 8500-2 is currently making it uneconomical for concrete producers to stockpile and use RA, and this is restricting the full potential of C&D wastes as aggregates being realised. The CTU is initiating a major project aimed at maximising the use and value of RA in concrete construction. The project seeks to develop a technically sound classification system for RA produced from C&D wastes based on aggregate properties, instead of composition as is currently the case in BS 8500-2. RA quality will then be related to concrete performance across the whole range of potential RA variability, independently of composition and source (Figure 3). The concept for this is that good quality aggregates (conforming to Class A) will be suitable for high performance applications, whilst lower classes of aggregate will be more appropriate for lower value applications, but all meeting the relevant standards and specifications.

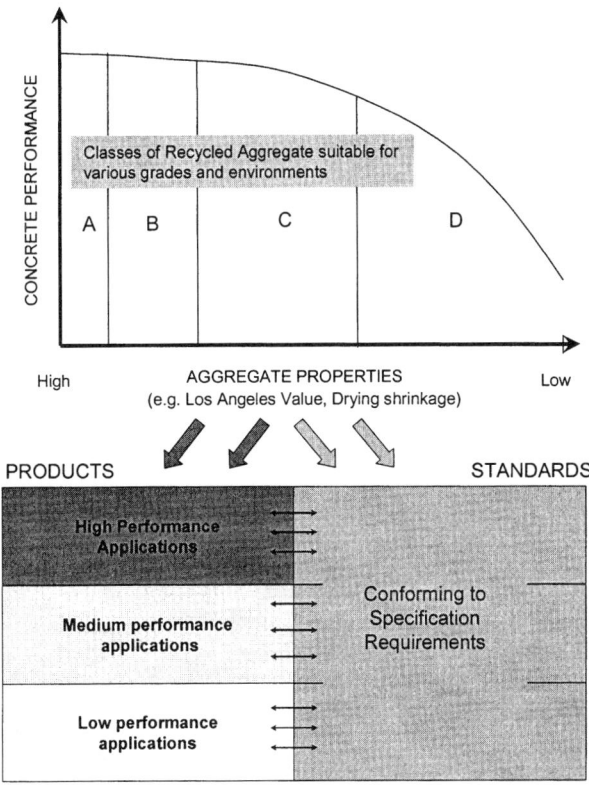

Figure 3 Performance based approach to use of recycled aggregates in concrete
that distinguishes recycled aggregate by its properties rather than by composition

Thus the project aims to remove the main technical barrier that is preventing uptake of RA in concrete, and lead to greater user confidence in specifying and using RA routinely as a valuable resource.

RECONSTITUTION OF CONSTRUCTION AND DEMOLITION WASTE

One of the most important ways in which the sustainability of products derived from C&D wastes can be ensured is through adding value. This 'added-value' approach to the processing and use of materials will always involve some form of optimisation.

One example of optimisation of waste-derived products is the a project currently being carried out by the CTU to reconstitute C&D wastes [4]. The aim of this project is to develop a methodology for processing these materials for use either as a component in reactive powder concrete or as a reactive material for use in cement combinations. Many of the constituents of C&D wastes (such as fired-clay ceramics, glass, and unreacted Portland cement) possess cementitious properties. Moreover, exposure to elevated temperatures will, in many cases, lead to the formation of phases capable of undergoing pozzolanic and hydraulic reactions.

Processing for a reactive powder concrete ingredient simply involves grinding the material to particle sizes comparable to those of quartz powders typically used in such materials. Where the materials are solely ground into a powdered form, there is likely to be a degree of cementitious character in C&D wastes in the form of pozzolanas such as glass [5] and fired clay [6]. Thermal processing is likely to render the material more active through the decomposition of clays [7] and the formation of cement clinker compounds [8].

The approach taken has been to use real C&D waste sources from around the UK. The chemical composition of these C&D wastes covered a fairly broad range, as can be seen in Figure 4, which shows the CaO-SiO$_2$-Al$_2$O$_3$ ternary system for the first ten materials studied by the project. The overall aim is to devise a methodology for combining and processing materials from a range of sources in such a way that a product of consistent performance can be obtained. Thus, it will be necessary to relate the chemical and mineralogical characteristics of each material to performance, and identify underlying trends.

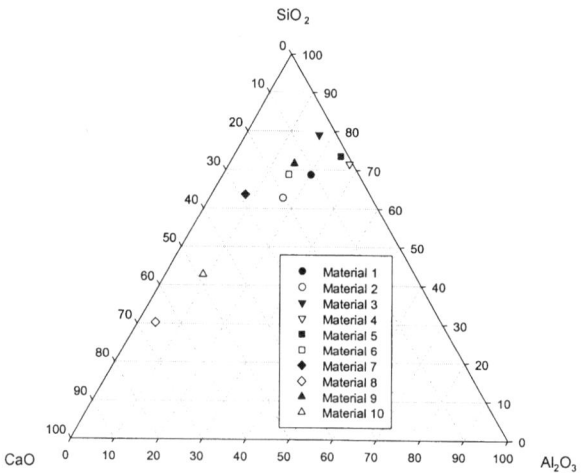

Figure 4 A CaO-SiO2-Al2O3 ternary diagram showing the compositions of 10 demolition wastes from around the UK

This variation in chemical composition (and mineralogy) means that the performance of cement combinations of these materials also varies significantly (Figure 5). It is anticipated that analysis of the data being generated by the project will reveal relationships between composition and performance. One of the aims of the project is to assess whether methodologies for combining C&D wastes can be formulated which allow a product of consistent performance to be produced routinely.

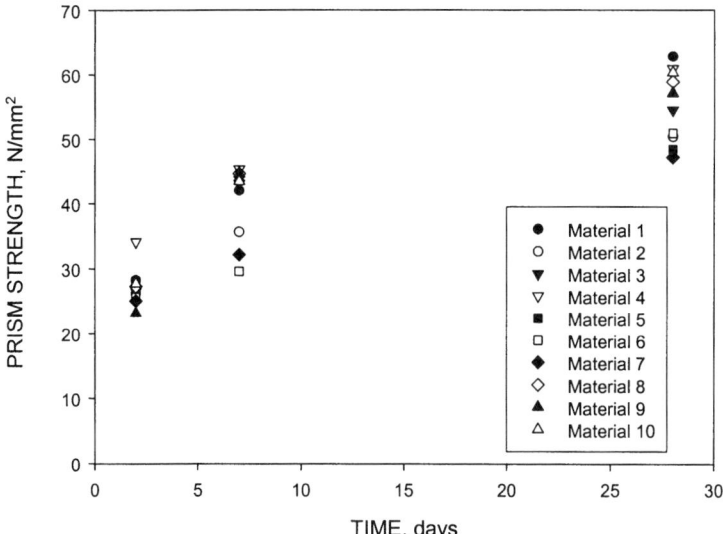

Figure 5 Strength development of mortars made with cement combinations containing 20% powdered demolition materials thermally processed at 900°C.

SUMMARY

The CTU has carried out considerable work in the field of utilisation of C&D wastes since 1996. This work has:

1. Shown that RCA can be used as a coarse aggregate at up to 30% by mass in both normal strength and high strength concrete without affecting mix proportions required for a given strength.

2. Shown that RCA can be used as a coarse aggregate at up to 100% by mass in both normal strength and high strength concrete provided mix proportions are adjusted slightly to achieve the design strength.

3. Determined that achievement of design strength will provide assurance in other aspects of concrete performance (e.g. durability) using up to 100% RCA as coarse aggregate .

4. Provided guidance and methods for dealing with concerns regarding alkali-silica reaction, and use of RCA containing high-alumina cement or gypsum powder.

5.   Demonstrated the use of RCA in aggressive environments through full-scale trials.

6.   Developed methods for reconstituting C&D waste to produce 'value-added' materials suitable for use as cement constituents.

ACKNOWLEDGEMENTS

The authors are grateful to all the organisations who have funded the various stages of research: Department of the Environment, Transport and the Regions, Department of Trade and Industry, RMC Environment Fund (RMCEF), Scottish Environment Protection Agency (SEPA), Quarry Products Association (QPA), British Cement Association (BCA), and industry (Aggregate Industries plc, AMEC Capital Projects, BAA plc,, The Business Environmental Partnership, Castle Cement Ltd., Costain Building Products Ltd., Dundee City Council, Envirocenter, Greater Manchester Waste Limited, John Doyle Construction Ltd., John Fyfe Ltd., MacWilliam Contract Ltd., RMC Aggregates (UK) Limited, Scott Wilson Kirkpatrick and Co., Tarmac Precast Concrete, Tayside Contracts, William Tracey Ltd., W S Atkins Consultants Ltd.)

A study of this size, scale and importance could not have been completed without the significant individual and collective contributions from the following experts: T Allan, J Barling, D Billington, B Brown, S Brown, C Clear, A Dryer, R Edwards, P Goring, D Grant, D Kay, K Laing, J Lay, J Lea, P Livesey, D MacLeod, G MacWilliam, A Melvin, B Menzies, S Neil, P Philips, G Raybould, D Sneddon, A Spalding, Z Stritton, M Tracey, J Tubman, C Waddell, J Webster, and R Wolstenholme.

REFERENCES

1.   DHIR, R K, PAINE, K.A., DYER, T.D. and TANG, M.C. Value-added recycling of domestic, industrial and construction arisings as concrete aggregate. Concrete Engineering International, Vol. 8, No. 1, 2004, pp 43-48.

2.   DHIR, R K, LIMBACHIYA, M C and BEGGS, A. Resolving application issues with the use of recycled concrete aggregate. Concrete Technology Unit, CTU/1601, 2001,171pp.

3.   DHIR, R K, PAINE, K A and O'LEARY, S. Use of recycled concrete aggregate in concrete pavement construction: A case study. Sustainable Waste Management, Proceedings of International Symposium, Dundee, September 2003, Thomas Telford Publishing, pp 373-382

4.   DHIR, R K and DYER, T D. Opportunities for creating medium-value construction products via demolition waste reconstitution, DTI/Industry Funded Project, Concrete Technology Unit (in progress).

5.   PATTENGILL, M and SHUTT, T C. Use of ground glass as a pozzolan, Symposium on Utilization of Waste Glass in Secondary Products, Alberquerque, New Mexico, 1973

6.   WILD, S. Observations on the use of ground waste clay brick as a cement replacement material, Building Research and Information, Vol.24, 1996, pp35-40

7.   TAYLOR, H F W. Cement Chemistry, Thomas Telford Publishing, London, 2nd ed., 1997, 459pp

8.   WOLTER, A. Influence of the Kiln System on the Clinker Properties, Zement-Kalk-Gipsum, Vol.38, 1985, pp612-614

DEVELOPING PRECAST CONCRETE PRODUCTS MADE WITH RECYCLED CONSTRUCTION AND DEMOLITION WASTE

N Jones, M N Soutsos, S G Millard, J H Bungey, R G Tickell

University of Liverpool

J Gradwell

Enviros Limited

United Kingdom

ABSTRACT. This paper outlines a new project, which has commenced at Liverpool University and is investigating the use of recycled general construction and demolition waste (C&DW) as aggregates in new concrete products. C&DW is currently recycled for high-volume, low-value end uses such as road sub-base construction. The project is investigating both the economic case and the technical feasibility of using both concrete and masonry recycled C&DW in higher specification applications such as precast concrete building blocks. The laboratory study will focus on the use of both concrete and masonry recycled C&DW as a replacement for either/or coarse and fine aggregate.

Preliminary results are presented on the development of a procedure for laboratory replication of the factory fabrication process and trials on the use of recycled masonry C&DW as a replacement aggregate in building blocks.

Keywords: Construction and demolition waste, Recycled aggregates, Precast concrete blocks, Sustainable construction.

Dr. N. Jones is a Research Associate at the Department of Civil Engineering, University of Liverpool.

Dr. S. G. Millard is a Senior Lecturer in Department of Civil Engineering at the University of Liverpool.

Dr. M. N. Soutsos is a Lecturer in the Department of Civil Engineering, University of Liverpool.

Professor J. H. Bungey is a Professor in the Civil Engineering department of the University of Liverpool and Head of the Structures and Materials Group.

Mr. R. G. Tickell is a Senior Lecturer in the Department of Civil Engineering, University of Liverpool.

Ms. J. Gradwell is an Environmental Consultant for Enviros, Manchester, specialising in waste and recycling.

INTRODUCTION

Construction and demolition waste (C&DW) accounts for approximately 17% of the annual UK waste stream, amounting to a total of 70 million tonnes of waste material [1], [2]. Across the EU, around 28% of C&DW is currently reused or recycled, with the remaining 72% going to landfill [3]. As C&DW is largely inert, consisting of soil, brick, glass, concrete and plaster etc., landfill charges for its disposal have traditionally been low. Additionally, a large proportion of C&DW is also used during landfill construction, enabling it to be disposed of free of charge.

Around 30% of the 70 million tonnes arising in the UK is reused in low-grade applications such as bulk fill, where some crushing and separation of materials such as metal and wood is required [2]. Higher-level use of the waste has not been thought possible previously because of the heterogeneous nature of the material compared with primary aggregates. .

Around 220 million tonnes of natural aggregates are quarried each year in the UK [4]. Because of the environmental impact of quarrying, there are currently initiatives in place - such as the Aggregates Levy, introduced in April 2002 - in the UK to minimise C&DW production and maximise the use of alternatives to primary aggregates [5].

Existing projects at the Universities of Nottingham [6] and Dundee [7] have found promising results with regard to incorporating C&DW from tightly controlled sources into fresh concrete. Reject concrete blocks, once crushed, have been investigated as a potential aggregate for fabricating new precast blocks, by the BRE with some success [8]. Poon et al. [9] found that reasonably high replacement levels of crushed block (of up to 50%) can be used with no detrimental effects on the block strength.

In this study, the technical aspects of using both concrete and masonry recycled C&DW as aggregates in the fabrication of new precast concrete blocks are investigated. The target strength of the blocks is $7N/mm^2$.

MATERIALS AND METHODS

Recycled Aggregates
The recycled aggregates were obtained from a local demolition contractor. The masonry contained mostly brick, with lesser amounts of plaster, mortar, tile, glass, metal, paper and wood. The concrete derived aggregate contained mostly concrete, with a small amount of red brick visible. Both were crushed to coarse (6-4mm) and fine (3mm-dust) fractions, to be as similar as possible to the natural limestone used in block production. Additionally, rapid hardening cement was used for all mixes.

Particle Density and Moisture Absorption
The aggregate particle density according to BS 812-2 [10] was determined for all aggregates used. The total sulphate content (SO_4) was also determined according to BS 812-118 [11] for the recycled aggregates. Literature values of total sulphate content were obtained for limestone. The test results, in Table 1, show that both the masonry and concrete derived aggregates have very high moisture absorption when compared to quarried limestone (7.5-9.5% for the coarse aggregate and 13.5-14.5% for the fine aggregate, compared to 0.65% 3.7% for coarse and fine limestone respectively).

Table 1. Aggregate Properties

AGGREGATE TYPE	PARTICLE DENSITY (Mg/m^3)		MOISTURE ABSORPTION	TOTAL SULPHATE CONTENT (as SO$_4$)
	SSD	Oven Dry	(% dry mass)	(% mass)
Fine Limestone (N)	2.41	2.24	3.70	<0.1
Coarse Limestone (N)	2.69	2.67	0.65	<0.1
Fine Masonry (R)	2.13	1.87	13.89	1.02
Coarse Masonry (R)	2.00	1.84	9.12	0.66
Fine Concrete (R)	2.07	1.81	14.29	0.50
Coarse Concrete (R)	2.88	2.67	7.58	0.36

(N) Naturally quarried   (R) Recycled

The SSD (saturated, surface dry) particle density o f the c oarse concrete derived a ggregate was found to be similar to that of coarse limestone, being 2.88Mg/m^3 and 2.67Mg/m^3 respectively. The particle densities of the other recycled aggregates, fine concrete and both coarse and fine masonry, were much lower than those obtained for quarried limestone. The high water absorption and low particle density of the recycled aggregates is similar to that of a man-made lightweight aggregate. Therefore a suitable mixing procedure of pre-mixing the aggregate and half the water first, followed by the cement and then the remaining water was adopted. The lower density of the recycled aggregates also meant that replacements were calculated on a volumetric rather than mass basis.

Sulphate Content
The sulphate content of the recycled C&DW aggregates may be an issue for blocks containing them. The Thaumasite Expert Group [12] has observed thaumasite on the surface of blocks made from natural aggregate, when subjected to the ideal conditions for thaumasite to form. There is potential for two types of attack (which could include thaumasite sulphate attack) to occur. The first is from the inside of the block due to the sulphate bearing aggregate. The second possibility of attack is on surrounding mortar, due to sulphates leaching out of the blocks.

Table 1 shows the sulphate contents of the recycled aggregates. It can be seen that the results are rather high when compared to those for limestone, particularly regarding the coarse masonry (1.02%). A maximum sulphate content of 1% by mass has been suggested for lightweight aggregates used in the manufacture of masonry units [13]. The values presented in Table 1 are total sulphate content determined by acid extraction, the likelihood of sulphates leaching out of the blocks, and in what quantities, is not known. At the time of writing it is thought that this is an issue to be acknowledged that may warrant further investigation later, but is not within the remit of this study.

Replication of the Factory Block Making Procedure

Fabrication

Preliminary work was carried out using naturally quarried limestone aggregates to develop a laboratory procedure for the fabrication of concrete blocks to BS 6073 [14]. Custom half-length block moulds were designed and fabricated, to make blocks of dimensions 220 x 215 x 100mm. The moulds were oversized in height, to allow for the larger volume of material before compaction. The amount of material required to produce a half block of dry mass just under 10kg was added to the mould and then simultaneously vibrated and compacted to the required height of 215mm. A vibration/compaction hammer with a custom designed stand and slide, shown in Figure 1, was used to simulate the industrial block making procedure.

Figure 1. Block Fabrication Using a Vibration/Compaction Hammer

Several trial mixes were carried out to determine the amount of water needed for a suitable mix, and the block density required to reach the target strength of 7N/mm². The results of some preliminary mixes are shown in Figure 2. It was found that the optimum water to cement ratio (W/C), which enabled the amount of material used to be successfully compacted into a block and retain its shape after demoulding, was quite a large range (W/C = 0.3-0.6). A block density of 2025kg/m³ was found to be suitable for achieving the target strength for a 100% limestone block.

Curing and testing

The blocks were cured under a 'tent' of wet Hessian sacking covered with polythene for the first 24 hours at room temperature, followed by storage on open laboratory shelving until the time of testing. Compressive strength testing was carried out on dry specimens at 7 days using fibreboard capping. A conversion factor of 1.06 was determined experimentally to convert the 7-day result to that of 28 day, wet tested, mortar-capped specimens. All strengths quoted in this paper are the converted 28-day values. Details of the testing and capping procedures are outlined in BS 6073 [14]. This 7-day testing procedure replicated that used by the factory.

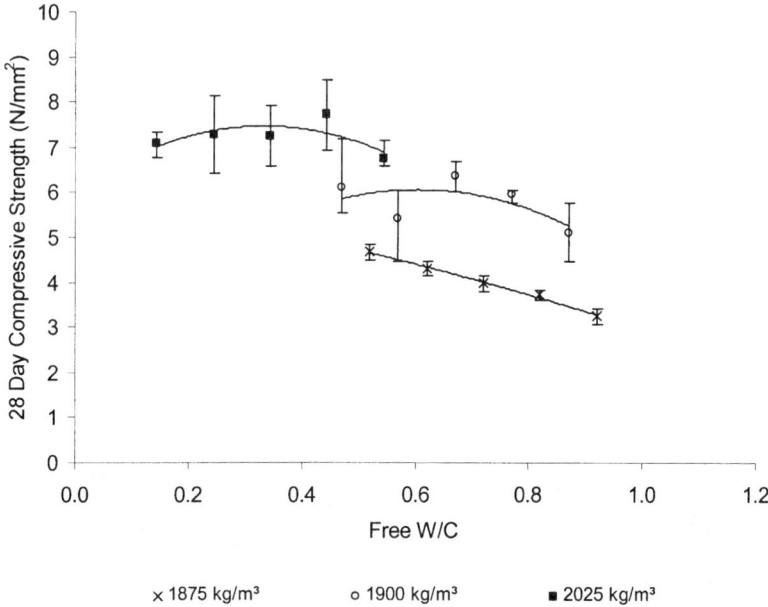

Figure 2. Strengths of Different Density Limestone Mixes Showing Free W/C

Replacement Mixes

Replacement mixes have included recycled masonry only at the time of writing. The recycled concrete aggregate mixes are to follow. Combined sieve grading curves were plotted for all mixes. The fine/coarse ratio for each mix was calculated by comparing it to the sieve grading for a 100% limestone mix. For the replacement mixes, the coarse/fine ratio used was that which provided the best fit to the limestone sieve grading. It is desirable, for economic reasons, to achieve the target block strength without the addition of extra cement, therefore the replacement mixes were designed with a cement content of 100kg/m³. The mix proportions and block densities of the low cement content mixes are shown in Table 2.

Table 2 shows that the water demand of the mix was not significantly influenced by the use of recycled aggregates as partial replacements. It was only when a mix of 100% coarse and fine masonry was used that the free water demand increased. It is thought that a free W/C of 0.5 should be sufficient for the future partial replacement mixes.

The replacement mixes were calculated on a volumetric basis due to the lower density of the recycled aggregates compared to limestone. The block density was ultimately dependent on the aggregate density and therefore the blocks made with recycled masonry as aggregate replacement had a lower density than those made with quarried limestone only. This reduced density had a direct effect on the block strength, as can be seen in Figure 3. It is therefore an aim to achieve the highest block density possible, whilst keeping the block weight below 10kg, in order to obtain the highest block strength possible.

Table 2. Mix Proportions for Low Cement Content Mixes

MIX PROPORTIONS	LEVEL OF MASONRY REPLACEMENT					
	Limestone	Coarse and Fine	Fine only	Coarse only		
kg/m³	100%	100%	100%	100%	60%	30%
Cement	100	100	100	100	100	100
Coarse LS	1000	-	892	-	385	683
Coarse MS	-	887	-	705	398	237
Fine LS	1250	-	-	1305	1288	1274
Fine MS	-	881	1243	-	-	-
Free W/C	0.3-0.6	0.7-1.3	0.3-0.6	0.4-0.6	0.4-0.6	0.4-0.6
Block Density	2025	1540	1770	1750	1820	1925

LS - Natural quarried limestone aggregate   MS - Crushed masonry derived aggregate

Figure 3 shows the results of the masonry replacement mixes. It can be seen that for a cement content of 100kg/m³ the highest block strengths were achieved at the lower replacement levels. For the 100% masonry (coarse only) mix it was necessary to increase the cement content to 200kg/m³ to achieve the target strength of 7N/mm². At the replacement level of 30% masonry (coarse only), a strength of almost 6N/mm² was achieved without adding extra cement. When 100% coarse and fine masonry was used the strength was very low, achieving only 5N/mm² with a cement content of 300kg/m³. It is intended that the target strength will be achieved at lower replacement levels without the addition of extra cement.

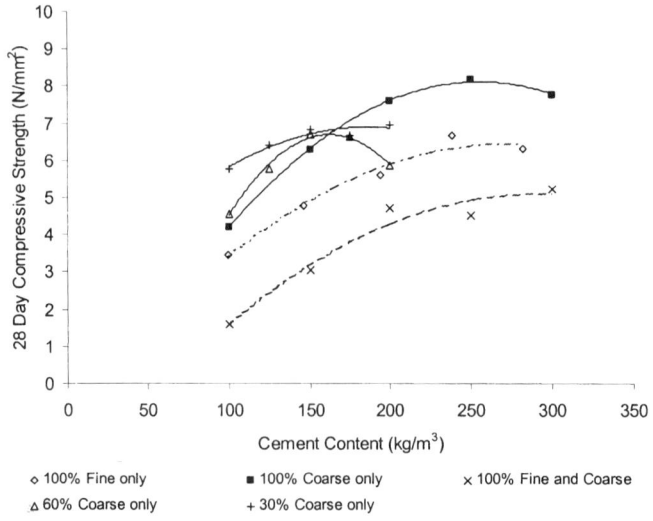

Figure 3. Strengths of the Masonry Replacement Mixes

CONCLUSIONS

A factory procedure for fabricating concrete blocks using natural aggregate has been successfully replicated. The physical characteristics of recycled aggregates have an adverse effect on the mechanical properties of blocks at high replacement levels. However, the results obtained so far show promise and it is expected that the target strength will be achieved by using lower replacement levels, whilst maintaining economical cement content.

ACKNOWLEDGEMENTS

The authors wish to thank the ONYX Environmental Trust and the Flintshire Community Trust Ltd. (AD Waste Ltd.) for funding this project. Thanks are also extended to the following industrial collaborators for their assistance: Clean Merseyside Centre, Marshalls Ltd., Forticrete Ltd., Liverpool City Council, Liverpool Housing Action Trust, RMC Readymix Ltd., W. F. Doyle & Co. Ltd., DSM Demolition Ltd., and Environmental Advisory Service – Merseyside.

REFERENCES

1. Estimated Annual Waste in the UK by Sector (Million Tonnes)., Environment Agency, (2000).

2. Williams, P. T. (1998) 'Waste Treatment and Disposal'. John Wiley & Sons Limited.

3. Symonds Group Ltd., ARGUS, COWI & PRC Bouwcentrum (Feb. 1999). 'Construction and Demolition Waste Management Practices, and Their Economic Impacts'. in *Final Report to DGXI, European Commission*. p. 208

4. Department of Environment, Transport and the Regions. (2000). 'Planning for the Supply of Aggregates in England - a Draft Consultation Paper'. p. 70.

5. HM Treasury. 'Consultation on the Objectives of Sustainability Fund under the Aggregates Levy Package'.

6. Johnston, A. G., Dawson, A. R., and Brown, G. J. (1998). 'Recycled Concrete Aggregate - an Alternative Concept'. in *Sustainable Construction: Use of Recycled Aggregate*. University of Dundee, 11-12 November 1998. p. 31-44

7. Dhir, R. K., Limbachiya, M. C., and Leelawat, T. (1999). 'Suitability of Recycled Concrete Aggregate for Use in BS5328 Designated Mixes'. *Proceedings of the Institution of Civil Engineers: Structures & Buildings*. August 1999: p. 257-274.

8. Collins, R. (2003). 'Recycled Concrete'. Advanced Concrete Technology - Processes, ed. J. Newman and B.S. Choo. Butterworth-Heinemann.

9. Poon, C. S. (2002). 'Use of Recycled Aggregates in Molded Concrete Bricks and Blocks'. *Construction and Building Materials*. 16: p. 281-289.

10. BS 812-2: Testing Aggregates - Methods of Determination of Density. (1995).

11. BS 812-118: Testing Aggregates - Methods for Determination of Sulphate Content. (1988).

12.   Thaumasite Expert Group Report: Review after Three Years Experience. Prepared by Professor L. A. Clark and BRE in Consultation with the Thaumasite Expert Group. (March 2002).

13.   BS 3937: Specification for Lightweight Aggregates for Masonry Units and Structural Concrete. (1990).

14.   BS 6073-1: Precast Concrete Masonry Units - Specification for Precast Concrete Masonry Units. (1981).

USING CONSTRUCTION DEMOLISHED WASTE IN CERAMIC

L B Svatovskaya

V Y Shangin

D N Buharina

A V Borodylja

L L Maslennikova

V D Martynova

A A Tenerjadko

E A Popova

St Petersburg State University of Transport

Russia

ABSTRACT. The new classifications of the solid-waste are suggested for ceramics technology. The base of the new classifications is the next – electron family in the Mendeleev's table and the surface of the solid-state waste. According to the new classifications the solid-state waste are being observed. A few examples of the new classification in ceramics using are being shown. It turned out, that ceramics materials obtained using waste not only utilized different nature solid waste but have very good exploitation properties.

Keywords: Ceramics, Classifications, Surface, Properties, Waste, Compressive strength, Flexural strength.

Professor, Dr L B Svatovskaja is the Head of Engineering Chemistry Department, Railway University, St Petersburg. Russia. She specializes in the chemistry and ecology.

Professor, Dr L L Maslennikova is Lecturer in Engineering Chemistry Department, Railway University, St Petersburg. Russia. Scientific interests are ecology and ceramics.

Dr V Y Shangin is Lecturer in Engineering Chemistry Department, Railway University, St Petersburg. Russia. Area of scientific interests is building materials.

Dr V D Martynova is Lecturer in Engineering Chemistry Department, Railway University, St Petersburg. Russia. Area of scientific interests is building materials.

Mrs D N Buharina is Post-graduate in Engineering Chemistry Department, Railway University, St Petersburg. Russia. Area of scientific interests is ecology.

Mrs A A Tenerjadko is Post-graduate in Engineering Chemistry Department, Railway University, St Petersburg. Russia. Area of scientific interests is ecology.

Mr A V Borodylja is Post-graduate in Engineering Chemistry Department, Railway University, St Petersburg. Russia. Area of scientific interests is building materials.

Mrs E A Popova is Post-graduate in Engineering Chemistry Department, Railway University, St Petersburg. Russia. Area of scientific interests is ecology.

INTRODUCTION

There are a lot of demolished constructions challenging for utilization. On the other hand, ceramics is known to be extremely substances absorbing air of technologies. In this paper the system of knowledge is suggested, that helps to use a lot of solid state-waste for ceramics technology material obtaining. The first position for that are classifications of the material. In the Table 1 is being shown the classification of the solid waste on the base of the Mendeleev's table. According to this classification there are s-p-and d-solid-waste materials that depend on the base cation in the waste.

The s-p and d-cations contented solid waste can be used for building ceramics with good properties. The second classification is believed to take into consideration composite nature of the building ceramics. The main notion is interaction between solid-state phase, one of that is clay and the other is a phase like filler. In the Table 2 is being shown reactions between two phases during firing till temperature 1000°C, and donor- and acceptors interactions forming the contact phase, that is shown on the Figure 1. According to the second classification the main practical properties of the ceramics materials are connected with phase that takes place between clay and fillers.

MATERIALS

In the Tables 2 and 3 is being shown the solid waste materials taken into consideration like filler. The mixes with clay and that filler from table 1 where firing, and then we have building ceramics material.

Table 1. Classification according to Mendeleev's table

ELECTRON - FAMILY	DEMOLISHED WASTE	IMPROVING MAIN PROPERTIES OF BRICKS
S - waste	$CaSO_4*2H_2O$ NaOH, NaCl Na, NaOH, Na_2CO_3-oil-demolished waste	Reduction heat conductivity, improving decorate properties
p-waste	SiO_2	Increasing flexural properties
s-,p-d - waste	C_2AS, CMS, C_3MS_2, β-C_2S, glass phase β-,Υ-C_2S, $2CaO*MgO_2SiO_2$glass phase	Increasing flexural properties
d-waste	FeO, Fe_2O_3, Fe_3O_4 $Zn(OH)_2$, $Cu(OH)_2$ $Pb(OH)_2$, $Cr(OH)_3$, $Ni(OH)_2$ $Fe(OH)_2$ $Fe(OH)_3$ Cr(III), Fe(III), Cu(II)	Improving both mechanical decorate and properties

Table 2. Forming of the composed material

TEMPERATURE OF THE PRESSES	MODEL OF THE REACTION	№
	☐ **Interaction with water**	
Standard temperature	Φ_1 A + $H_2O\downarrow\uparrow$ = Φ_1 A↓↑ OH$^-$ + H$^+$	1
	Φ_2 Д$\downarrow\uparrow$ + H$_2$O = Φ_2 Д$\downarrow\uparrow$H$^+$ + OH$^-$	2
	Discovery of the active centers	
Firing till 800°C	Φ_1 A↓↑OH$^-$ → Φ_1 A + OH$^-$ ⎫	3
	Φ_2 Д$\downarrow\uparrow$H$^+$ → Φ_2 Д$\downarrow\uparrow$ + H$^+$ ⎭ − H$_2$O	4
	Contact phase forming ☐	
Firing till 1000°C	Φ_1 A + $\downarrow\uparrow$Д Φ_2 → Φ_1 A$\downarrow\uparrow$Д Φ_2	5

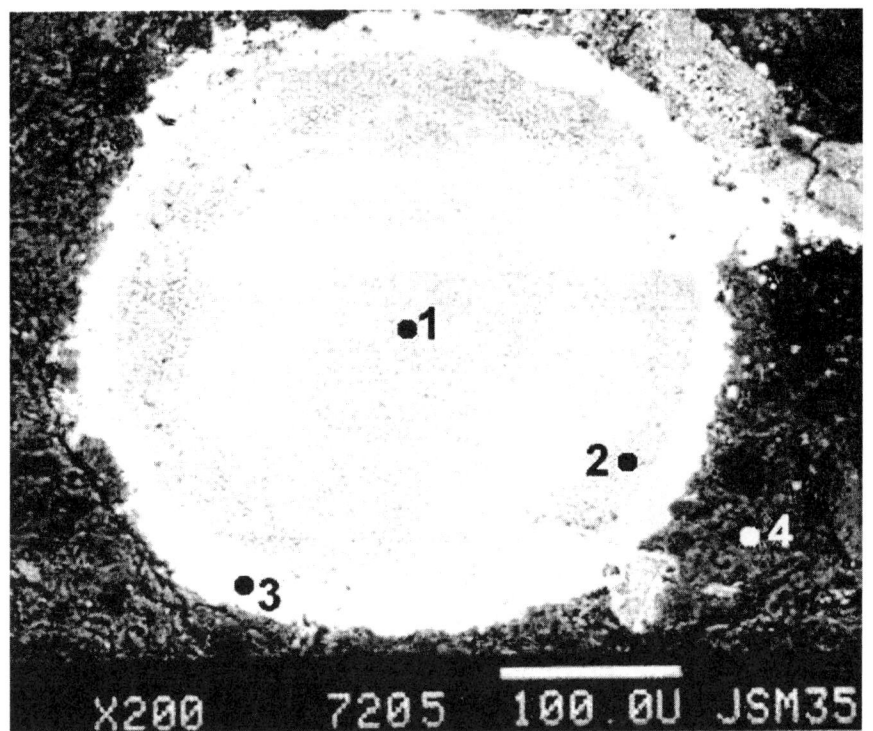

Figure 1. Fe-, Cu-waste (1), previous phase (2), contact-zone (3), matrix (4).

RESULTS AND DISCUSSION

In the Figures 2, 3 and 4 is being shown a few relationship and it can see that the best waste is d-one according to Table 1 (Figure 2, F-, Cu-waste) because of the flexure strength and than the railway waste; s-waste, according to Table 1, for example $CaSO_4 \cdot 2H_2O$ (demolished forms) need increasing firing temperature. The more s-waste, the higher the firing temperature (Figure 3), but only till 900°C instead of 1000°C, and cold resistance is very good in the all cases (Figure 4).

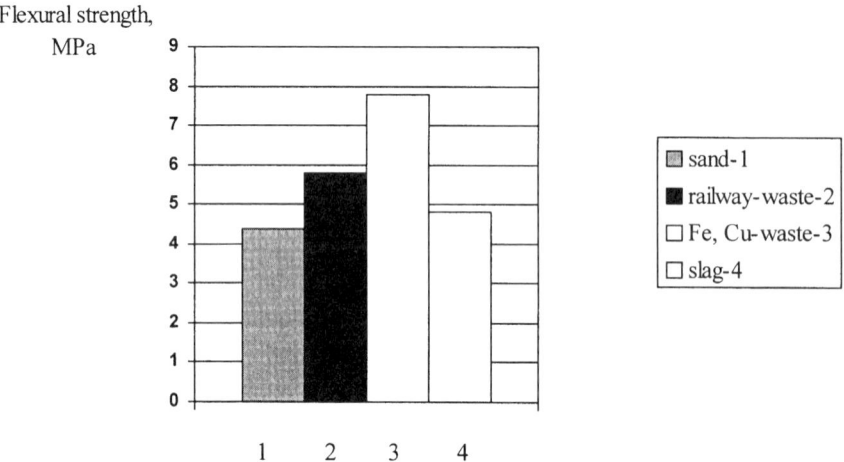

Figure 2. Relationship between flexural strength and natural waste

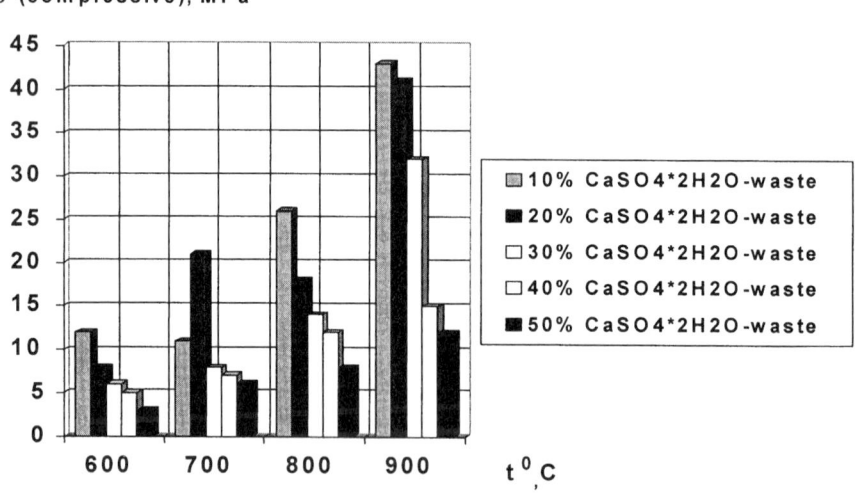

Figure 3. Relationship between firing-temperature and compressive strength

Cold-resistance, circles

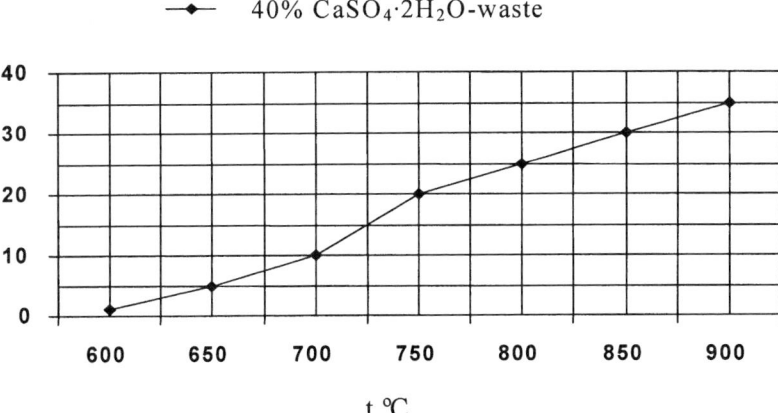

Figure 4. Relationship between cold resistant and firing-temperature

On the Figure 5 it can see the relationships between amount sand or s-waste ($CaSO_4*2H_2O$) and there is no difference between natural product like sand and CaSO4*2H2O-waste. In the tables 3,4,5,6,7 is been shown the other properties bricks with s-, p- and d-waste. And every time both compressive strength and flexural strength are higher for experimental bricks than manufactured ones. Apart from that it can see that bricks have high cold-resistance and the latter is especially impotent for cold country like Russia.

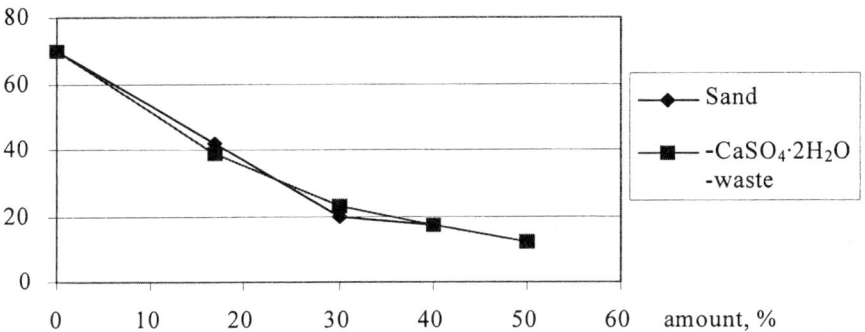

Figure 5. Relationship between compressive strength and amount of filler.

Table 3. Bricks with Fe- Cu-waste

CONTENT	COMPRESSIVE STRENGTH, MPa	FLEXURAL STRENGTH, MPa	CRACK RESISTANT	WATER ABSORPTION, %
Manufactured bricks	9,48	2,6	0,27	7,8
5% Fe-, Cu-waste bricks	12,1	4,4	0,36	6,7
7% Fe-, Cu-waste bricks	11,2	3,2	0,28	7,1

Table 4. Properties of the bricks with oxide-waste

CONTENT	COMP. STRENGTH, MPa	FLEXURAL STRENGTH, MPa	COLOR	WATER ABSORPTIO, %	t^0 firing
Clay- waste-100%	14,4	5,0	Scarlett	10,5	960
Clay – waste 90% MnO_2 -10%	20,5	9,6	Deep brown	7,5	960
Clay – waste –90%; (Fe_2O_3) - waste-10%	18,0	9,2	Deep - red	8,8	960
Clay – waste –90% TiO_2 -10%	16,9	8,9	Yellow	10,0	960
Clay – waste –90% CuO – 10%	17,2	10,1	Black	7,1	960

Table 5. Properties of the bricks with slip

CONSIST OF THE SLIP	COLOUR	COLD RESISTANT, circles
Clay - 90% TiO_2 – 10%	Light-yellow	25
Clay - 95% TiO_2 – 5%	Yellow	25
Clay - 90% (Fe_2O_3)-waste - 10%	Red-brown	25
Clay - 90% Cr_2O_3 - 10%	Deep-brown	25

Table 6. Properties of the bricks with railway-waste

PROPERTIES OF THE BRICKS	COMPRESSIVE STRENGTH, MPa	FLEXURAL STRENGTH, MPa
Bricks with sand (20%)	. 10,8	3.0
Bricks with railway-waste	14,8	5,2

Table 7. Bricks with Fe- Cu-waste

CONTAIN	COMP. STRENGTH, MPa	FLEXURAL STRENGTH MPa	WATER ABSORPTION, %	t^0, firing	COLD RESISTANT, circles
Clay-80% Sand-20%	11,1	4,5	18,0	980	20
Clay -80% Fe, Cu-waste-20%	16,9	7,6	11,3	980	29

CONCLUSIONS

1. Two classifications are suggested to fore cost an ability sold-waste to be useful for bricks – obtaining technology. One is according to the position in the Mendeleev's table, and the other according to the regularity contact – zone forming.

2. The improving building properties with solid-waste ceramics are being shown.

ACKNOWLEDGMENTS

The authors would like to acknowledge the scientists of the department for discussing problems.

REFERENCES

1. SVATOVSKAJA, L.B.: Engineering and chemical problems of the foam materials of the third millennium, St. Petersburg, 1999, pp32-46.

2. SVATOVSKAJA, L.B.: Modern natural scientific bases in the science of the constructive materials and ecology, St. Petersburg, 2000.

3. SVATOVSKAJA, L.B., MASLENNIKOVA, L.L., SOLOVYEVA V.Y.: Basic approaches for creation of the new complex technologies for biosphere purification, St. Petersburg state university of Transport, St. Petersburg, 2003.

4. BAIDARASHVILI, M.M.: Adaptation of the indicating method for biosphere purification from heavy metal ions, St. Petersburg state university of Transport, St. Petersburg, 2001.

GEOTECHNICAL ASPECTS OF RE-CYCLED CONSTRUCTION WASTES

V Sivakumar

Queen's University of Belfast

D Glynn

Central Procurement Directorate in Northern Ireland

United Kingdom

ABSTRACT. P erformance of various construction waste materials in relation to possible geotechnical applications is examined in this paper. Four materials were considered: Freshly quarried basalt (for comparison purpose), quarry waste, building debris and crushed concrete. These m aterials w ere e xamined u nder v arious t est c onditions; n amely dry, w et a nd m ixed with 10% and 20% clay slurry. Tests were also carried out to examine how the performance of these recycled waste materials could be enhanced using geo-grids. Laboratory tests were carried out in a 305mm×305mm direct shear box. The results show the performance of the re-cycled materials is not significantly different to that of freshly quarried basalt, although the conditions under which the products were tested (ie dry, wet and smeared with clay slurry) appeared to influence their performances in a significant way. Recycled construction wastes are known to be susceptible to crushing under repeated loading. Accordingly, recycled construction wastes were subjected to repeated loading under three different test conditions: (a) freshly recycled wastes (b) soaked in clean water for a long period and tested under wet conditions and (c) soaked in contaminated water (containing concentrations of sulphates) and subsequently tested under wet conditions. The results show the recycled waste material exposed to water contaminated with sulphate exhibited a considerable a mount of c rushing during repeated loading but not particularly different from dry or wet materials.

Keywords: Construction demolition waste, Sustainable construction, Recycled aggregates, geo-grids, ground improvement.

Dr V Sivakumar is a Lecturer in the School of Civil Engineering, Queen's University Belfast, His research interests include the recycling and reuse of waste materials, modelling of saturated and unsaturated soils and ground improvement.

Mr D Glynn is a C hartered C ivil E ngineer a nd P rincipal G eotechnical Engineer w ith t he Department of Finance and Personnel, Central Procurement Directorate in Northern Ireland.

INTRODUCTION

Granular aggregates are widely used for concrete production, backfilling, highway/embankment construction, and sub-bases/ballast for roads and railway tracks, etc. They are also used for ground improvement purposes such as vibrated stone columns, gabion walls, slope stabilisation, load transfer platforms, etc. Large quantities of freshly quarried stones are normally used in these applications. However, it is possible for recycled construction waste products to meet the necessary design requirements of such applications and, assuming they are a practical alternative, a considerable amount of savings could result. This paper focuses on the performance of various recycled construction wastes in relation to possible geotechnical applications.

The UK construction industry faces significant concerns about material supply in the future. Natural mineral resources are not unlimited and the extraction causes increasingly unacceptable environmental damage and raises health and safety concerns. The industry also produces a large amount of waste material (concrete, brickwork, cement mortar), at about four times the normal rate of household waste production (Coventry and Guthrie, 1998), generated from activities such as demolition, excavation, site preparation and a range of other activities. Waste from construction sites in the United Kingdom totals some 70 million tonnes per annum. A report by Sir John Egan (1998), on the improvement of the construction industry, highlighted the need for eliminating or recycling waste materials and reducing construction costs.

There is great potential for minimising waste without compromising the quality of the design or the functionality of the project. The construction industry can make savings in material purchasing and waste disposal costs through an increased emphasis on waste reduction, reuse and recycling. The landfill levy in the UK will also alter the economics of choice between new and reclaimed materials because reused materials are exempt from this landfill 'tax'- they are not strictly classified as waste. Savings from waste minimisation should be welcomed in view of the survey published in 1994 (Latham, 1994), which highlighted the scope for increasing the efficiency of UK construction practices.

NATURE OF CONSTRUCTION WASTE

Construction wastes are mainly the result of demolition, rehabilitation of railway tracks and highway construction processes. These materials typically comprise heterogeneous materials with particle sizes ranging from gravel to boulders. In some cases unused construction materials are also categorised as construction wastes. They include harden 'Ready Mix' concrete, aggregates, unsuitable bricks and pavement blocks. As a consequence, there is considerable variation in mechanical strength both for each of the categories and for individual particles within each group. Such materials can be recycled and reused as replacement for more expensive traditionally-used coarse granular materials such as crushed rock.

The performance of all construction materials is strongly influenced by the conditions in which they are used. For example, the strength of dry stone may be different from wet stone. Stone mixed with fine material may behave differently from aggregates in wet or dry conditions. Smearing or clay contamination of stones can be a particular problem in many civil engineering applications eg, filter drains, stone columns, sub-bases for roads and railway track ballast, etc. (Bell et al., 1981). In addition, if the reuse of waste material is considered a viable alternative, these materials are often found in a poor condition at the site. For example, they can be covered with dust/mud or experienced softening due to prolonged exposure to rain/water. In addition, the performance of these materials is affected by repeated loading (Sivakumar at al 2004). The recycled construction wastes are particularly p rone to particle crushing during shearing. The amount of crushing experienced by recycled wastes is also affected by the condition in which they are used, namely when soaked in water and/or when they come into contact with deleterious chemicals or agents. The purpose of this paper is to report some of the work carried out at Queen's University Belfast in relation to the reuse of construction waste for geotechnical applications.

LABORATORY EXPERIMENTS

The laboratory work consisted of testing various coarse-grained or granular materials which included: (a) 40 mm size uniformly graded crushed basalt (b) crushed concrete prepared in the laboratory (c) building debris and (d) quarry waste. The materials were tested under various conditions: (a) dry (b) wet (c) smeared with clay slurry (d) long term soaking in water (e) long term soaking in water contaminated with sulphate. In addition, tests were also performed to examine the effects of introducing geo-grids in an attempt to enhance the performance of the recycled materials.

Material Types
The 40mm uniformly crushed rock was obtained from a local supplier (Whitemountain Quarries, Belfast). The crushed concrete was generated from concrete cubes, after they had been tested in the Material Testing Station at Queen's University Belfast. The cubes had an average compressive strength of 35 MN/m^2. The cubes were crushed and the resulting material was sieved. Particles having sizes between 20 and 40 mm were considered to be suitable for testing. The building debris was obtained from a convenient demolition site. The debris contained predominantly bricks (at least 95% by weight) and mortar. Wood content was removed and the debris was crushed and then sieved. Particle sizes between 20-40 mm were again considered suitable for testing. Quarry waste was supplied by John McCleery Quarries, a local quarry near Ballymactaggart, Co. Down. The material is commercially referred to as quarry dust.

Equipment
Tests on crushed rock, crushed concrete and building debris were carried out in a large direct shear box having dimensions of 305×305 mm. The large box is suitable for materials containing particles up to 50 mm (Head, 1994). Tests on quarry waste were carried out in a 100×100 mm shear box, which is better suited to materials containing smaller particles. The tests used four different vertical pressures. In the case of crushed rock, concrete and building debris, the materials were compacted in the box in three layers. Each layer was compacted using a P roctor h ammer. S amples w ere s heared at 1 .5 m m/min. D uring s hearing t he s hear displacement, vertical displacement and the shear load applied to the sample were carefully recorded.

Material Preparation
Tests were carried out in four stages. In the first stage, the materials were tested under different conditions namely: dry, wet and mixed with 10% and 20% kaolin slurry. For tests on dry materials, the products were placed in the direct shear apparatus without any preparation. For 'wet' conditions, the specimens were allowed to soak in water overnight before testing. For tests on smeared materials, the following procedure was adopted: The required amount of

dry kaolin powder was determined by the dry mass of the material to be tested in a mix ratio of 10:2. For example, 100 kg of dry material required 20 kg of dry kaolin powder in order to achieve 20% smearing. The powdered kaolin initially mixed with water to form a slurry 1.5 times the liquid limit. The slurry and the dry material were thoroughly combined in a concrete mixer. In the second stage of the programme, the above mentioned waste materials were tested under similar initial conditions but layers of geo-grid were included in the sample makeup to provide added strength to the test specimen. In the third stage of the testing, freshly crushed basalt, crushed concrete and building debris were tested in the large direct shear box in order to examine the performance of these materials under repeated loading. Samples were loaded and unloaded up to 8 times (single loading, 2, 4 and 8 loading cycles). When the cyclic loading tests were completed particle size analysis was carried out to determine the amount of particle crushing induced. In the forth and the final stage of the testing programme, crushed concrete were subjected to repeated loading up to 4 times. These samples were tested under three different conditions: (a) dry (b) long-term soaking in water and (c) long-term soaking in water contaminated with sulphate.

TEST RESULTS AND EVALUATION

For simplicity only a limited number of results are reported here. For further information refer to publications: McKelvey at al. (2002), Touhamia et al. (2002) and Sivakumar at al. (2004). During the shearing process, the samples were taken to the maximum shear displacement which could be applied in one direction in the large direct shear box apparatus. This corresponds to a shear strain of about 10% in all the tests. Most of the samples tested reached peak shear stresses at this strain. Samples tested in the wet and dry conditions exhibited the typical response of dense granular materials; initially showing a slight compression followed by a significant dilation. The applied vertical pressures did not significantly affect the amount of dilation of the samples. However, in the case of samples mixed with clay slurry, the volume change characteristics were untypical. The samples tested at low vertical pressures showed a tendency for dilation whereas the samples tested at high vertical pressures showed little or no volume change in the case of 10% mixing and a significant amount of compression in the case of 20% mixing. This may have been due to the lubricant effect provided by clay slurry. The maximum shear stresses achieved during the tests associated with vertical pressures were used to determine the strength parameters for each material.

Figure 1 shows the relationship between the maximum shear stresses plotted against the maximum vertical pressures for crushed rock and concrete tested with the four different initial conditions (results obtained on crushed building debris is not shown here). In the case of crushed rock, it appears that the slope of the failure envelope, $\square$', is quite significantly influenced by the testconditions. The $\square$' for dry material is approximately 51°. This soil parameter reduced to 48° when wetted,. This further reduced to 42° was observed when mixed with 10% clay slurry. However, the amount of slurry had a less significant effect on the frictional angle- with the 20% clay slurry content - only causing a further small reduction in the $\square$' value to reach 40°. It also appears that the stones mixed with 20% clay slurry exhibited an apparent small amount of cohesion with c' = 7 kN/m^2 being measured. This phenomenon can possibly be explained by the strength of the slurry, where it is either trapped between particles or contained within the void spaces. The slurry was applied to the dry crushed rock at 1.5 times the liquid limit. Since the crushed rock was dry when the slurry was applied, this may have resulted in some water being absorbed from the slurry causing a marginal gain in strength in the slurry paste itself. Similar behaviours were observed with all other materials.

The best performance in relation to shear strength of the various materials tested was obtained for the traditionally quarried crushed basalt, which showed an angle of internal friction of 51° in the range of pressures used in the tests. This value reduced to 40° when smeared with 20% kaolin slurry. These findings are comparable with results from a previous study (Bell et al., 1981) where tests were carried out to examination the effect of clay contamination in pavement sub-bases. Recycled crushed concrete yielded a value of 39° in the dry state, reducing to 30° when smeared with kaolin slurry. The building debris gave a result of 37° in the dry state but this reduced to 35° when tested in the wet state. The crushed concrete exhibited a reduction of about 10° in the angle of internal friction compared with that of high quality crushed rock. The reduction for the quarry waste was 5° to 7° depending on test conditions. For building debris, the reduction may be as large as 13° though no tests were actually carried out on smeared building debris. A reduction in □' amounting to 10° will reduce the overall shear resistance by 60%. Designers should also consider that recycled materials may experience particle crushing when subjected to shearing. The amount of particle crushing is most significant in situations where these materials are used under repeated loading.

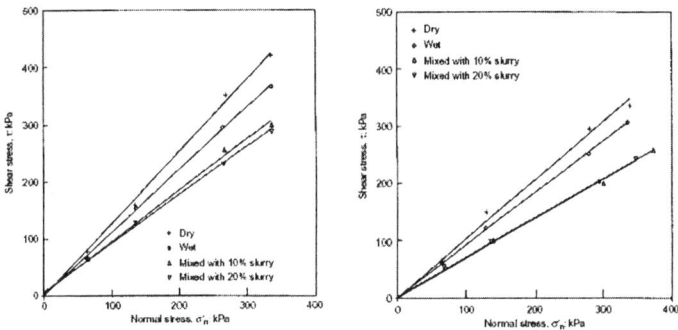

Figure 1 Failure envelopes for crushed rock and concrete

Figures 2 shows the relationship between the maximum shear stress and the maximum vertical pressure for all three materials tested under 1,2,4 and 8 loading cycles. An estimated angle of internal friction ϕ' for basalt under one loading is approximately 47 degrees but this is reduces to 45 degrees after 8 repeated loading (Note- the material in this testing programme is different from the material reported previously). The reduction in the friction angle is small and negligible for practical applications. In the case of crushed concrete, the angle of internal friction, ϕ', is approximately 43 degrees based on the first loading cycle, which reduces to 38 degrees after 8 loading cycles. The reduction in the angle of internal friction due to repeated loading is significant and cannot be ignored.

A similar performance was observed when building debris was taken through the sequence of loading cycles. The angle of internal friction ϕ' was approximately 43 degrees when the analysis was based on the first loading cycle but it reduced to 39 degrees when the sample was taken through 8 loading cycles. The reduction in the angle of internal friction of these materials was largely due to particle crushing caused by the repeated loading. Immediately following the test the specimen in the shear box was carefully examined to assess the amount of particle crushing. Three distinct layers were identified when crushed concrete and building debris were subjected to repeated loading: (a) the top layer generally contained intact material and it showed no significant evidence of particle crushing. (b) the middle layer contained distinctively different material from the original material used, particularly in the case of crushed concrete. The intense shearing process stripped off the mortar (hardened cement paste) surrounding the 12mm aggregates, which w ere originally used in m anufacturing the concrete. This formed a thin lens of 12 mm aggregate at mid-specimen level and (c) the bottom layer contained smaller particles and the stripped hardened mortar from the concrete along with the original material used in preparing the specimens.

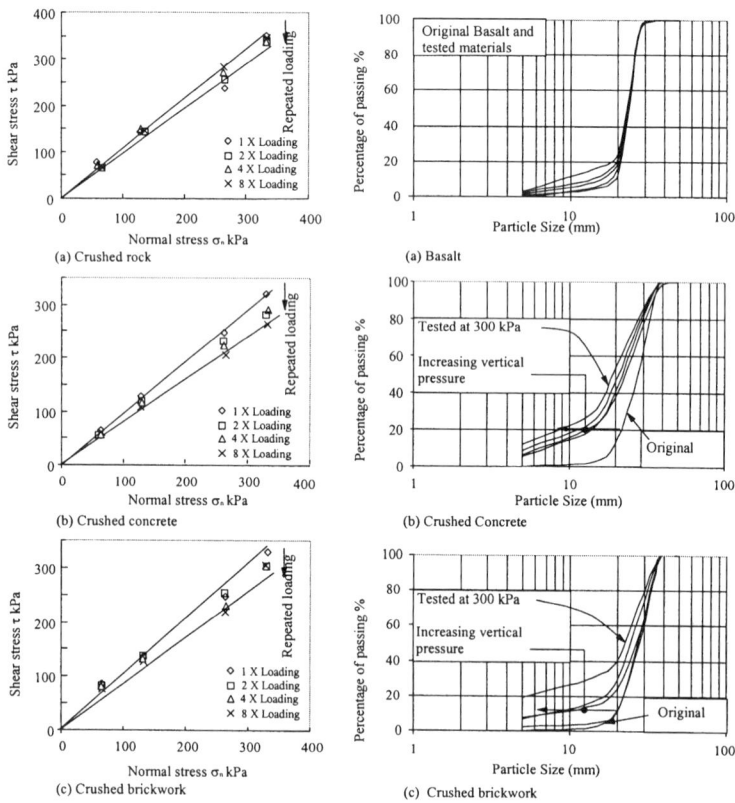

Figure 2 Failure envelopes during repeated loading. Figure 3 Particle size analysis on tested material

After each test and upon completion of the particular loading cycle, the tested material was sieved in order to establish the particle size distribution. Figure 3 shows the particle size distribution for basalt, crushed concrete and crushed brickwork subjected to a number of loading cycles under various vertical pressures. As can be seen, basalt was more resistance to particle crushing. The coefficient of uniformity C_u ($C_u = D_{60}/D_{10}$) of the original material was approximately 1.15. This increased slightly to 2.3 after 8 loading cycles when sheared under 300 kPa of vertical stress. It appears then that crushed concrete and building debris are very susceptible to particle crushing. The amount of crushing experienced by the re-cycled products appears to have increased with the magnitude of vertical pressure and the number of loading cycles. The coefficient of uniformity C_u of the original crushed concrete and building debris were 1.15 and 1.6 respectively. Following the shearing process through 8 loading cycles under 300 kPa of vertical pressure, the respective uniformity coefficients increased to 4.5 and 7 for crushed concrete and brickwork.

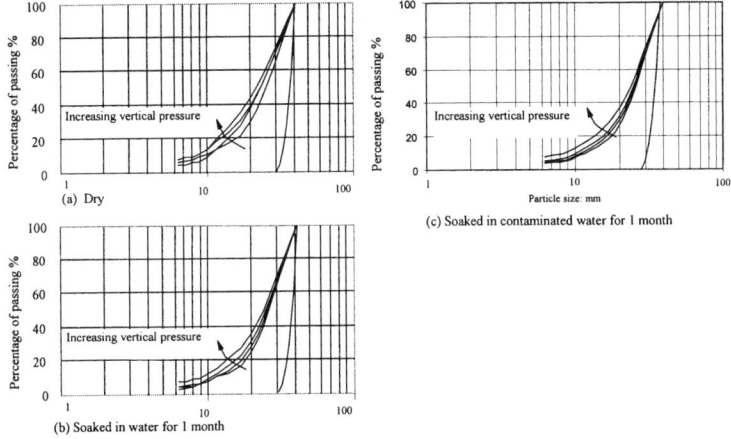

Figure 4 Particle size analysis on dry, soaked and soaked with contaminated water

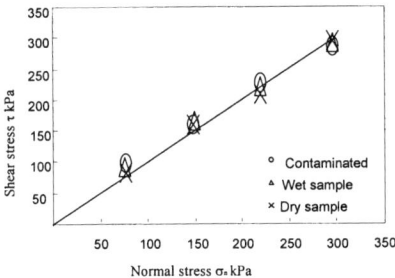

Figure 5 Failure envelope on dry, soaked and soaked in contaminated water
(after 4 loading cycles)

It is believed that the performance of recycled materials may deteriorate with time if they are exposed to harsh environs, such as the chemically contaminated water used in this testing regime. This will be a particular problem when using recycled concrete and building debris. Ingress of water into the crushed concrete and building debris can lead to an increased susceptibility to particle crushing under repeated loading. The presence of any contaminants in the water, particularly sulphates, can accelerate the weakening of these recycled materials. A series of tests were performed to examine the performance of crushed concrete during long term storage or exposure to water and sulphate-contaminated water. Tests were performed under repeated loading conditions in which the samples were subjected to four loading cycles under various vertical pressures. Particle size analysis was carried out to evaluate the degree of crushing in each case. Figure 4 shows the particle size distribution of the tested materials after four loading cycles, whilst Figure 5 shows the strength envelopes of the crushed concrete after the same loading cycles. It appears that there is no significant difference in the angle of internal friction of the crushed concrete (at least after four loading cycles) regardless of the testing conditions. However the recycled concrete underwent a considerable amount of particle crushing. The amount of crushing is not particularly affected by the testing conditions (ie dry and soaked in water and contaminated water). Further tests were carried out to examine ways in which the performance of the recycled materials can be improved. Initial consideration was given to include geo-grids in the sample materials tested in the large shear box Touhamia et al. (2002). The results have shown that the angles of internal friction of the recycled materials can be increased by at least 10° by incorporating geo-grids in the backfill operation using recycled construction wastes.

CONCLUSIONS

It is known that the availability and ready supply of high-quality natural aggregates poses stiff competition for the reclaimed aggregates industry. However, the market is large and there are opportunities to increase the use of reclaimed aggregates. Concrete, particularly high-grade crushed concrete and building debris such as brickwork, have considerable potential for use in general bulk backfill operations/applications. The strength requirement in such applications is generally modest and deformations are not considered to be critical. In addition, these materials can be used in various ground improvement applications without necessarily compromising the quality of the design. The work reported here examined the performance of various recycled construction wastes. The results show that their mechanical performance is satisfactory for general geotechnical applications envisaged. However, one of the major problems or drawbacks associated with these recycled waste products is their susceptibility to crushing or 'crushability' and particulate breakdown under intense repeated loading regimes.

ACKNOWLEDGEMENTS

Funding for the research was provided by P.J Carey Contractors Ltd, Wembley, Middlesex, London.

REFERENCE

1. BELL, A.L., MCCULLOUGH, L.M. and GREGORY, B.J. (1981) Clay contamination in crushed rock highway sub-bases. Proc. of the 2nd Australian Conference on Engineering Materials. Sydney.

2. COVENTRY, S. and GUTHRIE, P.M. (1998) Waste minimisation and recycling in construction – design manual. CIRIA SP134.

3. EGAN Sir J. Rethinking construction: the report of the Construction Task Force to the Department of the Environment, Transport and the Regions. DETR, London, 1998.

4. HEAD KH. Manual of soil laboratory testing – volume 2, 2nd edn. John Wiley & Sons, Inc. 1994.

5. LATHAM M. Constructing the team: joint review of procurement and contractual arrangements in the United Kingdom. BRE Construction Industry Final Report 1994.

6. McKELVEY, D., SIVAKUMAR. V, BELL, A. and McLAVERTY. G.(2002). Shear strength of recycled construction materials intended for use in vibro ground improvement. Ground Improvement Journal, Vol 6, pp 59-68.

7. TOUHAMIA. M., SIVAKUMAR., V and McKELVEY, 2002. Application of Geo-grids in re-use of waste materials. Construction materials, Vol. 16

8. SIVAKUMAR V., McKINLEY J. D. and FERGUSON.D. Reuse of construction waste: performance under repeated loading. Geotechnical Engineering, Vo. 157, No. 2 pp 91-96.

CONCRETE AND MORTAR PERFORMANCE BY USING RECYCLED AGGREGATES

V Corinaldesi

G Moriconi

Marche Polytechnical University of Ancona

Italy

ABSTRACT. In the light of the wide availability of rubble from building demolition to be recycled, several fields of employment other than roadbeds or floor foundations have been examined in this experimental activity. In order to achieve 100 % recycling of all materials produced by a real recycling plant, not only recycled aggregates were used for manufacturing concrete but also for preparing mortar and self-compacting concrete. The experimental results showed that recycled aggregate fractions up to 15 mm size, although containing masonry rubble at a level of 25-30 %, proved to be suitable for manufacturing structural concrete. For this, they could also be employed as a total substitution of the fine and coarse natural aggregate fractions. Moreover, an attempt was made to improve the quality of the recycled aggregates for concrete, by reusing in an alternative way the most detrimental fractions, that are the materials coming from masonry rubble and the fine recycled aggregate fraction. Finally, even the finest fraction produced during the recycling process, that is the rubble powder, was reused as a filler for producing self-compacting concrete. By means of this approach surprising and unexpected performance could be achieved, particularly for mortar, which showed excellent bond strength with bricks, and for self-compacting concrete showing an optimized rheological behaviour.

Keywords: Construction and demolition waste, Recycled-aggregate concrete, Recycled-aggregate mortar, Rubble powder, Self-compacting concrete.

V Corinaldesi is a Civil Engineer and has a Ph.D. in Materials Engineering; she is developing a research project on recycling of building materials at the Department of Materials and Environment Engineering and Physics at the Marche Polytechnical University of Ancona, Italy.

G Moriconi is Professor of Materials Science and Technology at the Marche Polytechnical University of Ancona, Italy and is the Head of the Department of Materials and Environment Engineering and Physics. He is the author and co-author of numerous papers in the field of cement and concrete technology, building materials performance and durability. He has been recently awarded by ACI for his contribution to the research development in the general field of concrete durability.

INTRODUCTION

In the recent past, several authors [1-3] studied the possibility of using recycled aggregates to prepare structural concrete and they showed that, in recycled-aggregate concrete, the fine recycled aggregate fraction seems to be particularly detrimental to both mechanical performance and durability of concrete.

In this work, the possibility of reusing the fine aggregate fraction in other ways was examined. The paper is a final report of a wide experimental program aimed at evaluating the possibility of 100 % recycling of rubble from building demolition by reusing them as aggregates for either structural concrete or mortar. Moreover, the possibility of properly treating and grinding these materials, in order to produce a mineral addition for preparing self-compacting concrete, was evaluated too.

RECYCLED-AGGREGATE CONCRETE

The mechanical behaviour of recycled-aggregate concrete was tested out on specimens manufactured by fully replacing natural with recycled aggregates. A coarse fraction (15 mm maximum size) and a fine fraction (6 mm maximum size) of recycled aggregate were used. These fractions were directly supplied by an industrial crushing plant in Villa Musone, Italy, in which debris from building demolition are currently disposed and suitably selected, ground, cleaned and sieved. The average weight composition of this recycled aggregate, resulting from three-year monitoring, consisted of concrete (70%), masonry (27%), bitumen (2%) and miscellaneous (1%). The specific gravity of the recycled aggregates was 2320 and 2150 kg/m^3 for the coarse and the fine recycled fractions respectively and their water absorption was 8% and 10% respectively.

At the same time, concrete specimens were prepared as reference by using two ordinary natural aggregate fractions with the same grain size distribution of the coarse and fine recycled aggregate fractions. The influence on concrete performance of the different type of aggregate, recycled or natural, was assessed with parameters such as sieve distribution, maximum diameter, type of cement, workability, curing and test conditions, all kept constant. This evaluation was carried out by means of compression tests performed on specimens prepared by varying water/cement (0.40, 0.50 and 0.60) and curing time (1, 3, 7, 28 days).

Further evaluation concerned specimens manufactured with recycled aggregates in the presence of an acrylic-based superplasticizer, added to the mixtures at a dosage of 1.8% by weight of cement.

The mixture proportions are reported in Table 1. Fresh concrete always had a fluid consistency (slump between 16 and 20 cm). The cubic specimens were cast into polystyrene formworks and wet cured at 20°C.

Results and Discussion
The average values of triplets of compression tests are reported in Table 2.

Table 1. Mixture proportions for concretes with natural aggregate (NA), recycled aggregate (RA) and recycled aggregate plus superplasticizer (RAS)

| MIXTURE | W/C | MIXTURE PROPORTIONS, kg/m³ | | | | |
| | | Water | Cement | Aggregates | | Superpl. |
				coarse (5-15 mm)	fine (0-6 mm)	
NA04	0.40	255	635	1002	363	-
NA05	0.50	255	520	911	548	-
NA06	0.60	255	420	824	717	-
RA04	0.40	255	635	931	216	-
RA05	0.50	255	520	850	378	-
RA06	0.60	255	420	768	529	-
RAS04	0.40	170	420	882	667	7.6
RAS05	0.50	170	345	759	783	6.2
RAS06	0.60	170	275	751	839	4.9

Table 2. Compressive strengths of concretes with natural aggregate (NA), recycled aggregate (RA) and recycled aggregate plus superplasticizer (RAS)

| MIXTURE | VOLUMIC MASS, kg/m³ | AVERAGE COMPRESSIVE STRENGTH, MPa | | | | R_{ck}, MPa |
| | | Curing Time, days | | | | |
		1	3	7	28	
NA04	2250	25.6	34.6	36.6	46.1	40
NA05	2237	17.3	24.8	30.2	40.7	35
NA06	2216	9.6	20.5	23.6	34.3	30
RA04	2127	21.4	27.3	30.0	34.8	30
RA05	2065	12.8	20.0	21.0	30.8	25
RA06	2005	10.2	15.0	18.0	24.4	20
RAS04	2107	11.1	20.1	24.8	32.0	25
RAS05	2015	6.9	13.6	17.7	24.0	20
RAS06	1945	5.6	11.2	14.0	18.1	15

By comparing the mechanical performances after 28 days of 'NA06', 'RA06' and 'RAS04' mixtures, all prepared with the same cement dosage, it can be clearly noted that the strength loss due to the presence of recycled aggregates can be compensated by the W/C decrease obtained by adding the superplasticizer (RAS04).

In general, the strength loss of RA (without superplasticizer) with respect to NA concrete decreases for high W/C or early curing time. The reason could be that a weak cement matrix tends to attenuate the negative effect of a poor-quality aggregate. Another interesting comparison can be made between the 'RA06' and 'RAS06' mixtures with the same W/C, i.e. with the same cement matrix quality, but with a different dosage of the fine fraction (Table 1). The reason of the weakness of the concrete specimens prepared by adding the superplasticizer (Table 2) probably lies in the very large volume filled up by the fine recycled aggregates.

RECYCLED-AGGREGATE MORTARS

As shown by several authors [1-3], the presence of masonry in concrete rubble harms the mechanical performance of recycled-aggregate concrete and, as already shown, the same negative effect is detectable when natural sand is replaced by the fine recycled aggregate fraction. The strength loss can be counteracted by adopting appropriate measures, such as the reduction of water to cement ratio and the use of mineral additions [4]. However, all these actions lead to reduced use of fine recycled material, which comes to be only partially employable in concrete. An alternative use of both masonry rubble and surplus fine recycled material could be in mortars, in which recycled instead of natural sand or powder obtained by brick grinding could be used as a partial cement substitution. Both attempts were carried out within this experimental activity.

Several mortars were prepared; the proportions of their mixtures are given in Table 3. The cement to sand ratio was 1:3 (by mass); the water content of each mortar was set to achieve the same consistence of 110±5 mm, evaluated according to EN 1015-3. When recycled sand (RA) was used, a higher water dosage was necessary to achieve the same workability, because of the higher water absorption of the recycled sand with respect to the natural one.

Table 3. Mortar mixture proportions

MIXTURE	W/CM	MIXTURE PROPORTIONS, kg/m³			
		Water	Cement	Natural Sand	Recycled Aggregate	Brick Powder
Ref	0.50	225	450	1350	-	-
BP	0.50	250	315	1350	-	135
RA	0.67	300	450	-	1350	-

Results and Discussion
Prismatic specimens (40 x 40 x 160 mm) were manufactured, cast and wet cured at 20°C. The compressive strength was evaluated according to EN 196-1. The results obtained are reported in Fig. 1. Mortars containing brick powder (BP) and recycled aggregate (RA) have lower compressive strengths with respect to the reference cementitious mortar (Ref).

In order to evaluate the bond strength developed during the shearing of a brick with respect to another brick along a mortar layer 10 mm thick, a test method, derived from the draft European Standard prEN 1052-3 [5], was adopted. In this way the masonry behaviour in the absence of normal stress was investigated, corresponding to the constant term in the Mohr-Coulomb friction law.

The tested model, shown in Figure 2, is composed of three bricks; it has a symmetric structure thus avoiding eccentric loads. The applied load (L) was measured and, at the same time, the vertical displacement of the central brick (δ) was also monitored. Usually, at the end of the test only one joint was cracked, so the bond strength was calculated by dividing half maximum load by the fracture area where brick and mortar were in contact (approximately 120 x 200 mm); test results are also included in Fig. 2. In particular, very high bond strength was obtained by coupling red bricks, characterized by a higher value of I.R.A. (Initial Rate of Absorption), with recycled-aggregate mortar (RA). For further details concerning this part of the experimental activity, see [6, 7].

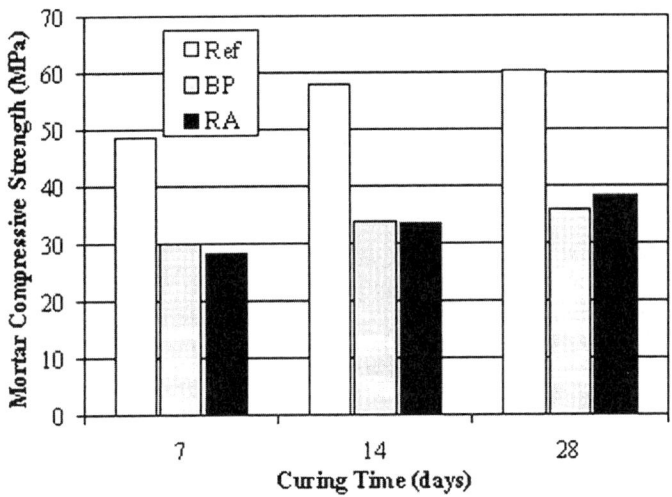

Figure 1. Compressive strength of the mortars as a function of curing time.

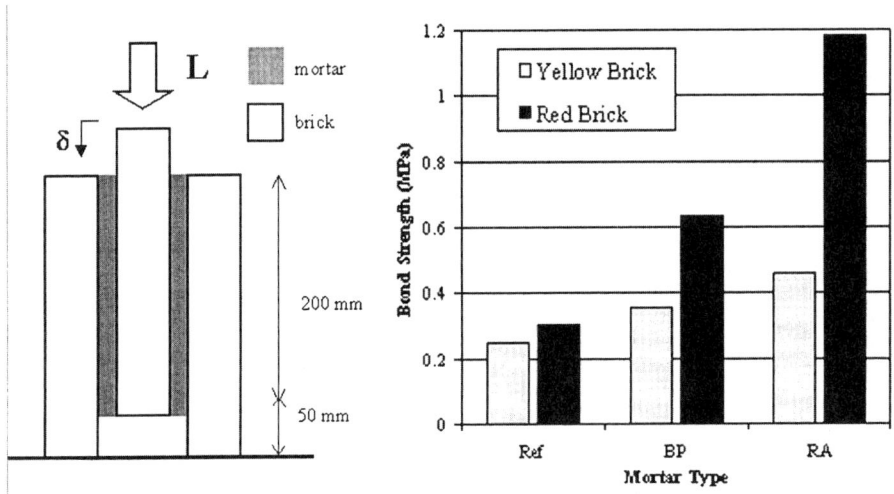

Figure 2. Masonry model and maximum values of the mortar-brick bond strength.

SELF-COMPACTING CONCRETE CONTAINING RUBBLE POWDER

Finally, a third experimental step was carried out concerning self-compacting concrete by using the dust produced during the rubble processing as a filler for manufacturing self-compacting concrete. In fact, this dust proved to be particularly detrimental for the mechanical performance of the recycled-aggregate concrete, due to its high water absorption, and the possibility of reducing its content in the fine recycled-aggregate fractions can improve concrete quality.

The mixture proportions of two self-compacting concretes are reported in Table 4. Both concretes were prepared with the same W/C of 0.45. In order to optimize the grain size distribution of solid particles in the concretes, the fine and the coarse aggregate fractions were combined at 28% and 72% by volume respectively, keeping into account also the suggestions reported in the literature concerning the mixture proportion of self-compacting concrete [8,9].

The volume of very fine particles was equal to 190 l/m^3. Two different mineral additions, rubble powder and limestone powder, were used at a dosage of 100 kg and 120 kg respectively. Rubble powder was obtained from the recycling process of construction and demolition waste, by collecting the material passing through the sieve ASTM n° 170 of 90 $\Box$m. This rubble powder had a Blaine fineness of 0.99 m^2/g and a specific gravity of 2.15 kg/m^3.

The limestone powder used was obtained as a by-product of marble working. Its Blaine fineness was 0.59 m^2/g and its specific gravity was 2.65 kg/m^3. An acrylic-based superplasticizer was employed at a dosage of 1.0% and 1.2% respectively in the case of limestone powder or rubble powder addition, due to their different fineness.

Table 4. Self-compacting concrete mixture proportions

MIXTURE	W/C	MIXTURE PROPORTIONS, kg/m³						
		water	Cement	Limestone powder	Rubble powder	Natural sand	Crushed aggregate	Superpl.
SCC+LP	0.45	200	440	120	-	1110	430	4.4
SCC+RP	0.45	200	400	-	100	1110	430	5.3

Results and Discussion

As a first step, properties of the fresh concrete other than slump were evaluated, since in this case the slump value is not relevant due to a very fluid concrete. Therefore, the attention was focused on the measurement of the slump flow and on the L-box test with horizontal steel bars. Compression tests according to Italian Standards UNI 6132-72 were carried out on cubic specimens, which were tested at right angles to the position of casting. The results obtained in terms of both fresh and hardened concrete performance are reported in Table 5.

Table 5. Performances of the fresh and hardened self-compacting concretes

MIXTURE		SCC+LP	SCC+RP
Slump flow test	Φ_{fin} [1] (mm)	750	700
	t_{500} [2] (s)	1	4
	t_{fin} [3] (s)	1	11
L-box test	ΔH_{fin} [4] (mm)	30	65
	t_{edge} [5] (s)	5	1
	t_{stop} [6] (s)	12	6
Compressive strengths (MPa)	curing time (days): 1	17.0	17.4
	3	25.4	27.8
	7	32.1	32.9
	28	40.1	40.9

(1) mean diameter of the slumped concrete;
(2) elapsed time to gain the mean diameter of 500 mm;
(3) elapsed time to gain the final configuration;
(4) difference in the concrete level in the opposite ends of the box;
(5) elapsed time to reach the opposite edge of the box;
(6) elapsed time to establish the final configuration.

In relation to the slump flow test, both concretes showed enough fluidity to be self-compactable; but a certain flow-segregation, with the presence of a halo of cement paste around the slumped concrete, was observed for the 'SCC+LP' concrete while the 'SCC+RP' concrete seemed to behave as a quite viscous system.

In relation to the L-box test, both concretes showed good results in terms of mobility through narrow sections. Concerning the flow-segregation, a certain separation between the coarse aggregate particles and the surrounding cement paste was again observed only in the case of the 'SCC+LP' concrete.

In terms of mechanical performance, the concretes prepared with either limestone powder or rubble powder performed similarly, with a 28-day compressive strength of 40 MPa.

CONCLUSIONS

Recycled-aggregate fractions up to 15 mm, although containing 25-30% masonry rubble, proved to be suitable for manufacturing structural concrete even if employed as a total substitution of the fine and coarse natural aggregate fractions.

Moreover, the exceeding fine fraction with particle size up to 5 mm, if reused as aggregate for mortars, allowed excellent bond strengths between mortar and bricks, in spite of a lower mechanical performance of the mortar itself. Also the masonry rubble can be profitably treated and reused for preparing mortars.

Finally, even for the finest fraction produced during the recycling process of construction and demolition waste, that is the rubble powder, an excellent way of reuse was found, that is as filler in self-compacting concrete.

In conclusion, the attempt to improve the quality of the recycled aggregates for concrete by reusing in alternative ways the most detrimental fractions, that are the material coming from masonry rubble and the finest recycled materials, allowed to achieve surprising and unexpected performance for mortar and self-compacting concrete.

REFERENCES

1. KASAI, Y., Reuse of Demolition Waste, Proc. of the 2nd Int. RILEM Symp. on "Demolition and Reuse of Concrete and Masonry", Chapman and Hall, London, U.K., 1988.

2. HANSEN, T.C., Recycling of Demolished Concrete and Masonry, Report of Technical Committee 37-DRC on "Demolition and Reuse of Concrete", RILEM Report 6, E & FN Spon, London, U.K., 1992.

3. DHIR, R.K., HENDERSON, N.A., LIMBACHIYA, M.C., Use of Recycled Concrete Aggregate, Proc. of the Int. Symp. on "Sustainable Construction", Thomas Telford Publishing, London, U.K., 1998.

4. CORINALDESI, V. AND MORICONI, G., Role of Chemical and Mineral Admixtures on Performance and Economics of Recycled Aggregate Concrete, Proc. of the 7th CANMET/ACI Int. Conf. on "Fly Ash, Silica Fume, Slag and Natural Pozzolans in Concrete", (Ed. V.M. Malhotra), ACI, Farmington Hills, MI, U.S.A., 2001, SP 199-50, 869-884.

5. prEN 1052-3: 1996, Methods of test for masonry; Part 3: Determination of initial shear strength. Draft for public comment, 1996.

6. CORINALDESI, V., GIUGGIOLINI, M., MORICONI, G., Use of Rubble from Building Demolition in Mortars, Waste Management, 22(8), 2002, pp. 893-899.

7. MORICONI, G., CORINALDESI, V., ANTONUCCI, R., Environmentally-friendly mortars: a way to improve bond between mortar and brick, Materials & Structures, 36(264), 2003, pp. 702-708.

8. BUI, V.K., MONTGOMERY, D., Mixture Proportioning Method for Self-compacting High Performance Concrete with Minimum Paste Volume, Proc. of the 1st Int. RILEM Symp. on "Self-Compacting Concrete", (Eds. A. Skarendahl & O. Petersson), Stockholm, Sweden, 1999, pp. 373-384.

9. JACOBS, F., HUNKELER, F., Design of self-compacting concrete for durable concrete structures, Proc. of the 1st Int. RILEM Symp. on "Self-Compacting Concrete", (Eds. Skarendahl & Petersson), Stockholm, Sweden, 1999, pp. 397-407.

THE COMPRESSIVE STRENGTH OF CONCRETE CONTAINING TILE CHIPS, CRUSHED SCALLOP SHELLS, OR CRUSHED ROOFING TILES

M Sugiyama

Hokkai Gakuen University

Japan

ABSTRACT. This paper addresses waste disposal through recycling; tile chips, crushed scallop shells, or crushed roofing tiles were mixed into concrete, and compressive strength was tested. The experiment was carried out in three parts. Series 1 measured the compressive strength and the static elastic modulus of concrete containing tile chips. That replaced 30% to 100% of the volume of the coarse and fine aggregate. The results showed that compressive strength increased as the proportion of tile chips was increased.

Series 2 measured the compressive strength of p orous concrete containing crushed scallop shells that replaced 50% to 100% of the volume of the coarse aggregate. The results showed that compressive strength decreased as the percentage of crushed scallop shells was increased. Series 3 measured the compressive strength and the static elastic modulus of concrete containing crushed roofing tile material that replaced 20% to 100% of the volume of the coarse aggregate. The results showed that compressive strength was increased when the percentage of roofing tile material was 20% to 60% of the volume of the coarse aggregate.

Keywords: Compressive strength, Crushed roofing tile, Crushed scallop shell, Static elastic modulus, Tile chips.

Dr. SUGIYAMA Masashi is a professor of the Faculty of Engineering, Hokkai Gakuen University, Sapporo, Japan. He received his doctorate in engineering from Hokkaido Univ. in 1981. His doctoral dissertation was "The Effect of Drying on the Physical Properties of Concrete." His specialty is research and development to improve the durability of concrete. He received a prize from the Japan Concrete Institute in 1988 for research into super-high-durability concrete. His interests include chemical admixtures, compressive strength, elastic modulus, freezing and thawing resistances and neutralization etc. He is a member of RILEM, ACI, JCI, AIJ and JSCE.

INTRODUCTION

The annual investment in construction in Japan amounts to about 70 billion yen, corresponding to about 14% of the gross domestic product. This activity produces about one hundred million tons of waste from construction sites, about 25% of the total amount of industrial waste in Japan [1, 2]. It is important to promote the recycling of construction materials as well as other commercial by-product in order to suppress the generation of waste. In this paper, tile chips, crushed roofing tiles, or crushed scallop shells were mixed into concrete, and compressive strength and static elastic modulus were tested.

SERIES 1: TILE CHIPS

Design of Experiments

Series 1 measured the compressive strength and the static elastic modulus of concrete containing tile chips. The mixing ratio of the tile chip was varied in the following manner;
  1) A test series in which the tile chips replaced 30% to 100% of the volume of coarse aggregate.
  2) A test series in which the tile chips replaced a fixed 30% of the coarse aggregate and 30% to 100% of the fine aggregate.
  3) A test in which the tile chips replaced 100% of both the coarse and fine aggregate.

Mix Proportions

Mix proportions are shown in Table 1.

Table 1. Mix proportions

SPECIMEN	TILE CHIP AGGREGATE		W/C (%)	RIVER GRAVEL (L/m³)	TILE COARSE Agg.	RIVER SAND (L/m³)	TILE FINE Agg	Sl. (cm)	AIR (%)
	Coarse	Fine							
G0 S0	0%	0%	50	394	0	274	0	19.5	0.8
G30 S0	30%	0%	50	276	118	274	0	19.0	0.9
G60 S0	60%	0%	50	158	236	274	0	18.5	1.0
G100 S0	100%	0%	50	0	394	274	0	18.0	1.0
G30 S30	30%	30%	50	276	118	191	83	18.5	0.9
G30 S60	30%	60%	50	276	118	110	164	18.0	0.9
G30 S100	30%	100%	50	276	118	0	274	19.0	1.0
G100 S100	100%	100%	50	0	394	0	274	18.5	0.9

The mix proportions all contain a 50% water cement ratio. Plain concrete was selected in order to minimize the effect of variations in entrained air quantity on compression strength. The coarse aggregate river gravel (2.75 density, 1.08% water absorption rate) was from Shizunai town, Hokkaido. The fine aggregate river sand (2.69 density, 1.42% water absorption rate) was from Mukawa town, Hokkaido. The tile was originally used as decorative for reinforced concrete external wall. By crushing, the tile was divided in coarse and fine aggregates: the coarse tile aggregate sized between 20mm and 5mm (6.52 fineness modulus, 2.30 density, 0.11% water absorption rate) and fine tile aggregate sized 5mm or less (3.83 fineness modulus, 2.30 density, 0.11% water absorption rate). The target slump was 18cm. The curing of the concrete was carried out in water at 20 degrees centigrade.

Compressive Strength Results (Figure 1)

The concrete containing no tile chips is designated G0S0, the control samples. G30S0, the concrete with tile chip coarse aggregate replacing 30% of the river gravel, exhibited a compressive strength larger than that of G0S0. This result was also found for the G60S0 (tile chip coarse aggregate replacing 60% of the river gravel) and G100S0 (tile chip coarse aggregate replacing 100% of the river gravel). In fact, the compressive strength increased as the proportion of the tile chips replacing the river gravel was increased.

G30S30, tile chip coarse aggregate replacing 30% of the river gravel and tile chip fine aggregate replacing 30% of the river sand, also exhibited a compressive strength larger than G0S0. As the proportion of tile fine chip aggregate replacing river sand was increased from 30 to 100% with the proportion of crushed tile coarse aggregate fixed at 30% (G30S30, G30S60 andG30S100, respectively) the compressive strength increased. The concrete containing coarse and fine aggregates composed only of tile chips, G100S100, exhibited the largest compressive strength of all the concrete specimens in Series 1.

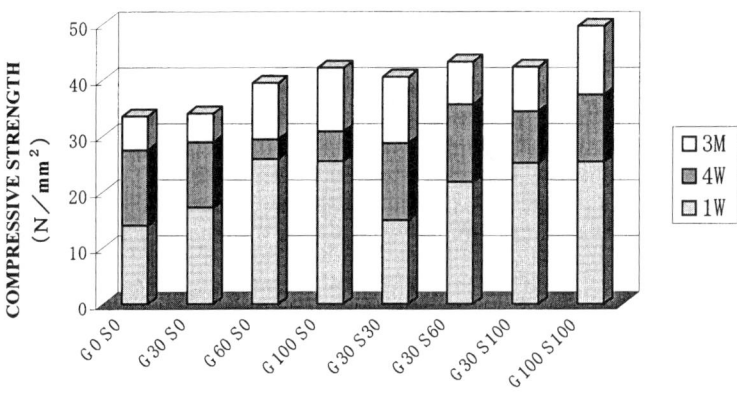

TEST SPECIMEN

Figure 1. Compressive strength of concrete containing tile chips

Static Elastic Modulus Result (Figure 2)

Static elastic modulus was calculated from the 1/3 stress-strain relationship of the maximum load. The static elastic modulus of concrete containing any level of tile chip aggregate is larger than that of the control concrete, G0S0.

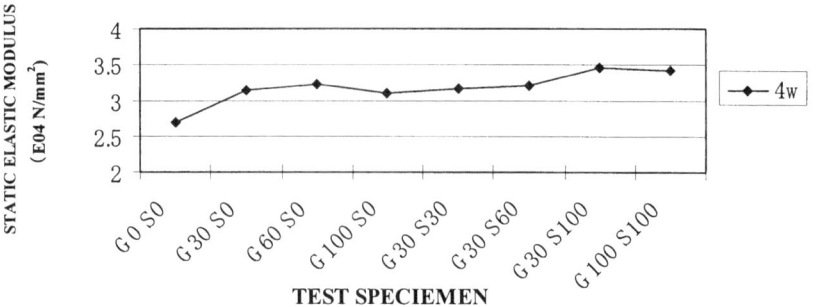

Figure 2. Static elastic modulus of concrete containing tile chips

SERIES 2: CRUSHED SCALLOP SHELLS

Design of Experiments

It has been shown that crushed scallop shell material can filter muddy water, effectively cleaning it [3]. If this beneficial effect can be obtained from porous concrete containing this crushed shell material, the environment will be improved with the ability to recycle both the shells as well as the resulting concrete when it becomes waste. In this test series, crushed scallop shell material was mixed into porous concrete, and the compressive strength was tested.

Mix Proportions

The mix proportions are shown in Table 2. The water-cement ratio and the target air void for all test specimens were 22% and 30%, respectively. As mentioned above, the crushed shell material was used to replace the aggregate at volumetric mixing ratios of 0%, 50%, and 100%. The materials used in the concrete are as follows: ordinary cement (density 3.16), aggregate (crushed stone, sieve size 5-15 mm, density 2.84), crushed shell material (sieve size 5-25 mm, density 2.50), and high-range water reducing admixture (1.68%). The porous concrete was cured in water at 20°C for four weeks. The concrete was manually placed into forms using table and hand vibrators. The concrete specimens were removed from the forms the next day and then cured in water for four weeks at 20°C. The compressive strength tests were conducted after 28 days of curing. The dimensions of the specimens were 10 cm ϕx20cm.

Table 2. Mix proportions

CRUSHED SHELL (%)	W/C (%)	WATER (kg)	CEMENT (kg)	AGGREGATE (kg)	CRUSHED SHELL (kg)	ADMIXTURE
0	22	53	240	1622	0	High-range Water Reducing Admixture
50	22	53	240	811	714	
100	22	53	240	0	1428	

Air Void Percentage Measurement Method for Porous Concrete

The air void percentage was measured using Japan Concrete Institute Methods: volumetric and weight [6]. The dimensions of the specimens were 10ϕx20cm.

1. Volumetric method

$$A= (1-(W_2-W_1)/V) \times 100$$

Where   A= Air void percentage of porous concrete (%)
W_2= Weight in air
W_1= Weight in water
V= Volume.

2. Weight method

$$A= (T-W)/T \times 100$$

Where   A= Air void percentage of porous concrete (%)
T= Weight, per unit volume, of concrete from which air was removed
W= Weight, per unit volume, of the concrete specimen

Compressive Strength and Air Void Percentage Results

The results of compressive strength and air void percentage tests are shown in Table 3. The compressive strength measured after four weeks of curing ranged from 10.6 to 0.3N/mm^2. The compressive strength at 0% crushed shell, i.e., no shell material added, is sufficient for porous concrete. It should be noted that the compressive strength decreased as the percentage of crushed shell material was increased. The target void percentage was 30%. The actual air void percentage determined by the volumetric method ranged from 28.1 to 55.7%; the weight method showed this parameter ranging from 28.3 to 51.1%. Thus, both test methods produced similar results. The air void percentage at 0% crushed shell content is close to the target value. However, the air void percentage increased significantly as the percentage of crushed shell material was increased.

Table 3. Compressive strength and air void percentage

CRUSHED SHELL (%)	AIR VOID PERCENTAGE (%)			COMPRESSIVE STRENGTH, After 4 Weeks of Curing (N/mm^2)
	Target of Air Void	Volumetric Method	Weight Method	
0	30	28.1	28.3	10.6
50	30	43.1	40.0	1.9
100	30	55.7	51.1	0.3

SERIES 3: CRUSHED ROOFING TILE MATERIAL

Design of Experiments

Series 3 measured the compressive strength and the static elastic modulus of high volume fly ash concrete containing crushed roofing tile material.

Mix Proportions

The mix proportions are shown in Table 5. The water-cement ratio for all test specimens was 40%. The crushed roofing tile material was added as 0%, 20%, 40%, 60%, 80%, and 100% of the volume of the coarse aggregate.

The m aterials u sed i n t he c oncrete a re a s f ollows: o rdinary c ement (density 3 .16), fly ash (density 2.16, loss on ignition 1.7%, specific surface area 3780cm^2/g, methylene blue absorption 0.55mg/g), fine aggregate (river sand, density 2.69), coarse aggregate (river gravel, density 2.80), and crushed roofing tile material (density 1.30). The concrete was cured in water at 20°C. The Slump was from 15 to 19cm. The compressive strength tests were conducted after 7, 28, and 91 days of curing. The dimensions of the specimens were 10φx20cm.

Table 5. Mix proportions

ROOFING TILE (%)	W/C+F (%)	WATER (kg)	CEMENT (kg)	FLY ASH (kg)	FINE AGG. (kg)	COARSE AGG.(kg)	ROOFING TILE (kg)
0	40	158	197	197	764	1112	0
20	40	158	197	197	764	890	103
40	40	158	197	197	764	666	207
60	40	158	197	197	764	445	309
80	40	158	197	197	764	221	413
100	40	158	197	197	764	0	516

Compressive Strength Results

The results of compressive strength tests are shown in Figure 3. The results show that compressive strength was increased when approximately 20% to 60% of the volume of the coarse aggregate was replaced by roofing tile material.

Static Elastic Modulus Results

The results of static elastic modulus tests are shown in Figure 4. Even if the mix rate of roofing tile material is increased to 80%, the static elastic modulus does not change significantly. However, when the mix rate of roofing tile material is 100%, the static elastic modulus is somewhat decreased.

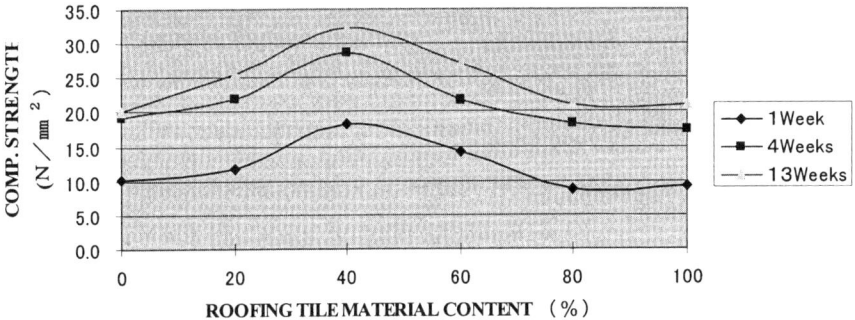

Figure 3. Compressive strength of concrete containing roofing tile material

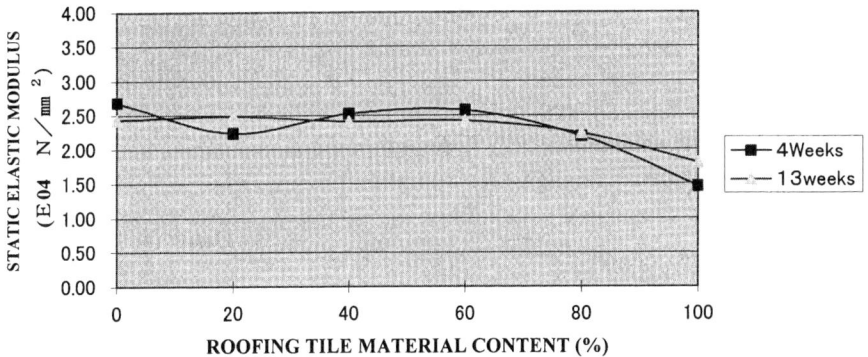

Figure 4. Static elastic modulus of concrete containing roofing tile material

CONCLUSIONS

This study examined the compressive strength and the static elastic modulus of concrete containing tile chips, crushed scallop shells, or crushed roofing tiles substituting for a portion or all of the aggregate. The conclusions are listed in the following;

1. Both the compressive strength and the static elastic modulus increase as the proportion of tile chips utilized as aggregate is increased.

2. The compressive strength of porous concrete decreases as the percentage of aggregate replaced by crushed scallop shell material is increased. Future studies will center on improving this result.

The compression strength can be increased by substituting roofing tile material for 20% to 60% of the volume of the coarse aggregate. A mix rate of crushed roofing tile material of up to approximately 80% has little effect on the static elastic modulus; however, when the mix rate is 100%, the static elastic modulus is somewhat decreased.

REFERENCES

1.   Construction Material Subcommittee: Present state and future trends of reuse technology concerning construction by-products, Architectural Inst. of Japan Kanto branch, March 1996.

2.   Construction Material Subcommittee: Present state and future view of zero emission in architectural projects, Architectural Inst. of Japan, Kinki branch, Aug. 1999.

3.   KOYAMA, N.: Scallop shell: bionic design (in Japanese), (website: Koyama Institute of Technology) http://www.pow.hi-tech.ac.jp/kenkyu/koyama/index.html

4.   SUGIYAMA M.: The experiment on compressive strength and freeze-thaw resistance of concrete containing tile chip, Recycling and Reuse of Glass Cullet: Proceedings of International Symposium, Dundee, Thomas Telford, March 2001.

5.   SUGIYAMA M.: The freeze-thaw test results of porous concrete with crushed scallop shell material added, Recycling and Reuse of Waste Materials: Proceedings of International Symposium, Dundee, Thomas Telford, March 2003.

IMMOBILISATION OF CALCIUM SULPHATE IN DEMOLITION WASTE

J Ambroise
J Pera

Unité de Recherche Génie Civil, INSA LYON
France

ABSTRACT. The results of a laboratory programme undertaken to examine the treatment of demolition waste containing calcium sulphate by calcium sulpho-aluminate clinker (CSA) are described. The quantity of CSA necessary to entirely consume calcium sulphate was determined. By means of infrared spectrometry analysis and X-ray diffraction, it was shown that calcium sulphate was entirely consumed when the ratio between CSA and calcium sulphate was 4. Standard sand was polluted by 4% calcium sulphate. Two solutions were investigated:-either global treatment of sand by CSA,-or immobilisation of calcium sulphate by CSA, followed by the introduction of this milled mixture in standard sand. Regardless of the type of treatment, swelling was almost stabilised after 28 days of immersion in water.

Keywords: Calcium sulpho-aluminate cement, Ettringite, Sand, Swelling, Treatment.

Dr Jean Ambroise is an Assistant Professor at the National Institute of Applied Sciences of Lyon. He is engaged in applied research, consulting and training activities in the fields of cementitious materials and concrete technology.

Professor Jean Péra is a Director of Civil Engineering Department at the National Institute of Applied Sciences of Lyon, from which he received his Dr Eng and Doctor of Sciences degrees. His research interests include the design of innovative materials for civil engineering and recycling and reuse of waste materials.

INTRODUCTION

The gypsum content of recycled aggregates varies from 0.03% in concrete rubble to more than 6% in some sieving sands. Gypsum occurs in finely dispersed form and originates mainly from plaster work [1]. The presence of gypsum has a negative effect on the material quality for reasons of solubility and low hardness and density. These properties prevent the use of gypsum containing products as unbound elevation or foundation material. Additionally gypsum affects the hardening reactions during concrete production. In cement bound applications, an excess of sulphates causes loss of strength, expansion and disintegration of concrete. Therefore, the sulphate content is a major criterion in the valorisation of the construction and demolition waste as a secondary aggregate.

Immobilisation of sulphate can be achieved by chemical binding in the form of ettringite and other calcium sulpho-aluminates. Mixture compositions based on Portland, high alumina and blast furnace slag cements have been defined [2].

This paper examines the feasibility of treating gypsum containing demolition waste by calcium sulpho-aluminate clinker (CSA), in order to immobilise calcium sulphate. Sulpho-belite cements, also called sulpho-aluminate-belite cements, contain the phases belite (C_2S) and yeelimite, or tetracalcium trialuminate sulfate ($C_4A_3\overline{S}$), and gypsum ($C\overline{S}H_2$) as their main constituents. When CSA clinker hydrates in the presence of gypsum, ettringite ($C_6A\overline{S}H_{32}$) is formed according to the following reactions [3]:

$$C_4A_3\overline{S} + 2C\overline{S}H_2 + 36H \Rightarrow C_6A\overline{S}H_{32} + 2AH_3, \text{ in absence of calcium hydroxide,}$$

$$C_4A_3\overline{S} + 8C\overline{S}H_2 + 6CH + 74H \Rightarrow 3C_6A\overline{S}H_{32}, \text{ in presence of calcium hydroxide.}$$

The properties of these cements have been described in the literature [4-7].

The first investigation was carried out on plain pastes. It consisted in optimising the quantity of CSA necessary to entirely consume calcium sulphate. In a second step, standard sand was polluted by 4% calcium sulphate. Two solutions were investigated: either global treatment of sand by CSA, or immobilisation of calcium sulphate by CSA, followed by the introduction of this milled mixture in standard sand. The performances of these polluted mortars were assessed.

MATERIALS

The CSA used in this study, presented the following composition:
- $C_4A_3\overline{S}$ = 66%,
- C_2S = 17%,
- Perovskite = 9.9%,
- $C_{12}A_7$ = 7.1%.

The source of calcium sulphate was plaster: $CaSO_4, 0.5 H_2O$. Standard siliceous sand (EN 196-1) was used as aggregate. Normal portland cement (NPC), class CEM I 52.5, according to the European Standard EN 197-1, was utilised to prepare mortars.

INVESTIGATION ON PLAIN PASTES

Plain pastes containing different amounts of plaster and CSA were hydrated at water to solids ratio of 0.35. The composition of pastes is given in Table 1.

Table 1. Composition of plain pastes (w_t%)

PLASTER	CSA
40	60
30	70
20	80

These pastes were stored in Plexiglas moulds (ϕ = 20 mm, h= 40 mm) for 7 and 28 days, respectively, and then analysed by means of infra-red spectrometry and X-ray diffraction. The IR spectra are presented in Figures 1 to 3.

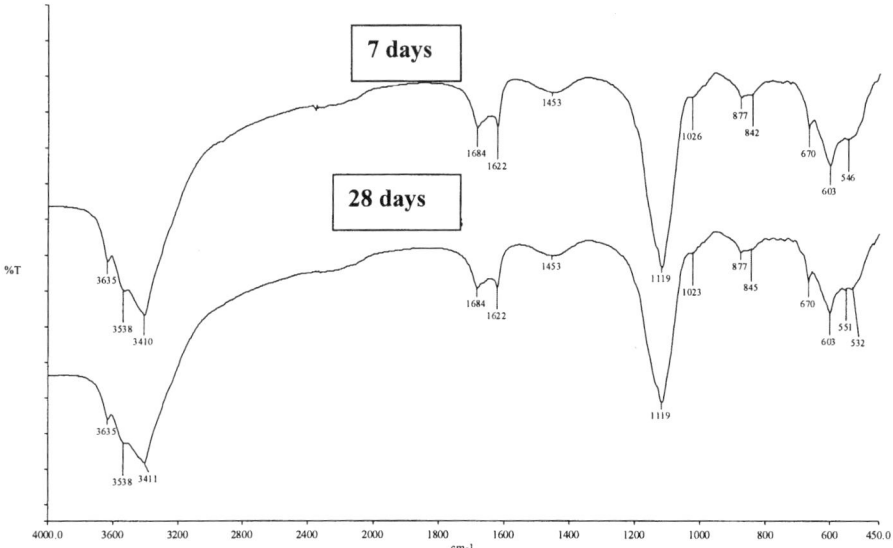

Figure 1. IR spectra of the mixture composed of 40% plaster and 60% CSA.

Figures 1 and 2 point out the presence of bands at 603 and 670 cm-1, characterising the presence of residual plaster. The amount of CSA is not sufficient to consume all the plaster present in these mixtures. Figure 3 shows that plaster is entirely consumed in the mixture containing 20% plaster and 80% CSA. This result was confirmed by X-ray diffraction. This technique also pointed out that the main product of hydration was ettringite. To entirely consume the plaster present in the mixture, the following condition has to be met: CSA/Plaster ≥ 4.

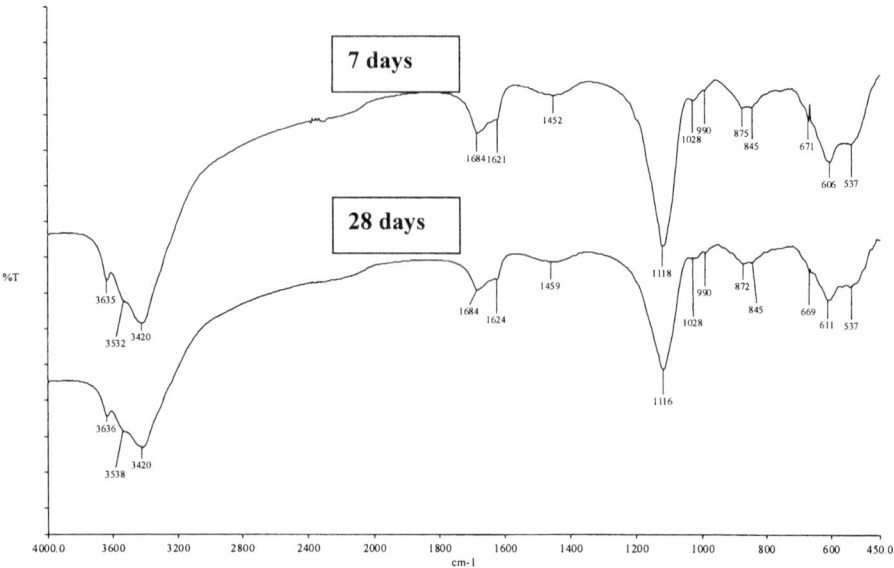

Figure 2. IR spectra of the mixture composed of 30% plaster and 70% CSA.

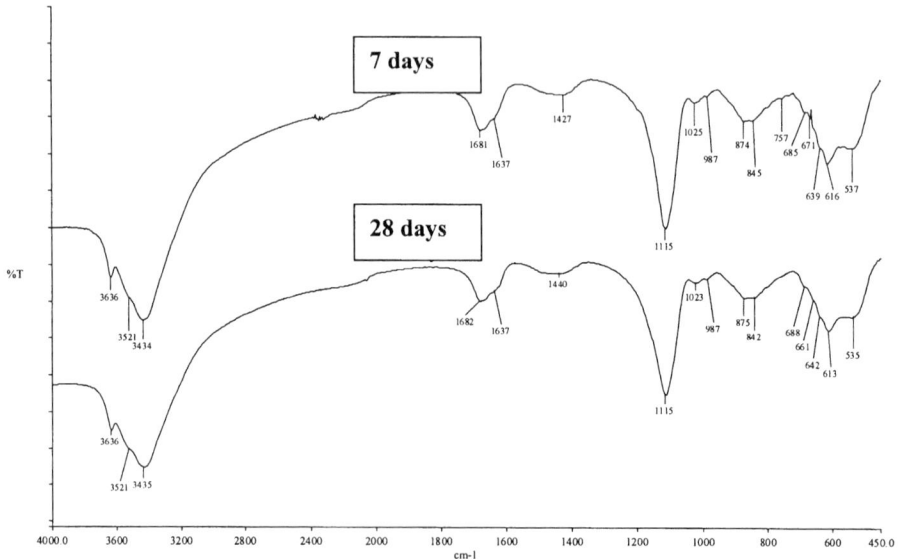

Figure 3. IR spectra of the mixture composed of 20% plaster and 80% CSA.

INVESTIGATION ON MORTARS

Two approaches have been undertaken:
- in the first one, standard sand was first polluted by plaster and then treated by CSA,
- in the second one, plaster was first mixed with CSA and hydrated for 7 days, then milled and introduced in standard sand.

Standard mortars were cast using NPC and these sands. Their dimensional variations were compared to those of control mortar cast with pure sand, when immersed in water. The amount of plaster present in the sand was 4%, by mass.

Results of the First Approach

In this approach, only two CSA/Plaster ratios have been investigated: 1.5 (60% CSA + 40% Plaster) and 2.33 (70% CSA + 30% Plaster). The compositions of treated sands are presented in Table 2.

Table 2. Compositions of treated sand (in gm).

COMPONENT	CSA/PLASTER	
	1.5	**2.33**
Standard sand	1296	1296
Plaster	54	54
CSA	81	126
Water	220	220

After 7 days of hydration in sealed bags, these sands were crushed to get particles smaller than 3 mm. The crushing was difficult due to the hardness of the treated sand, and therefore, this phenomenon has to be taken into account in an industrial process.

Standard mortars were cast using NPC and either pure, or polluted, or treated sand, in the following proportions:
- Sand: 1350 g,
- NPC: 450 g,
- Water: 225 g (pure or polluted sand) or 285 g (treated sand).

The quantity of water was adjusted to get the same workability. The samples were demoulded after 24 hours and immersed in water at 20°C. The dimensional variations of the different mortars are shown in Figure 4.

Figure 4 points out that the mortar containing polluted sand (4% plaster) developed continuous swelling, reaching 915 µm/m at 90 days of age. For pure or treated sand, swelling was stabilised after 7 days of immersion in water. The swelling obtained at 90 days for the different mortars is presented in Table 3.

Table 3. Swelling of mortars after 90 days of immersion in water.

TYPE OF SAND	SWELLING, at 90 days (μm/m)
Pure	69
Polluted (4% plaster)	915
Treated CSA/Plaster = 2.33	158
Treated CSA/Plaster = 1.5	208

Table 3 proves the efficiency of the treatment. Adding CSA to plaster reduces swelling of the mortars by about 80%.

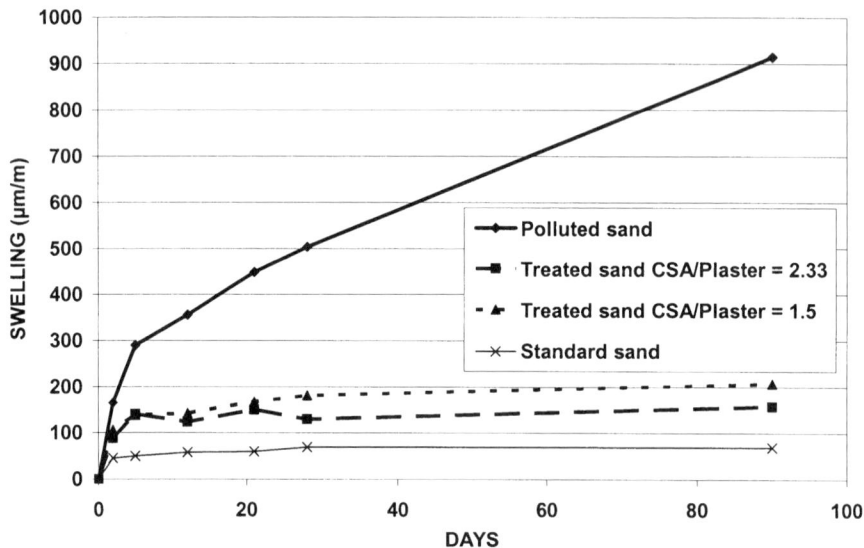

Figure 4. Swelling of mortars cast with treated sand versus time.

Results of the Second Approach

Plaster was hydrated in the presence of CSA at W/S = 0.35 in the following proportions: CSA/Plaster = 1.5, 2.33 and 4. After 7 days of hydration in plastic bags, pastes were milled to get particles smaller than 200 μm. These fine particles were introduced in sand to prepare standard mortars with NPC. The composition of mortars is shown in Table 4. The quantity of plaster present in each mortar was 4%.

The swelling of mortars immersed in water is presented in Figure 5. The swelling observed with such mortars is in the same range as that observed for mortars cast with treated sand as shown in Table 5. The two types of treatment are equivalent.

Table 4. Composition of mortars prepared with treated plaster (in g).

COMPONENT	CSA/PLASTER		
	1.5	2.33	4
Standard sand	1222	1185	1117
Treated plaster	128	164	233
NPC CEM I 52.5	450	450	450
Water	225	225	225

Table 5. Swelling observed at 90 days with the two types of treatment.

TYPE OF TREATMENT	CSA/PLASTER = 1.5	CSA/PLASTER = 2.33	CSA/PLASTER = 4
Treated sand	208	158	n.a.
Treated plaster	210	195	170

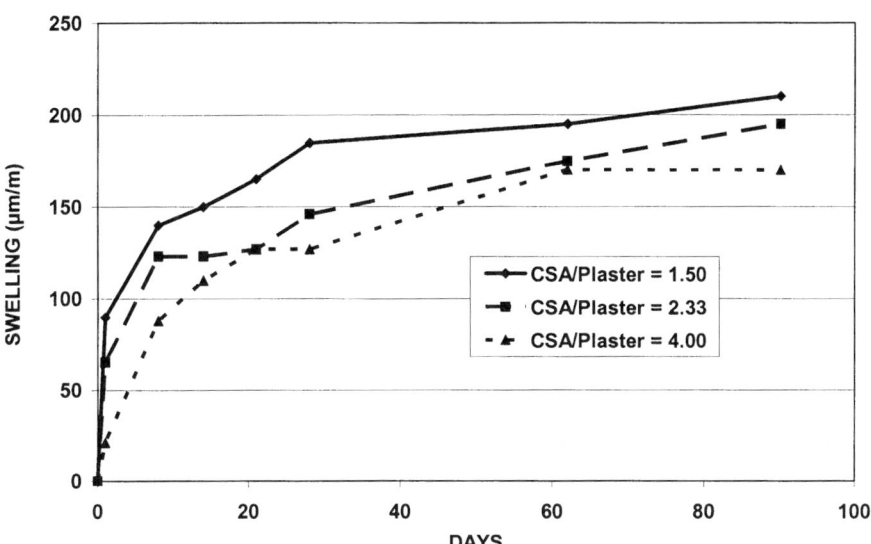

Figure 5. Swelling of mortars cast with treated plaster versus time.

CONCLUSIONS

The use of calcium sulpho-aluminate clinker (CSA) is efficient to stabilize the swelling of mortars containing sand polluted by plaster. The consumption of plaster is completed when the ratio between CSA and plaster reaches 4. The reduction of swelling is effective for CSA/Plaster values higher than 1.5: the swelling drops by 80%. Two ways of treatment are possible: either treating directly the polluted sand or separating plaster from the sand and treating this plaster by CSA before re-introducing it in mortar.

ACKNOWLEDGMENTS

The authors would like to acknowledge the financial support provided for the project by Carrières du Boulonnais.

REFERENCES

1.   VRANCKEN, K.C. AND LAETHEM, B.: Recycling options for gypsum from construction and demolition waste, Proceedings of the International Conference on the Science and Engineering of Recycling for Environmental Protection, Harrogate, Pergamon, June 2000, pp325-331.

2.   ODLER, I.: Cements containing calcium sulfoaluminate, Special Inorganic Cements, London, E & FN Spon, 2000, pp69-87.

3.   ZHANG, J.L. AND GLASSER, F.P.: New concretes based on calcium sulfoaluminate cement, Proceedings of the International Conference on Modern Concrete Materials: Binders, Additions and Admixtures, London, Thomas Telford, 1999, pp.261-274.

4.   A.K. CHATTERJEE: Special cements, Structure and Performance of Cements, London, E & FN Spon, 2002, pp226-231.

5.   SU,M., WANG Y., ZHANG, L., AND LI, D.: Preliminary study on the durability of sulfo/ferro-aluminate cements, Proceedings. of the10[th] International. Congress on the Chemistry of Cement, Gothenburg, Sweden, June 1997, IV, p.4iv029, 12 p.

6.   PERA, J., AMBROISE, J. AND CHABANNET, M: Valorization of automotive shredder residue in building materials, Cement and Concrete Research, Vol.34, N° 4, 2004, pp557-562.

7.   PERA, J.AND AMBROISE, J.: New applications of calcium sulfoaluminate cement, Cement and Concrete Research, Vol.34, N° 4, 2004, pp671-676

INVESTIGATION OF THE USE OF WASTE PLASTIC AS AGGREGATE FOR CONCRETE

R Boutemeur

M Taibi

N Ouhib

A Bali

Ecole polytechnique

Algiers

ABSTRACT. Important quantities of plastic waste issued from multiple sources such as wrappings and packings and waste demolition are stockpilled in discharges despite the new Algerian laws requiring the recycling and the reuse of these wastes in order to protect the environment. In the first stage, the objective of our study is to analyze the possible valorisation of plastic waste issued from mainly bottles of polyethylene terephtalate (PET), as agregates for concrete. Two types of recycled agregates with different shapes (cuttings and grains) have been used in the experimental study.A partial substitution o f fine gravel and sand respectively by cuttings and grains at different percentages was sudied.

The p roperties o f t he c oncrete w ith p lastic i n f resh a nd h ardened s tates a re e xamined a nd compared with those of an ordinary concrete containing natural aggregates.

The encouraging results of this study showed that it is possible to manufacture a concrete, based on plastic aggregate of acceptable mechanical characteristics (strength and workability).

This investigation could be extented to the valorisation of other types of plastic wastes recovered from discharges.

Keywords: Plastic waste, Recycling, Concrete, Reuse, Aggregates, Cuttings, Grains

R Boutemeur is a Doctor Lecturer in Civil Engineering Departement at Ecole polytechnique of Algiers. He is working on recycling and valorisation of waste materials.

M Taibi and **N Ouhib** are students in Civil Enginering Departement of Ecole Polytechnique of Algiers.

A Bali is Professor in th Civil Engineering Departement, Head of Materials Laboratory at Ecole Polytechnique of Algiers.

INTRODUCTION

Algeria has been facing serious problems of environmental degradation together with natural ressources lost. The situation is very critical. Important actions have to be taken immediately in order to restablish the natural equilibria to avoid an ecological disaster. The environment protection has to be considered seriously and should be integrated within the economical development.

Treatment and valorisation of wastes could positively contribute to the protection of the environment and to the reduction of the waste quantities evacuated towards public discharges increasing the duration of their exploitation. This must help Algeria:

- To have a substantial economy of natural matters;
- To recover a better balance of one's ecosystems;
- To preserve the natural resources;
- To have a healthy environment.

The objective of our study is a contribution to the valorisation, in civil engineering field, of an urban plastic waste which is harmful, taking up too much space and aesthetically not pleasing. Despite the important quantities which are rejected, these plastic wastes are not recycled in Algeria. Our study consists of introducing the plastic waste in concrete and evaluating its performance in order to reach acceptable strengths with the new formulated concrete.

EXPERIMENTAL METHOD AND EQUIPEMENT

Waste Presentation

The studied waste is an urban plastic waste issued from packing materials in particular from manufactured polyethylene terephtaltic bottles commonly known as P.E.T. In this study the recycling of this waste consists of integrating the solid mater in order to obtain a new elaborated material.

Two shapes of the integrated waste are presented:
- grains of (1/3) diameter obtained after grinding followed by a washing
- cuttings of diameter (2/5) obtained after grinding, washing and extrusion

Other Materials

The other materials included in the formulas for concrete are:
- sea sand (0/3) from Zemmouri (Boumerdes)
- gravel (5/15 and 15/25) from Cap Djinet (Boumerdes)
- cement OPC from Meftah

The characteristics of the overall used materials have been identified using tests according to the french standards AFNOR.

Composition

In the purpose of assessing the performances of the elaborated concrete, namely the mechanical ones, a number of formulas (established up according to Faury method) have been studdied while varying the percentage of the incorporated plastic waste. The plastic waste has been substituted by taking into account the granulometry of each shape, with respect to the granules for the cuttings and to the sand for the grains, within the various compositions, consistently with the percentages ranging from 0% to 50 % with an increment of 10.

Workability

The workability of manufactured mixtures has been measured by the Slump-test consistently with NF standard P18-451 (AFNOR). All the concretes have been manufactured through constant workability (plasticity range), so that the comparison between the obtained results are facilitated.

Strengths

The performances of the elaborated material have been evaluated, for each concrete and each age, using compression and flexion tests, carried out on cubics (7cm x 7cm x 7cm) and prismatic (7cmx7cmx28cm) specimen, respectively. The tensile strength is measured through the direct flexion test.

Temperature Effect

To evaluate the elaborated material performances with regard to heat, mechanical compression tests have been carried out on specimen's mainted in an oven during 48 hours, at a steady temperature. Four different values of temperature with a 50° increment have been fixed to work out the resistances evolution.

RESULTS AND DISCUSSION

Fresh Concrete

The workability of fresh elaborated concrete with different percentages of incorporated plastic waste (shaped as grains or cuttings) given in table 1.

We can notice that the use of the amount of water obtained by the Faury formula for the ordinary concrete (without waste) provides a hard concrete (slump = 3.3 cm), requiring therefore an increase of that amount to have a plastic concrete (5 cm<Slump<9 cm).

While incorporating the plastic waste (either cuttings or grains) in the concrete, we find out a workability decrease with the increase of the waste percentage, especially when the percentage ranges between 40 and 50° (when the concrete becomes hard). This occurrence has led us to add an additional amount of water to come back to the suitable workability.

Table 1. Measurement of the slump and the E/C ratio for the different percentages of waste shaped as cuttings and grains.

WASTE, %	PLASTIC QUANTITIES (kg) for 1m³ of Concrete		WATER (l)		MEASURED SLUMP (Cm)		E/C RATIO	
	Cuttings	Grains	Cuttings	Grains	Cuttings	Grains	Cuttings	Grains
0%	0	0	204	204	3.3	3.3	0.58	0.58
			217	217	8.5	7.75	0.62	0.62
10%	13.98	30.82	217	217	8	6.7	0.62	0.62
20%	27.95	61.65	217	217	6	6	0.62	0.62
30%	42.93	92.5	217	217	5.3	5.3	0.62	0.62
40%	55.9	123.29	217	217	3	3	0.62	0.62
			226	226	6	5.1	0.645	0.645
50%	69.88	154.16	226	226	4.2	3.2	0.645	0.645
			235	235	5	5	0.67	0.67

Hardened Concrete

Tensile strength

The obtained strength results for the concrete formulas with the different waste percentages are shown in Figure 1. When dealing with the waste in cuttings, we note a slight improvement of the tensile strength with regards to the witness (ranging 1 Mpa for 30% of substitution), which could be explained by a possible sewing of the microcracks and therefore, the fracture is somewhat delayed.

Beyond a substittution of 30%, the resistance slightly decreases as far as the witness is concerned, probably, because of the mixture porosity. As to the waste in grain shape, we notice a slight reduction (about o.5 Mpa for a 40% substitution) the shape of the grains being not an obstacle to the spreading of the cracks; however even with a 50% substitution, the resistance that prevails is sastisfying.

Compressive strength

The obtained strength results for the concrete formulas with different percentages of waste are shown in Figure 2. As to the waste in cuttings shape, a slight improvement of the resistance is recorded, whereas the optimum is reached with a 30% substitution (ranging 4 MPa) ; the cuttings ensuring a better adhesion. When the substitution percentage increases, the material becomes more porous and hence the relative reduction of the resistance can be explained. However, even with a 50% substitution the gained resistance is satisfying (27.5 MPa by 28 days).

As to the waste in grain shape, the 28 daycompressive strength increases with the percentage of incorporated waste between 10 and 20 %. The optimum of strength is gained for 10%. Beyond 20% the resistance with regard to the witness, somewhat decreases but remains satisfying even for a 50% substitution (27.5 MPa by 28 days).

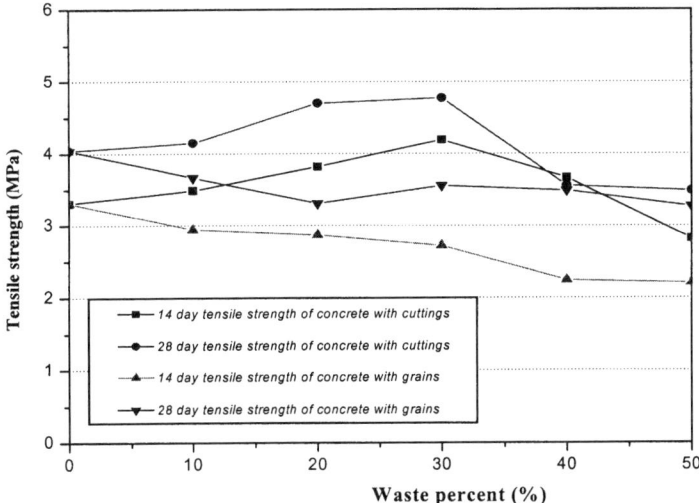

Fig. 1. Tensile strength variation according to waste % incorporated.

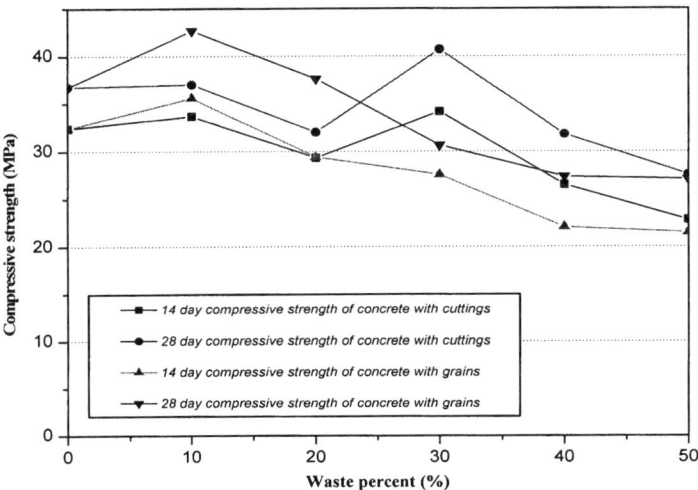

Fig. 2. Compressive strength variation according to waste % incorporated.

Temperature Effect

The variations of compressive strength based on the established formulas, under the temperature effect are shown in Figures 3 and 4.

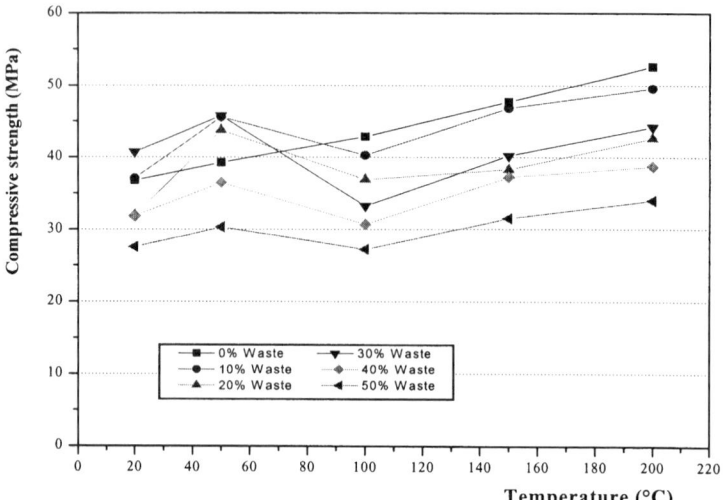

Fig.3 Variation of compressive strength vs tempetrature
for concrete with plastic cuttings.

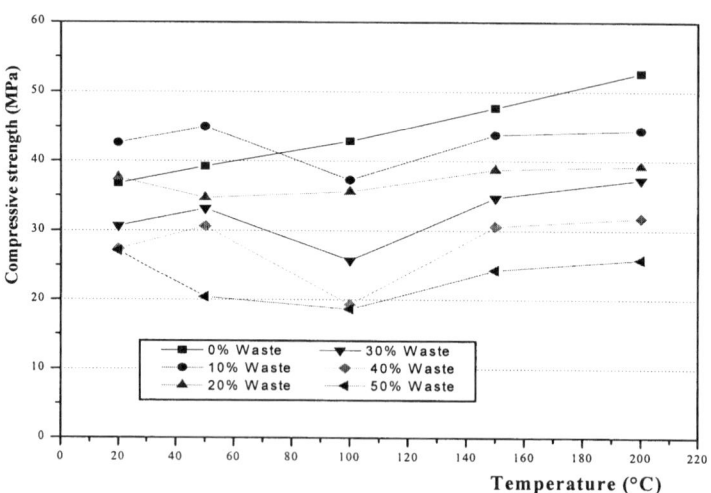

Fig.4 Variation of compressive strength vs temperature
for concrete with plastic grains.

As to the witness concrete, we notice an increase of the compressive strength dealing with the temperature rise, because of stresses occuring inside the concrete, considering that the components have different volume expansion ratios (11). One part of the external load will be balanced by these stresses and the other one will induce the fracture.

As to the concrete including a substitute, we notice that the curve features show a similar variation. The recorded compressive strength increase to compression between 0 and 50°C can be explained in the same way that it is done for the witness concrete, while we can notice a slight reduction of resistance when experiencing the 100°C level.
In fact, this is due to the plastic softening, for, the glassy transition temperature Tv (ranging 75°C) has been overstepped.

Between 100 and 200°C, we notice an increase of strength because the temperature has widely overstepped Tv, thus creating volume expansions which will stress a part of the external load. We must note that the strengths decrease with the adding of waste (especially for the grain concrete), still, they remain acceptable.

CONCLUSIONS

The present experimental survey has allowed us to confirm that the valorisation of plastic waste, interfering with a partial substitution to the natural aggregates, is technically possible. Indeed, the studdy and the analysis of the main mechanical characteristics of the concrete in fresh state and hard state, developed with the waste, allow the display of the following results.

As to the concrete containing an amount of plastic cuttings granules, it has been noticed an improvement of the strength (18% for tensile and 11% for compression) related to a substitution which can go up to 30%, with workability close to that of the witness concrete. Beyond a 30% substitution, the strength slightly decreases but remain satisfying even with a 50% substitution.

As to the concrete containing an amount of plastic grains, we can record an improvement of the compressive strength (16%) related to a 10% substitution with a workability close to that of the witness one.

This resistance slightly decreases but remains acceptable even with a 50% substitution. The tensile strength decreases by 20% with regard to a 50% substitution.

For both developed concretes, acceptable strengths have been obtained with temperatures reaching up to 200%C.

The encouraging results of this study showed that it is possible to manufacture a concrete, based on plastic aggregate of acceptable mechanical characteristics (strength and workability).

This investigation could be extended to the valorisation of other types of plastic wastes recovered from discharges in order to protect the environment.

.

REFERENCES

1. A.NAVARRO : Gestion et traitement des déchets . Technique de l'ingénieur, traités généralités et construction, A8 660 (2001).

2. S.ABDERREZAK : Gestion des déchets solides en Algérie. Séminaire sur la gestion des déchets solides en Algérie 2000.

3 - K.KOIDE, M.TOMON et T.SASAKI : Investigation of the use of waste plastic as an aggregate for light weight concrete (Japan 2000) . Proceedings international congress. Challenge of concrete Construction. 5 -11 Sep 2002, Dundee, Scotland U.K.

4 - S.BEDJOU: Contribution à la valorisation des déchets de construction . Thèse de magister en génie civil E.N.P. Alger 2002.

5 - N.BENREKAA : Contribution à l'étude de propriétés physico thermiques du polyéthylène téréphtalique par mesures de courants thermiquement stimulés (CTS). Thèse demagister en physique U.S.T.H.B. Alger 1999.

6 - M.A.CHABOU : Etude de la valorisation du polyéthylène basse densité régénéré par recyclage mécanique . Thèse de magister de l'environnement E.N.P. Alger 2000.

7 - M.HELMUT KRIST : Aspects du recyclage des matières plastiques. Séminaire sur la gestion des déchets solides en Algérie 2000.

8 - B.BARTHELEMY et J.KRUPPA : Résistance au feu des structures béton – acier – bois. Editions Eyrolles 1978.

9 - Laboratoire Central des Ponts et Chaussées (LCPC) : Colloque international sur l'utilisation des sous-produits en génie civil . Paris 28-29-30 Novembre 1978.

HARDENED PROPERTIES OF RECYCLED AGGREGATE CONCRETE PREPARED WITH FLY ASH

S. C Kou

C S Poon

L Lam

D Chan

The Hong Kong Polytechnic University

Hong Kong- China

ABSTRACT. The effects of the incorporation of Class F fly ash on the hardened properties of recycled aggregate concrete were determined. The concrete mixtures were prepared with a water-to-binder (W/B) ratio of 0.55. The recycled aggregate was used as 0, 20, 50 and 100 % by weight replacements of natural aggregate. In addition, fly ash was used as 0, 25 and 35 % by weight replacements of cement. The results showed that the compressive strengths, the tensile strengths and the static modulus of elasticity values of the concrete at all ages decreased as the recycled aggregate and the fly ash contents increased. Furthermore, an increase in the recycled aggregate content decreased the resistance against chloride-ion penetration and increased the drying shrinkage of concrete. The use of fly ash as a substitution of cement improved the resistance against chloride-ion penetration and decreased the d rying s hrinkage o f the r ecycled a ggregate c oncrete. T he r esults s how t hat o ne o f t he practical ways to utilize a higher percentage of recycled aggregates in structural concrete is by incorporating 25-35 percents of flyash.

Keywords: Hardened properties; Recycled aggregate concrete; Fly ash.

Dr C.S. Poon is a Professor and Leader of the *Research Group on the Reuse of Waste Materials for Construction* at the Hong Kong Polytechnic University. His research interests include construction and demolition waste management, concrete technology and recycling of waste materials for construction.

S.C. Kou is a PhD candidate; **L Lam.** is a Research F ellow; and **D Chan**. is a Research Assistant respectively of the *Research Group on the Reuse of Waste Materials for Construction* at the Hong Kong Polytechnic University.

INTRODUCTION

Recycled aggregate has been used as a partial replacement of the natural aggregate for a number of years in many different countries. The potential benefits and limitations of using recycled aggregate in concrete are well understood and extensively documented [1-5]. The use of recycled aggregate generally increases the drying shrinkage and creep of concrete compared to those of natural aggregate concrete. Hasaba et al. [6] reported that the drying shrinkage of concrete prepared with coarse recycled aggregate and natural sand was 50 % higher than that of conventional concrete. When both coarse and fine recycled aggregates were used, the drying shrinkage of recycled aggregate concrete was as much as 70 % higher than that of conventional concrete.

Hansn and BØEGH [7] reported that recycled aggregate concrete had 15 to 30 % lower modulus of elasticity and 40 to 60 % higher shrinkage than those of conventional concrete. Olorunsogo and Padayachee [8] found that the water sorptivity of concrete prepared with 100 % recycled aggregate was about 39 % higher than that of natural aggregate concrete at the curing age of 28 days. Dhir et al. [9] showed that the compressive strength of concrete with 100 % coarse and 50 % fine recycled aggregates was between 20 – 30 % lower than that of the corresponding natural concrete.

It has been shown from previous research studies that one of the main potential problems of recycled aggregate concrete is the high drying shrinkage value compared to that of conventional concrete. But this shortcoming can be improved by incorporating a certain amount of fly ash into the concrete mixture since fly ash is known to be able to mitigate the drying shrinkage of concrete [10-15]. However, fly ash can be used as a replacement of cement or as additional cementitious materials in concrete. The different applications of fly ash produce concrete with totally different properties. In this study, fly ash was used as a cement replacement in proportioning recycled aggregate concrete. The effects of fly ash on the compressive strength, the tensile splitting strength, the static modulus of elasticity, the drying shrinkage and the resistance of concrete against chloride penetration on the recycled aggregate concrete were investigated.

EXPERIMENTAL DETAILS

Materials

ASTM Type I Portland cement and ASTM Class F low-calcium fly ash were used in the concrete mixtures. The chemical and physical properties of cement and fly ash are given in Tables 1 - 2.

Natural and recycled aggregates were used as the coarse aggregate in the concrete mixtures. In this study, crushed granite was used as the natural aggregate and recycled aggregate sourced from a recycling facility in Hong Kong was used. The nominal sizes of the natural and recycled coarse aggregates were 20 and 10 mm and their particle size distributions conformed to the requirements of BS 812 [16].

The physical and mechanical properties of the coarse aggregate are shown in Table 3. The porosity of the aggregates was determined using mercury intrusion porosimetry (MIP). River sand was used as the fine aggregate in the concrete mixtures.

Table 1 Chemical compositions of cement and fly ash

MATERIALS	COMPOSITION (%)						
	LOI	SiO$_2$	Fe$_2$O$_3$	Al$_2$O$_3$	CaO	MgO	SO$_3$
CEMENT	2.97	19.61	3.32	7.33	63.15	2.54	2.13
FLY ASH	3.90	56.79	5.31	28.21	<3	5.21	0.68

Table 2 Physical properties of cement and fly ash

PROPERTIES	MATERIALS	
	Cement	Fly Ash
DENSITY (g/cm^3)	3.16	2.31
SPECIFIC SURFACE AREA (cm^2/g)	3519.5	3960

Table 3 Properties of natural and recycled coarse aggregates

TYPE	NOMINAL SIZE	DENSITY (kg/m^3)	WATER ABSORPTION (%)	STRENGTH (10% fines) KN	MIP POROSITY (%)
CRUSHED GRANITE	10mm	2.62	1.12	159	1.62
	20mm	2.62	1.11		
RECYCLED AGGREGATE	10mm	2.49	4.26	126	8.69
	20mm	2.57	3.52		

Concrete mixtures

The concrete mixtures were prepared with a water-to-binder (W/B) ratio of 0.55. The fly ash was used as a partial replacement of cement and recycled aggregate was used as a replacement of the natural coarse aggregate.

In this study, fly ash was used as 0, 25 and 35% by weight replacements of cement and recycled aggregate was used as 0, 20, 50 and 100% by weight replacements of natural coarse aggregate. The absolute volume method was adopted to design the mix proportions of the concrete mixtures as shown in Table 4. In the concrete mixture, the 10 and 20 mm coarse aggregates were used in a ratio of 1:2.

Specimens Casting and Curing

For each concrete mix, 100 mm cubes, 70 x 70 x 285 mm prisms and 200 x 100 mm diameter cylindrical specimens were cast. The cubes and the rectangular prisms were used to test the compressive strength and the drying shrinkage respectively. The cylinders were used to evaluate the splitting tensile strength, the static modulus of elasticity and the chloride-ion resistance of the concrete. All the specimens were cast in steel moulds and compacted using a vibrating table. Three cubes and three cylinders were immediately used after demolding to test the 1-day compressive and splitting tensile strengths, and the rest of the specimens were cured in a water-curing tank at $27 \pm 1°C$ until the day of testing.

Table 4 Proportioning of concrete mixtures

NOTATION	FLY ASH (%)	RECYCLED AGGREGATE (%)	CONSTITUTES (kg/m^3)					
			Water	Total cementitious material	Sand	Granite	Recycled aggregate
R0	0	0	225	410	642	1048	0
R20	0	20	225	410	642	840	204
R50	0	50	225	410	642	524	506
R100	0	100	225	410	642	0	1017
R0 F25	25	0	225	410	611	1048	0
R20F25	25	20	225	410	611	840	204
R50F25	25	50	225	410	611	524	506
R100F25	25	100	225	410	611	0	1017
R0F35	35	0	225	410	598	1048	0
R20F35	35	20	225	410	598	840	204
R50F35	35	50	225	410	598	524	506
R100F35	35	100	225	410	598	0	1017

EXPERIMENTAL

Compressive and Splitting Tensile Strengths

The compressive and splitting tensile strengths of concrete were determined using a Denison compression machine with a loading capacity of 3000 kN. The loading rates applied in the compressive and splitting tensile tests were 0.2 - 0.4 N/mm^2.sec and 57 kN/min respectively. The compressive and splitting tensile strength tests were carried out on concrete specimens at the ages of 1, 4, 7, 28 and 90 days.

Static Modulus of Elasticity

The static modulus of elasticity of concrete was determined in accordance with ASTM C 469-65. This test was carried out on concrete specimens at the ages of 28 and 90 days.

Drying Shrinkage Test

A modified British Method (BS1881, part 5: 1970) was used for the test. After removing the concrete prisms from the curing tank, the initial length of each specimen was measured. The specimens were then stored in an environmental chamber with a temperature of 55°C and a relative humidity of 95 % until the next measurements at 1, 4, 7, 28, 56, 90 and 112 days. Before each measurement was taken on the scheduled day, the specimens were first removed from the environmental chamber and conveyed to a second cooling chamber for about 4 hours at a controlled temperature of 25°C and a relative humidity of 75 %.

The length of each specimen was then measured within 15 minutes before delivering the specimens back to the environmental chamber for the subsequence drying process. The procedure of drying, cooling and measuring continued until the final length measurement at 112 day was recorded.

Determination of Chloride Penetrability

The chloride penetrability of concrete was determined in accordance with ASTM C1202-94 using a 50 mm thick x 100 diameter concrete disc cut from the 100 x 200 mm concrete cylinder. The resistance of concrete against chloride ion penetration is represented by the total charge passed in coulombs during a test period of 6 hour. In this study, the chloride penetrability test was carried out on concrete specimens at the ages of 28 and 90 days.

RESULTS AND DISCUSSIONS

Compressive and splitting tensile strength

The compressive strengths results are presented in Tables 5. Each presented value is the average of three measurements. It can be observed that the compressive strengths of all concrete specimens decreased as the recycled aggregate content increased. Furthermore, the early day (1, 4 and 7-day) compressive strengths of concrete were significantly reduced as the fly ash content increased. However, at 90 days, the beneficial effects of the incorporation of fly ash became evident. The 90-day compressive strengths of concrete containing fly ash as a 25 % replacement of cement were comparable to or higher than those of the concrete prepared with no fly ash.

The results of the splitting tensile strength of concrete are presented in Tables 6. Each presented value is the average of three measurements. The results showed that the tensile splitting strength decreased as the recycled aggregate content increased. At the same recycled aggregate replacement level, the use of fly ash reduced the tensile splitting strengths of the concrete.

Static Modulus of Elasticity

The static modulus of elasticity values of all concrete mixtures are presented in Table 7. Each presented value is the average of two measurements. At the same fly ash content, the static modulus of elasticity values of concrete decreased as the recycled aggregate content increased. The 28-day static modulus of elasticity of concrete prepared with 100% recycled aggregate was about 20% lower than that of natural aggregate concrete. Furthermore, at the same recycled aggregate content, the static modulus of elasticity values of the concrete decreased as the fly ash content increased. However, the influence of fly ash on the static modulus of elasticity of concrete prepared with 100 % recycled aggregate was not significant.

Table 5 Compressive strength of concrete mixtures

NOTATION	FLY ASH (%)	RECYCLED AGGREGATE (%)	COMPRESSIVE STRENGTH (MPa)				
			1-day	4-day	7-day	28-day	90-day
R0	0	0	12.8	23.3	30.2	48.6	52.7
R20	0	20	11.9	22.4	29.1	45.3	50.8
R50	0	50	11.6	21.8	27.6	42.5	49.5
R100	0	100	10.2	18.6	24.4	38.1	45.5
R0 F25	25	0	12.1	22.8	28.6	43.6	57.9
R20 F25	25	20	11.5	24.3	32.8	42.8	57.3
R50 F25	25	50	11.1	22.9	30.4	41.7	53.4
R100F25	25	100	9.4	19.1	25.1	36.8	50.1
R0 F35	35	0	7.7	16.6	22.5	40.7	47.8
R20 F35	35	20	6.6	16.4	20.9	41.0	46.6
R50 F35	35	50	5.9	15.2	20.4	37.1	43.2
R100F35	35	100	4.8	14.6	19.4	25.2	37.4

Table 6 Splitting tensile strength of concrete mixtures

NOTATION	FLY ASH (%)	RECYCLED AGGREGATE (%)	COMPRESSIVE STRENGTH (MPa)				
			1-day	4-day	7-day	28-day	90-day
R0	0	0	1.30	2.32	2.62	3.32	3.68
R20	0	20	1.27	2.30	2.53	3.21	3.66
R50	0	50	1.25	2.28	2.51	3.16	3.48
R100	0	100	1.21	2.20	2.38	3.06	3.23
R0 F25	25	0	1.14	1.99	2.40	3.28	3.42
R20 F25	25	20	1.01	1.88	2.28	3.13	3.28
R50 F25	25	50	1.00	1.82	2.23	3.09	3.22
R100F25	25	100	0.75	1.75	2.15	2.26	2.79
R0 F35	35	0	0.67	1.53	1.85	2.90	3.11
R20 F35	35	20	0.63	1.43	1.70	2.76	3.08
R50 F35	35	50	0.52	1.34	1.66	2.58	2.85
R100F35	35	100	0.49	1.27	1.61	2.56	2.72

Table 7 Static elasticity modulus of concrete mixtures

NOTATION	FLY ASH (%)	RECYCLED AGGREGATE (%)	ELASTICITY MODULUS (GPa)	
			28-day	90-day
R0	0	0	30.0	31.2
R20	0	20	28.8	29.7
R50	0	50	26.3	27.8
R100	0	100	21.7	24.7
R0F25	25	0	29.0	31.7
R20F25	25	20	28.9	29.9
R50F25	25	50	28.7	29.1
R100F25	25	100	23.2	23.4
R0F35	35	0	28.5	30.1
R20F35	35	20	26.3	27.8
R50F35	35	50	24.8	25.2
R100F35	35	100	21.6	23.6

Table 8 Chloride-ion penetration and drying shrinkage of concrete mixtures

NOTATION	FLY ASH (%)	RECYCLED AGGREGATE (%)	TOTAL CHARGE PASSED (Coulombs)		DRYING SHRINKAGE $(X10^{-6})$
			28-day	90-day	112-day
R0	0	0	6291	4940	405
R20	0	20	6572	5194	447
R50	0	50	6687	5425	476
R100	0	100	6905	5750	540
R0 F25	25	0	4169	2923	365
R20 F25	25	20	4430	3031	398
R50 F25	25	50	4703	3126	419
R100F25	25	100	4972	3265	459
R0 F35	35	0	2920	2542	341
R20 F35	35	20	3344	2683	382
R50 F35	35	50	3596	2825	405
R100F35	35	100	3683	3036	460

Drying Shrinkage

The drying shrinkage values (tested at 112 days) of the concrete mixtures are shown in Table 8. The shrinkage values of recycled aggregate concrete were higher than those of natural aggregate concrete. The incorporation of fly ash as a partial replacement of cement reduced the drying shrinkage values of the concrete.

Chloride Penetrability

The resistances of concrete against chloride-ion penetration are shown in Table 8. It was found that the use of fly ash as a partial replacement of cement increased the resistance against chloride ion penetration. However, the resistance decreased as the recycled aggregate content increased.

SUMMARY AND CONCLUSIONS

The test results showed that the compressive and splitting tensile strengths of concrete decreased with an increase in recycled aggregate content. The early aged compressive and splitting tensile strengths of concrete significantly decreased as the fly ash content increased. However, at longer curing ages, the beneficial effects of fly ash became evident.

The static modulus of elasticity values of concrete decreased as the recycled aggregate content increased. The modulus of elasticity decreased with an increase in the fly ash content. However, the influence of fly ash on the static modulus of elasticity of concrete prepared with 100 % recycled aggregate was not significant.

The drying shrinkage value of concrete increased with an increase in the recycled aggregate content. But the shrinkage of concrete decreased with an increase in the fly ash content.

The resistance of concrete against chloride ion penetration decreased as the recycled aggregate content increased. However, this reduction could be counteracted by the use of fly ash.

The results show that one of the practical ways to utilize a higher percentage of recycled aggregates in structural concrete is by incorporating 25-35 percents of flyash.

ACKNOWLEDGEMENTS

The authors would like to thank the Environment and Conservation Fund and the Hong Kong Polytechnic University for funding support.

REFERENCES

1. R.K. DHIR AND T.G. JAPPY (edit), 1999, Proceedings of the International Conference on Exploiting Wastes in Concrete, Thomas Telford, UK.

2. HANSEN, T. C., RILEM REPORT 6.: Recycling of Demolished Concrete and Masonry, published by E&FN Spon, Bodmin, UK, 199

3. COLLINS, R.J. 1994. The Use of Recycled Aggregates in Concrete. BRE Information Paper IP 5/94, Building Research Establishment, Watford, UK.

4. R.K. DHIR, N.A. HENDERSON AND M.C. LIMBACHIYA (edit) 1998, Proceedings of the International Conference on the Use of Recycled Concrete Aggregates, Thomas Telford, UK.

5. RILEM, Proceedings of the 1st ETNRecy.net/RILEM workshop on Use of Recycled Materials as Aggregates in the Construction Industry, 11-12 September, 2000, Paris.

6. HASABA, S., KAWAMURA, M., TORLI, K. AND TAKEMOTO, K.: Drying shrinkage and durability of concrete made from recycled concrete aggregates, Transactions of Japan Concrete Institute, Tokyo, V, 3, 1981, pp.55-60.

7. HANSEN, T. C. AND BØEGH, E.: Elasticity and drying shrinkage of recycled aggregate concrete, Journal of the American Concrete Institute, 82 (5), pp. 648-652, 1985.

8. OLORUNSOGO, F. T. AND PADAYACHEE, N.: Performance of recycled aggregate concrete monitored by durability indexes, Cement and Concrete Research, 32 (2002)179-185.

9. DHIR, R. K., LIMBACHIYA, M. and LEELAWAT, C. T.: Suitability of recycled concrete aggregate for use BS 5328 designated mixes. Proc. Inst. Civ. Engrs & Bldgs, 1999, 134, Aug., 257-274.

10. HAQUE, M. N., LANGAN, B. W. AND WARD, M. A.: High fly ash concrete. ACI Mater. J., 81 (1), 54-60.

11. HANSEN, K. D., AND REINHARDT, W. G.: Roller-compacted concrete dams, McGraw-Hill, New York.

12. NANNI, A., LUDWIG, D., AND SHOENBERGER, J.: Roller compacted concrete for highway pavements. Concr. Int., 18(5) 33-39.

13. DELAGRAVE, A., MARCHAND, J., PIGEON, M., AND BOISVERT, J.: Deicer salt scaling resistance of roller compacted concrete pavements. ACI Mater. J., 96 (2), 164-169.

14. PITTMAN, D. W., AND RAGAN, S. A.: Drying shrinkage of roller compacted concrete for pavement applications. ACI Mater. J., 95 (1), 19-25.

15. COSTABILE, S.: Recycled aggregate concrete with fly ash: A preliminary study on the feasibility of a sustainable structural material. The first international conference on ecological building structure, San Rafael California, July, 5-9, 2001.

16. BRITISH STANDARD INSTITUTE, 1985. BS 812: Part 103.1: Methods for determination of particle size distribution.

FOAM CONCRETE CONSTRUCTION DEMOLISHED WASTE

L B Svatovskaya

M M Baidarashvili

T S Titova

M V Shershneva

A M Sychova

N I Yakimova

A V Khitrov

St Petersburg State University of Transport

Russia

ABSTRACT. Both, properties foam concrete demolished waste (named product №1) and foam concrete construction with product №1 are being shown. The latter in product №2. It was turned out, that the foam concrete waste surface is good enough for heavy metal ions adsorbing that are being shown by using a new method of surface researching. The details of that are determined. The product №1 is very useful for biosphere purification from heavy metal ions. The foam concrete with heavy metals ions adsorbed is good enough for the new foam concrete material like filler. The properties a foam concrete with filles like that is being shown. In the end we have two useful waste productions with both, the biosphere purification and improved building material properties.

Keywords: Foam concrete, Waste, Heavy metals ions, Surface, New method, Adsorbing, Filler.

Dr, professor L B Svatovskaja is Head of Engineering Chemistry and natural science Department St. Petersburg university of Transport. Her research interests include both, material for building and biosphere purification problems.

Dr M V Shershneva - lecturer of the St. Petersburg University of Transport, her research interest are in the ecology area.

Dr M M Baidarashvili - lecturer of the St. Petersburg University of Transport, her research interests are in the ecology area including the sold state surface.

Dr A M Sychova - lecturer of the St. Petersburg University of Transport, research interests include new building materials like foam concrete.

Dr T S Titova - lecturer of the St. Petersburg University of Transport, research interests are connected with environment condition.

Dr N I Yakimova - lecturer of the St. Petersburg University of Transport, research interests are connected with new technology of the environment purification.

Dr A V Khitrov - lecturer of the St. Petersburg University of transport, his research interests are new foam concrete material technologies.

INTRODUCTION

A new material for constructions is foam concrete and especially good is an autoclave foam material. [1,2,3]. But there are lots of the foam concrete wastes during obtaining that concrete. [4]. This paper examines properties of the surface of foam concrete waste with the new method - ACD (adsorbing centers distribution). This is the first using of the foam concrete waste, that we named the product №1.

This method was offered by Nechiporenko [5] and was described for the other aims. But this method could be good enough for ecology researching as well, because properties of the surface can be observed in the notion donors and acceptors. But heavy metal ions are good acceptors and that is why the information of the ACD – method could be useful for forecast adsorbing ability foam concrete. But after such kind of using we have the other waste – foam concrete demolished waste with heavy metal ions on its surface, the second waste. Such kind of product examines like filler in the foam concrete, and the latter is product №2. On the figure 1 the chain of the using of the foam concrete demolished waste is presented.

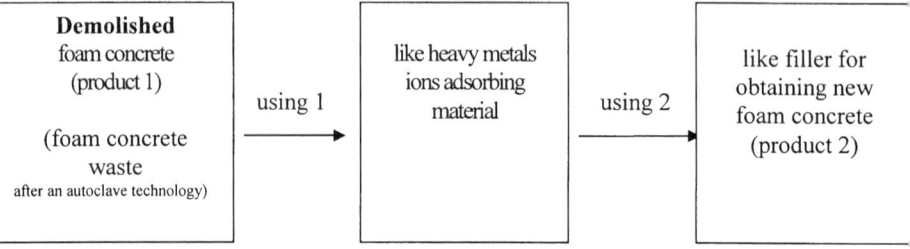

Figure 1. The chain demolished foam concrete using.

MATERIALS

The materials and adsorbing centers distributions is shown on the Table 1 and on the Figure2. Demolished foam concrete waste are seen to have different intense of ADC (fig.2), but according to our research especially interest is in the field of 0<pKa<12.

Table 1. Demolished foam concrete

DENSITY, kg/m³	CONTENT, %			
	Cement	Sand	Water	Foam
300	17,07	2,40	8,90	71,50
500	31,0	3,97	11,9	53,0
700	33,50	10,45	15,68	40,4
800	34,6	13,6	11,86	40,0
1000	25,3	37,5	9,6	27,5
1200	25,0	44,5	9,4	20,9

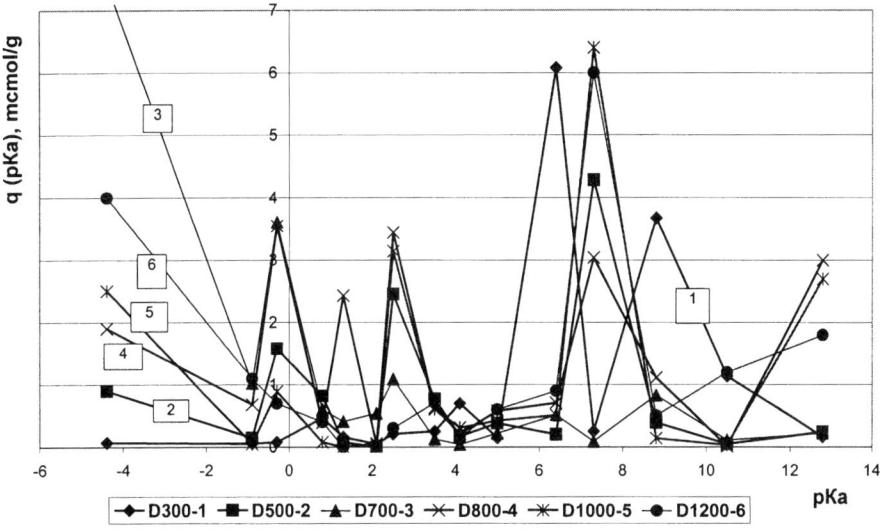

Figure 2. Adsorbing centers distribution on the foam concrete surface

DISCUSSION OF THE EXPERIMENT

Ions of heavy metals (Fe (III) and Mn (II)) is proved to adsorb on the surface foam concrete demolished waste. For example, for average densities 300 and 500 the best field of adsorbing is 5<pKa<11 (the Figure 3 and 4) and adsorbing ability of concrete is shown in the Table 2.

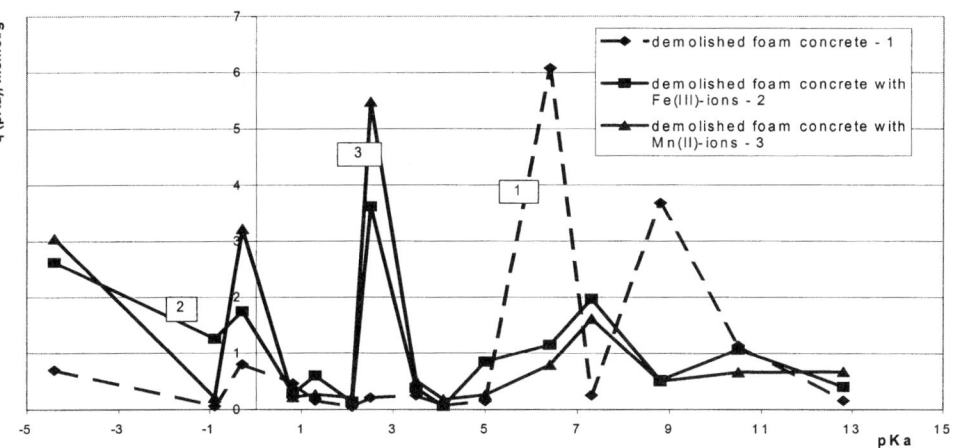

Figure 3. ACD on the foam concrete surface, density 300 kg/m^3

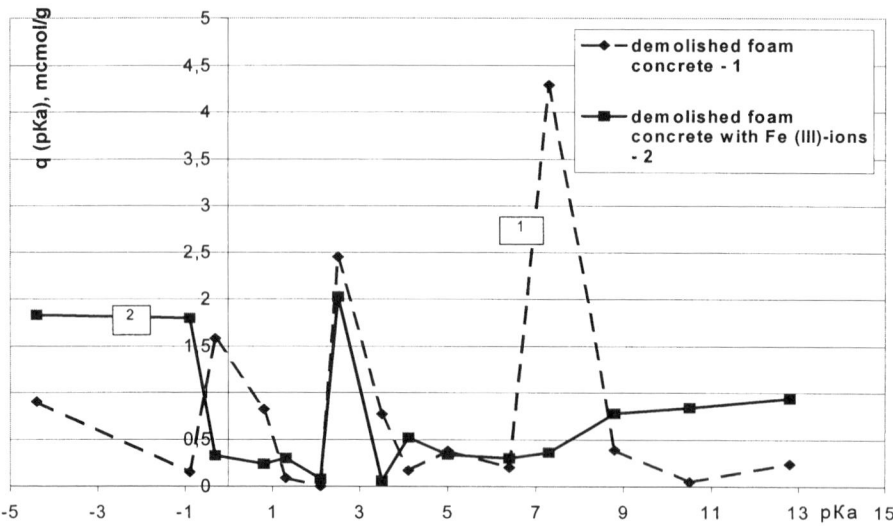

Figure 4. ACD on the foam concrete surface, density 500 kg/m^3

Table 2. Adsorbing ability foam concrete.

AVERAGE DENSITY, кg/m³	300	500	700
Adsorbing Ability for Fe (III), мg/g	0,70	0,55	0,22
Adsorbing Ability for Mn (II), мg/g	2,70	2,40	1,50

The foam concrete with heavy metal ions need in using. According to author [6] such kind of waste is good enough like filler. As a matter of fact there is a problem of fillers for foam concrete and the waste with the same nature like a base material is good for properties. In the Table 3 and on the figure 5 the heat conductivity of the foam concrete with new filler is seen to reduce (Figure 5), but and the other properties are good enough according to Russian standard (Table 3). That new foam concrete was named product 2.

Table 3. Properties of the foam concrete with demolished foam concrete filler

MASS % of the fillers	Density, kg/m³ (dried example)	FLEXURAL STRENGTH, MPa, 28 days	COMPRESSIVE STRENGTH, MPa, 28 days
5	500	10,2	15,2
10	500	9,8	15,0
15	500	7,2	15,0

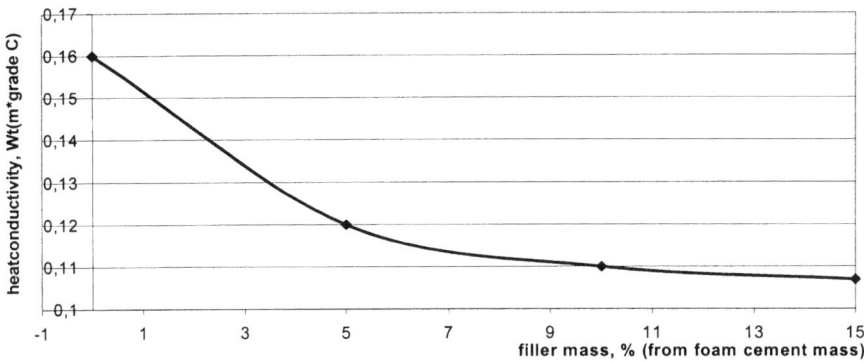

Figure 5.   Relationship between filler content and heat conductivity new foam concrete, density 600 kg/m³

PRACTICAL IMPLICATIONS

The warm monolith foam concrete floors were obtained using filler from product 1, nearly 2000m². Heat conductivity has been reduced more, than 20%.

CONCLUSIONS

- Surface of the foam concrete demolished waste can adsorb heavy metal ions and that is why that product is useful for biosphere purification.
- After that, the waste is good like filler for new foam concrete, especially for monolith foam concrete. The main property of that new foam concrete is reducing heat conductivity.

ACKNOWLEDGMENTS

The authors would like to acknowledge the scientists of the department for discussing problems.

REFERENCES

1. SVATOVSKAJA, L.B.: Engineering and chemical problems of the foam materials of the third millennium, St. Petersburg, 1999, pp32-46.
2. SVATOVSKAJA, L.B.: Modern natural scientific bases in the science of the constructive materials and ecology, St. Petersburg, 2000.
3. SVATOVSKAJA, L.B., MASLENNIKOVA, L.L., SOLOVYEVA V.Y.: Basic approaches for creation of the new complex technologies for biosphere purification, St. Petersburg state university of Transport, St. Petersburg, 2003.
4. BAIDARASHVILI, M.M.: Adaptation of the indicating method for biosphere purification from heavy metal ions, St. Petersburg state university of Transport, St. Petersburg, 2001.
5. NECHEPORENCO A.P.: Properties of the donors and acceptors of the solid stuffs surface, St. Petersburg, 1995.
6. SVATOVSKAJA, L.B.: Dry building mixes and new technologies in construction, St. Petersburg, 2003.

SHEAR BEHAVIOUR OF CRUSHED CONCRETE AND BRICKS

I Chidiroglou

E A Laycock

P S Mangat

Sheffield Hallam University

A K Goodwin

Mott MacDonald Ltd

United Kingdom

ABSTRACT. The results of an investigation into the shear strength properties of demolition waste materials are presented. Two crushed concrete materials and one crushed brick material were each tested in a large shear box at similar moisture contents and dry densities. Based on the results to date, it is conjectured that recycled concrete aggregates exhibit some non-linearity in their failure envelopes, whereas brick fills show substantial non-linearity across a wide range of stresses. It is thought that the non-linearity is due to crushing of the recycled aggregate at higher stresses, but further work is required to verify this.

Keywords: Recycled aggregates, Shear strength, Construction demolition waste, Crushed concrete, Crushed bricks, Properties of construction demolition waste

I Chidiroglou is a doctoral candidate at Sheffield Hallam University researching the physical, mechanical and chemical properties of commercially crushed demolition waste, specifically crushed concrete and bricks. He received both his BEng in Civil Engineering and his MSc in Integrated Environmental Control from Nottingham Trent University.

Dr A K Goodwin is an Associate Director of Mott MacDonald Ltd, and leads the geotechnical and geo-environmental team in Sheffield. In addition, he is the lead supervisor for two research projects into the behaviour of engineered fills. He gained his first degree from the University of Surrey, and his PhD from the University of Sheffield.

Dr E A Laycock received her Doctorate from the University of Sheffield in 1997. She is now a Senior Lecturer within the School of Environment and Development. Current teaching and research activities focus on the subjects of geology and construction materials.

Professor P S Mangat novel construction material based on industrial waste, binders and non-cement activators, repair technology and performance of corroding and repaired concrete element.

INTRODUCTION

The UK produces approximately 90 million tonnes of construction and demolition waste every year [1], but the construction industry has traditionally relied upon the use of primary aggregates rather than recycling the waste. Latham [2] and Egan [3] both considered that the industry would benefit from the increased use of recycled materials, and in 1994 the UK government set out its objective to reduce the quantity of primary aggregates used by 68% [4]. In line with this objective, economic pressure is being applied to the industry through an increase in the landfill tax of £1 per tonne per year since 1999, with a review of this policy due in 2004.

A large percentage of construction and demolition waste comprises bricks and concrete, and there is a growing need for research into the re-use of these materials as aggregates [5]. Little research has been undertaken into what is perceived as the low level re-use of the materials as bulk fills, though some recent work [6] indicates that there is increasing interest in this aspect. It has generally been assumed in practice that the behaviour of such fills would be similar to that of natural aggregates, and therefore accumulated data on the properties of such aggregates has been extrapolated and applied to the recycled materials.

This investigation is addressing the use of these types of materials for engineering fill applications. It compares the properties of three different types of materials and the effect that different degrees of processing has on their shear strength. The results presented are part of a wider investigation into the physical, mechanical and chemical properties of construction demolition waste.

MATERIALS

Sources of Materials

Three material types, referenced A to C, have been used to date in the experimental work. Each material was purposefully selected as being representative of commercial operational and crushing procedures, in preference to crushing of the materials in the laboratory using artificial procedures.

Material A was produced local to the university from the demolition of a late 20th century concrete building. The demolition waste was crushed on site using heavy demolition plant, and in excess of two tonnes of the material transported direct to the university laboratories. The material mainly comprised crushed concrete, as the demolition and processing method adopted on site had lead to the removal of the vast majority of other structural components.

Laboratory inspection procedures revealed that impurities such as steel reinforcement accounted for only 1% by mass of the sample delivered, and all such remaining materials were manually removed in the laboratory prior to testing.

Materials B and C were of crushed concrete and crushed brick respectively, and were both obtained from a demolition waste processing site in Hull operated by Sam Allon (Contracts) Ltd. Both the concrete and brick had been sourced originally from unknown developments, and subsequently transported to and processed at the crushing site.

A range of commercial products are prepared at the site, and the products used in this research project were selected as being compliant with industry standard specifications. Material B was selected as meeting the grading requirements of Clause 803 of the Department of Transport Specification for Highway Works [7] for Type 1 sub-base, whilst Material C was chosen to meet the grading requirements of Class 6F2 (selected granular materials) in accordance with Table 6/1 of the Department of Transport Specification for Highway Works.

Preparation of Sub-Samples

The complete testing programme required a very large quantity of each material to be prepared, but due to storage limitations it was not feasible to obtain sufficient of the material from the suppliers to allow quartering of the samples strictly in accordance with BS1377 [8] requirements. Rather, the entire 2 tonne mass of each material was initially sieved to determine the grading of the full sample. The fraction coarser than 37.5mm ?? was then removed to reduce the ratio between the minimum dimension of the test specimen and the maximum particle size to within acceptable limits, and the remainder mixed thoroughly on a clean concrete floor. The mixed material was sub-divided to prepare each sub-sample in turn via a process of quartering, with the remaining sample being retained each time for further sub-division into other sub-samples of approximately equal mass throughout the test programme.

Grading tests were undertaken on the prepared sub-samples throughout the shear box testing programme to verify the efficacy of the preparation procedure, and the mean grading curves for each test material are presented in Figure 1. The grading envelopes for all tests have been omitted for clarity, as all grading curves lay ±3% of the mean curve and confirmed that a high degree of repeatability had been obtained.

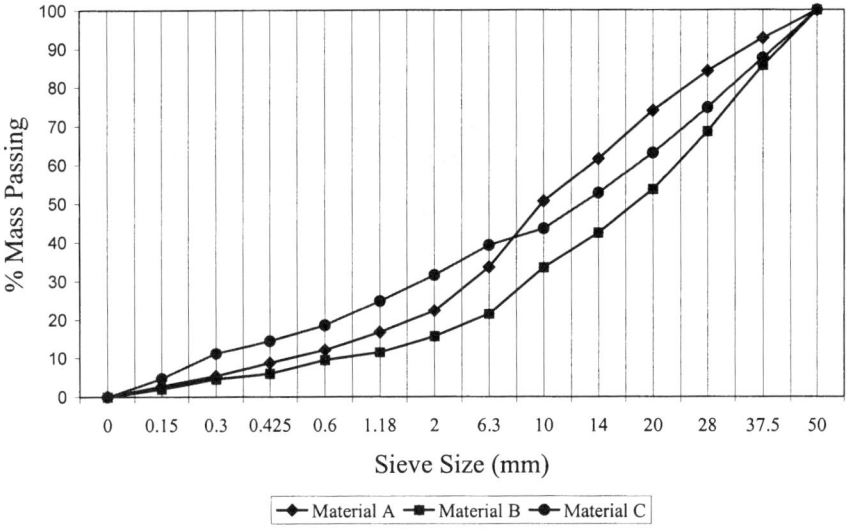

Figure 1. Mean Grading Curves for Test Materials change y axis – it must stop at 100%!!

TESTING PROGRAMME AND SPECIMEN PREPARATION

Test Programme

The recently completed full shear test programme was designed to determine the shear strength properties of the three materials at a range of densities and moisture contents. The results presented herein relate only to the first series of tests, at a dry density of 1.80 ± 0.08 Mg/m^3 and a moisture content of 2.0 ± 0.2%. A total of five different confining pressures between 95kPa and 317kPa were used in the work. A total of 100 samples were tested to deal with potential variability of large specimens.

Details of Shear Box

The shear box equipment used in the test programme is shown in Figure 2. The equipment is free-standing, and consists of a rigid base frame supporting a central horizontal loading jack. The thrust from the jack is applied to the lower half of the ball track mounted 316mm square by 160mm high shear box. The upper half of the shear box is attached to a yoke which is connected in turn to a 50 kN load cell.

The horizontal loading jack is driven via a multi-speed gear box mounted within the base of the machine. In all tests, a displacement rate of 0.125mm/minute was adopted. Vertical load is applied to each specimen through a rigid loading plate and a vertical loading frame. The lower crosshead of the loading frame passes beneath a horizontal loading beam, which transfers dead load from the end of the beam at a ratio of 20:1.

Figure 2. Large Shear Box and Computer Logging System

Specimen Preparation

To ease specimen preparation and avoid damage to the shear box frame, compaction of the materials in the shear box was undertaken with the shear box removed from its frame and placed on the concrete floor. Both halves of the shear box were secured together with screws, and a vibrating hammer used to compact a total of 8.4kg of material into the box in three layers to achieve the target dry density for each test specimen of 1.8Mg/m^3. Three layers were adopted to avoid coexistence of the shear plane with layer interfaces, and to promote a uniform density throughout the specimens. After the required density was achieved the top plate was placed in position and the whole box placed within the test frame.

RESULTS OF SHEAR TESTING

Shear Stress - Displacement and Volumetric Behaviour During Shear

Detailed review for the results for all three materials indicated very similar behaviour during shear. Figures 3 and 4 present the mean shear stress – displacement curves, and mean volumetric changes against displacement, for Material B at a range of normal stresses. The results indicate the following:

- strain to failure increases with the normal stress level
- a small volumetric reduction occurs during shear

The observed behaviour is typical of that expected for normally to lightly over-consolidated natural aggregates.

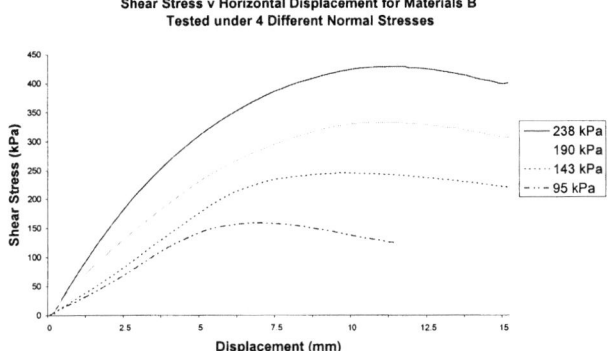

Figure 3. Mean Shear Stress Response of Material B to Shear

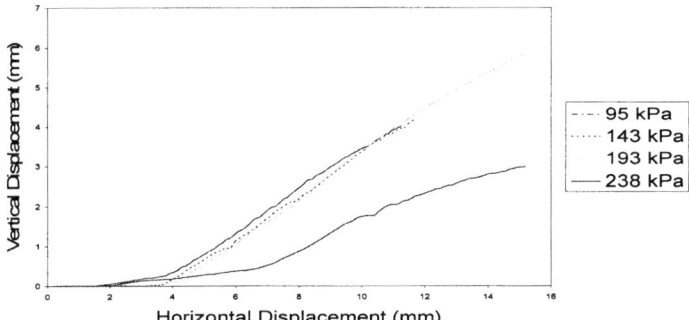

Figure 4. Mean Volumetric Response of Material B to Shear

Shear Strength Parameters

Figure 5 presents plots of shear stress at failure (σ_{sf}) against normal stress (σ_n) for all three material types. Each data point represent the mean of 5 test results. The repeatability of the results was satisfactory and the largest difference recorded was 37 kPa (from largest value to smallest for the same type of test)

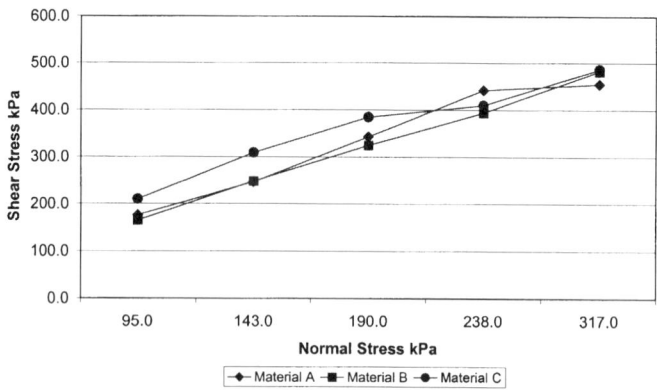

Figure 5. Shear stress against normal stress relations for Material A, B and C

Table 1 summarises the effective angles of friction ϕ' and effective cohesions determined from the tests for each material assuming a linear correlation.

Table 1. Linear regression analysis of failure envelopes for all normal stresses

MATERIAL	ϕ' (°) assuming c' = 0	VALUES BASED ON LINEAR REGRESSION Analysis of All Tests	
		ϕ' (°)	c' (kPa)
A	58.7	53.4	67.8
B	58.2	55.0	41.5
C	60.2	40.0	125.9

Closer inspection of the data indicates that there is some evidence of curvature of the failure envelope at higher normal stresses, and this is reinforced by the extremely high values for effective cohesion shown in Table 1. Table 2 summarises the shear strength parameters tests for each material assuming a linear correlation for the lowest three normal stresses alone. This shows that the values of effective cohesion for the concrete fills (Materials A and B) are substantially lower than that for the brick fill (Material C).

Comparison of the results in Table 2 with those in Table 1 for all normal stresses indicates that higher values of the shear strength parameters are valid at lower normal stresses. This effect is very marked for the brick fill (Material C), and may be evidence of crushing of this material in particular at higher stresses. This effect has previously been reported by others for recycled materials [6]. Further, others [9] [10] have conjectured that when materials with large particles are subjected to shear stresses their behaviour depends on a combination of dilatancy and degradation, and that the relative importance of these two factors varies as the normal stresses change. At low σ_{ns} dilatancy would normally be expected to dominate the behaviour pattern, whereas crushing and degradation would increasingly dominate as the normal stresses increase. Post-shearing grading curve analyses from this investigation, which are not reported herein due to space restrictions, support this contention and show significant particle breakage.

Table 2. Linear regression analysis of failure envelopes for normal stresses up to 190kPa

MATERIAL	ϕ' (°) assuming c' = 0	VALUES BASED ON LINEAR REGRESSION Analysis of All Tests	
		ϕ' (°)	c' (kPa)
A	60.8	60.4	4.5
B	59.8	59.3	5.7
C	64.5	61.5	38.5

CONCLUSIONS

Three types of commercially processed demolition waste have been tested during this investigation. The shear strength properties and behaviour of each material have been determined from an extensive programme of testing in a large shear box, the partial results of which have been reported in this paper.

The behavioural characteristics of the three materials tested were all similar, and the research has indicated that each waste material has substantial strength. Effective angles of friction of 59°, 58° and 60° for Materials A, B and C respectively have been determined assuming a linear failure envelope and zero cohesion for all normal stresses. If zero cohesion is not forced in the analysis, the effective angles of friction for Material A and B decrease slightly to 53° and 55° respectively, with apparently high values of effective cohesion due to the curvature of the failure envelope. In contrast, the effective angle of friction for Material C decreased substantially to 40°, with an even more marked increase in effective cohesion to 126kPa.

Further analyses allowing for the non-linearity in the failure envelope resulted in slightly higher effective angles of friction if zero cohesion is enforced, compared the equivalent values for the full range of normal stresses.

Removing the requirement for zero cohesion lead to only small changes in the shear properties for the concrete materials (Materials A and B), whereas there was a much more marked change for Material C, the brick fill.

Based on the results to date, it is conjectured that recycled concrete aggregates exhibit some non-linearity in their failure envelopes whereas brick fills show substantial non-linearity across a wide range of stresses. It is thought that the non-linearity is due to crushing of the recycled aggregate at higher stresses. However, further work is required to verify this.

ACKNOWLEDGEMENTS

The authors wish to thank Stephen L inskill, the technician responsible for the geotechnics laboratory at Sheffield Hallam University, for his valuable help in completing the testing for this investigation. Special thanks are also extended to Controlled Demolition Group Ltd for providing Material A, and to Sam Allon (Contracts) Ltd for providing Materials B and C.

REFERENCES

1. www.defra.gov.uk/environment/statistics/waste/wrkf09.htm

2. LATHAM, M. Constructing the Team: Joint Review of Procurement and Contractual Arrangements in the United Kingdom. Building Research Establishment, Watford, 1994, BRE Construction Industry Final Report

3. EGAN, Sir J. Rethinking construction: the report of the construction task force to the Department of the Environment, Transport & the Regions, DETR, London, 1998.

4. Department of the Environment, Transport & the Regions. Guidelines for aggregates provision in England. DETR, London, 1994, Minerals Planning Guidance MPG06

5. HURLEY, J.W., McGRATH, C., FLETCHER, .S.L. & BOWES, H.M., Deconstruction and reuse of construction materials, Building Research Establishment Ltd, 2001

6. SIVAKUMAR, V., McKINLEY J. D. & FERGUSON, D., Reuse of construction waste: performance under repeated loading. Journal of Geotechnical Engineering, Proceedings of the Institution of Civil Engineers, Vol 157, Issue GE2, pp91-96, 2004

7. Department of Transport, Specification for road and bridge works, HMSO, 1980

8. British Standards Institution, Methods of test for soils for civil engineering purposes, British Standard 1377, 1990

9. BISHOP, A.W., The strength of soils as engineering materials, Geotechnique, Vol. 16, 2, pp91-130, 1966

10. BILLAM, J., Some aspects of the behaviour of granular materials at high pressures. Proceedings of the Roscoe Memorial Symposium, 29th – 31st March 1971, Cambridge University, pp69-80, Henley-on-Thames, 1972

THE USE OF RECYCLED AGGREGATES IN VIBRO STONE COLUMN GROUND IMPROVEMENT TECHNIQUES

C J Serridge

Pennine Vibropiling Limited
United Kingdom

ABSTRACT. Increasing awareness for a sustainable environment is resulting in the greater use of recycled aggregates in vibro stone column ground improvement techniques. In the context of stone columns, crushed concrete and spent railway track ballast probably have the greatest potential for this application in the UK. Other potential sources of aggregate (recycled and secondary aggregates), include granular demolition rubble, waste rock and some steel slags. However, where such materials are considered for use, as an alternative to natural primary aggregates, it is important that there are appropriate quality control procedures/protocols in place, to ensure that these materials are "fit for purpose". Technical requirements for vibro stone column aggregate are discussed and some applications of recycled aggregates presented.

Of particular importance is the fines (clay/silt) content of the aggregate, as this can affect inter-granular contact within the completed columns, and have a significant effect on the angle of internal friction of the column material. This in turn has an influence on both the load carrying capacity of the stone column and the settlement characteristics of the stone column-soil composite.

Keywords: Vibro stone columns, Recycled aggregates, Sustainable construction, Recycled concrete, Recycled railway track ballast, Angle of internal friction, Quality control, Fit for purpose.

Mr C J Serridge is a Principal Geotechnical Engineer with Specialist Geotechnical Contractor - Pennine Vibropiling. He has over 15 years experience in the design and project management of ground improvement techniques (principally vibro stone column and dynamic compaction techniques), both in the UK and the Middle East. His research interests include the utilization of recycled aggregates in granular column ground improvement techniques and ground improvement in soft clay soils.

INTRODUCTION

Vibro stone columns have been utilized as a ground improvement/treatment technique in the United Kingdom (UK) since the 1960's, and are currently the most common form of ground improvement/treatment adopted in the UK. Whilst the use of recycled aggregates in stone columns is not an entirely new concept, increasing awareness of the principal of sustainable construction, is leading to the greater utilization of these materials.

VIBRO STONE COLUMN INSTALLATION

In simple terms, the stone column installation technique involves the insertion of a vibroflot (vibrating poker) into the ground to a suitable end bearing strata, or a pre-determined design depth. If the bore formed by the vibroflot is stable, and there is no significant interference from groundwater, the vibroflot is removed from the bore to permit introduction of a plug of stone aggregate, which is then compacted by re-insertion of the vibroflot. This cycle is repeated until a compact stone column is constructed to the surface. This is termed the dry top feed technique (Figure 1). In the event that the bore formed by the vibroflot, is not anticipated to remain stable during stone column installation, (particularly in cohesive strata), stone column installation would normally be undertaken by the bottom feed method (Figure 1), with the vibroflot remaining in the bore during stone column construction. Aggregate is fed directly to the tip of the vibroflot via a tremie pipe/stone delivery tube arrangement and subsequently compacted by the vibroflot. The bottom feed technique has largely superseded the wet top feed technique in the UK, on environmental grounds.

Further details on vibro stone column installation techniques, including the wet top feed technique (and the range of soils suitable for these techniques), can be found in Moseley and Priebe [1], BRE Specifying vibro stone columns [2] and Serridge [3], amongst others.

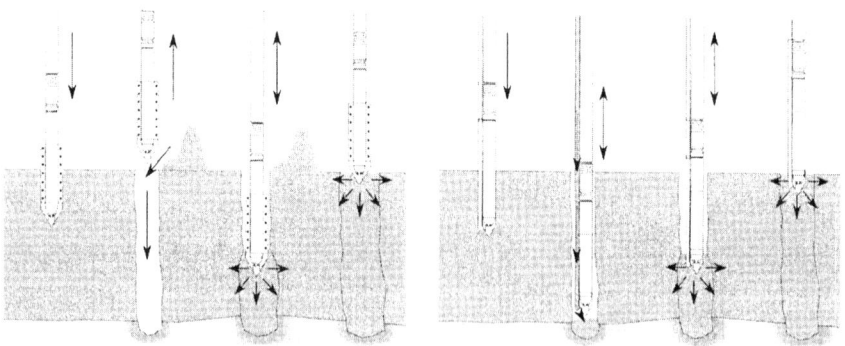

Figure 1.Vibro dry top feed technique   Figure 2.Vibro dry bottom feed technique

VIBRO STONE COLUMN DESIGN PHILOSOPHY

The objective of the vibro stone column technique is to provide a composite "ground structure" of in-situ soil and stone columns, which act as vertical reinforcement and which overall has lower compressibility and increased bearing capacity. In loose granular soils the vibratory action of the vibroflot will also densify material immediately surrounding the stone column, thus enhancing its geotechnical properties. A further advantage of the stone column technique, is that they provide very efficient drainage paths for pore water pressure dissipation (under applied surcharge), which in turn provides acceleration of consolidation settlements, the latter of which is particularly relevant when considering the support of highway or railway embankments. Stone columns can also provide vents for release of gas. Both the latter applications require use of high permeability and preferably single size aggregate to be most effective.

Design is typically based on Hughes and Withers [4] for determination of column length and load carrying capacity (Q_{ult}) of an individual stone column expressed as:

$$Q_{ult} = \frac{(1+\sin \varphi')}{(1-\sin \varphi')} (\gamma h + 4c_u + p)$$

where

φ' = the effective angle of friction of the stone column material.
γ = soil bulk density.
h   = depth at which failure of the stone column in bulging occurs (critical depth). This is usually taken as between 1 and 2 stone column diameters below formation.
c_u = shear strength of the soil at the critical depth.
p   = surcharge pressure.

Baumann and Bauer [5] is frequently adopted for analysis of stress distribution and hence the factor of safety against column overload. Foundation settlement performance is first estimated using available soil parameters and then appropriate settlement reduction factors applied, according to Priebe [6].

INFLUENCE OF STONE COLUMN MATERIAL ON DESIGN

Vibro stone columns represent piers of compacted granular material, and load carrying capacity is therefore a function of the angle of internal friction of the column material. The maximum adopted friction angle for stone columns in the UK is 45°. Laboratory determined values which are higher than this would normally be downgraded to 45°. Some further downgrading of friction angle may also be necessary when addressing stone columns in soft sensitive clays soils. The influence of friction angle on column carrying capacity is clearly demonstrated in Table 1.

Table 1. Ultimate carrying capacity of stone columns with different angles of internal friction (ϕ') [1] (normalised against 45° friction angle)

ϕ'	$\dfrac{(1 + \sin \phi')}{(1 - \sin \phi')}$	% (relative to 45°)
45°	5.828	100
40°	4.599	79
35°	3.690	63

Friction angle has a significant influence on settlement reduction factors, as described by Priebe [6]. Even a modest reduction in friction angle can result in a significant change in settlement reduction factor and therefore overall settlement of the treated ground, (Figure 3).

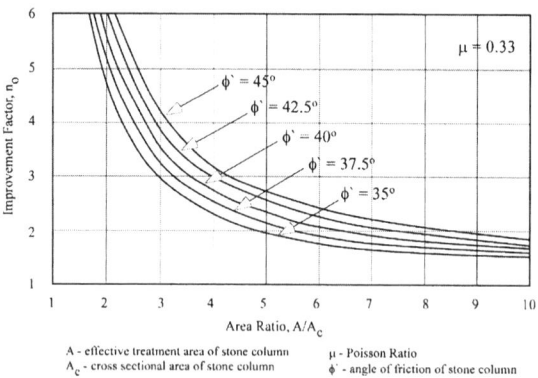

Figure 3. Basic improvement factor with respect to settlement, after Priebe [6]

Direct (plate) loading tests can be used to determine the in-situ measurement of stone column deformation modulus. Young's modulus (E), may be calculated from the following theoretical relationship:

$$E = \frac{I_q\, Q_h\, (1-v^2)}{v}$$

where

I_q	=	an influence factor ($I_q = \pi/4$ for a rigid circular plate)
Q_b	=	average plate bearing pressure
v	=	soil Poisson's ratio
v	=	average settlement

AGGREGATE REQUIREMENTS FOR STONE COLUMNS

The materials used to form stone columns should be "fit for purpose". Typically this requires them to be free draining, hard, inert and capable of forming stone columns with relatively high angles of friction, and to comply with acceptable criteria in terms of material type, grading, hardness, particle shape, flakiness and chemical stability.

Specifications for stone column material are included in the ICE Specification for Ground treatment [7] and BRE Specifying vibro stone columns [2].

The two most common vibro stone column techniques adopted in the UK and their aggregate property requirements are as follows:

(a) **Dry top feed technique:** a nominal single-sized aggregate within the range 40-75 mm or a graded material within these limits is appropriate with a maximum fines (clay/silt) content of 5% usually permitted.

(b) **Dry bottom feed technique:** a nominal single-sized material within the range 20-50 mm is normally considered appropriate, with a maximum fines content not exceeding 5%. It should be recognized, however, that rounded aggregate tends to travel down the stone delivery tube better than more rounded aggregates and the presence of excessive fines can tend to block the delivery systems.

In those circumstances, where the wet top feed technique is necessary (and permitted), a nominal single size material, within the range 20-50 mm and with less than 5% fines, would normally be appropriate.

Material selected to form stone columns should be able to withstand the impact forces of the vibrating poker and retain long-term stability under the applied foundation loads. Materials with an Aggregate Impact Value (AIV)- BS 812 Part 112 [8] or Aggregate Crushing Value (ACV)- BS 812 Part 110 [8] greater than 30% may not be suitable for stone column construction, as they are likely to crush and breakdown under the impact forces of the vibrating poker, particularly under water. The Ten per cent Fines Value (TFV)- BS 812 Part 111 [8], has also been used to define hardness (resistance to crushing/fragmentation) for stone column material.

Resistance to fragmentation is now determined by the Los Angeles Abrasion test in the new European Standards [9]. It is important to recognize, however, that the tests differ and cannot be compared directly. The National Guidance documents which accompany the European Standards provide some suggested limiting values for the Los Angeles test, that are regarded as comparable or equivalent to those which have been used for the TFV test.

Dependent upon the vibro technique adopted, acceptable particle shapes for stone column materials would include rounded; angular and irregular. Unacceptable materials would be flaky and elongated.

Appropriate quality control procedures are necessary; to ensure that there is consistent supply of material to the site. Storage conditions should be such that introduction of excessive fines is avoided, which would otherwise lead to poor inter-granular contact within the completed columns. It is important that these requirements are recognized when considering use of recycled/reclaimed aggregates in stone columns. Any contamination issues should also be addressed.

USE OF RECYCLED AGGREGATES IN VIBRO STONE COLUMNS

Within the UK construction industry, the following terminology has been developed for recycled/reclaimed aggregates:

(a) **Recycled aggregates (RA):** derived from re-processing materials previously used in construction. Examples include construction and demolition arisings (crushed concrete and brick) and spent railway ballast.

(b) **Secondary aggregates (SA):** Usually by-products of other industrial processes not previously used in construction. Examples include metallurgical slags, waste rock and foundry sand.

Recycled and secondary aggregates currently have around 23% (19% RA and 4% SA), of the aggregate market share in England.

Within the UK in general, several barriers have been cited, that may inhibit the full potential and utilization of these aggregates in construction, (and in turn for stone columns). They include:

(a) Lack of confidence and perceived risk with the product (i.e. a perception that they are inferior to natural primary aggregates)

(b) Lack of suitable specifications and testing protocols (quality control)

(c) Rigid specifications in contract documents

(d) Lack of awareness and industry culture

(e) Some Client's unwillingness to share risk

(f) Waste management licensing regulations/environmental issues

(g) Supply – demand issues and economics

In the UK, prior to the late 1990's, there was a general lack of specifications for recycled aggregates in construction, and consequently little basis for applying quality control. Most utilization of these materials was therefore in lower grade applications. However, developments in Europe and Japan in the drafting of specifications for recycled materials, prompted the publication of, for example, BRE Digest 433- Recycled Aggregates [10] and subsequently, BRE Quality Control- The Production of Recycled Aggregates [11]. The principal aim of these documents was to bridge the gap between UK practice at the time and the introduction of the new European Standards (EN's) for Aggregates in 2004,[12] which are generally more permissive with respect to recycled and secondary aggregates than the British Standards they replace. This is clearly a positive step for sustainability. However, the National Guidance documents, which are designed to accompany the new standards, whilst well developed for primary aggregates, are less well developed for recycled and secondary aggregates and require further improvement.

Until relatively recently, there has been limited information on the distribution, availability and quality of recycled materials in the UK. However, services such as AggRegain have been set up. Established by WRAP (Waste Resources Action Programme) and funded through the Aggregates Levy Sustainability Fund, the service is essentially designed to provide sources of practical information on the use of recycled and secondary aggregates, including "parity of performance with primary aggregates and potential cost benefits." In addition to the AggRegain web site [13] other useful web sites include the CIRIA Register of Construction Recycling Sites (GB) [14] and the Building Research Establishment BREMAP^{TM} web site [15]. Trade organizations such as the Quarry Products Association, may also provide useful information [16].

RECYCLED CRUSHED CONCRETE

Crushed concrete has perhaps the greatest potential of all recycled aggregates for use in vibro stone column construction, particularly in urban areas, where concrete structures are being demolished and floor slabs broken up, (Figure 4).

Figure 4. Production, grading and stockpiling of crushed concrete for use as recycled aggregate

However, crushed concrete is slightly weaker and has a lower angle of friction, than most natural primary aggregates utilized in stone column construction in the UK. Mc Kelvey et al. [17] recorded an angle of internal friction of 39°, using laboratory shear box testing for crushed concrete aggregate, with particle sizes in the range 20-40 mm. Whilst values of this magnitude are unlikely to result in significantly lesser bearing capacity or greater settlement of the improved ground, it is important to recognize that when this crushed concrete aggregate was smeared with up to 20% kaolin slurry in the laboratory, the friction angle reduced to around 30°. In addition, some particle crushing was observed at higher applied stresses.

Hence, in applications such as stone columns, it is important that designers and specifiers do consider these issues and ensure that the crushed concrete aggregate selected is "fit for purpose" and not contaminated with fines (silt/clay). Materials from bases of stockpiles therefore require to be selected with caution (and rejected where deemed necessary). Caution may also need to be exercised, particularly when higher foundation loads are required.

Use of Recycled Crushed Concrete Aggregate – Case History

One example, of an increasing number of vibro ground treatment projects in the UK, where crushed concrete has been utilized as a recycled aggregate for stone column construction, is at a site in Coatbridge in Scotland. Some 4000-4500 tonnes of recycled crushed concrete aggregate was used to construct stone columns beneath the main foundations and ground bearing floor slab of a proposed DIY retail unit.

Prior to the re-development of the site (which was a former ironworks), demolition of existing concrete structures and breaking of concrete floor slabs was necessary. In order to implement the principle of "sustainable" construction (and minimise the production of waste and the efficient use of materials and recycling of wastes), it was decided, by the relevant interested parties, to establish crushing and screening facilities on site, to allow production of crushed concrete which was suitable for general fill and higher quality material (i.e. with less fines), for use in vibro stone column construction.

Specifications and grading envelopes were provided by the Specialist Contractor, to confirm requirements for the stone column aggregate. Quality control procedures were also implemented on site incorporating regular inspection, sampling, testing and monitoring, to ensure consistency of supply (and quality) and therefore "fitness for purpose".

The ground treatment was designed to improve the bearing capacity and control the total and differential settlement characteristics of the underlying made ground, which comprised between 2.0 m and 6.0 m of ash and slag deposits (associated with the sites industrial legacy), to permit construction of shallow foundations (pad and strip foundations), and ground bearing floor slabs.

Stone columns were installed in closely spaced groups beneath pad and strip foundations, to provide allowable bearing pressures of 100 kN/m². Beneath ground bearing floor slab areas, the stone columns were installed on a square grid arrangement, to provide allowable bearing pressures in the range 30-50 kN/m². Total settlements were required to be restricted to less than 25 mm and differential settlements to better than 1 in 500. Stone column lengths varied from 2.5-6.0 m.

The crushed concrete aggregate proved very satisfactory for stone column construction, with an average deformation modulus of 40 MN/m², Table 2, determined from 600 mm diameter plate load tests. Primary aggregates did not need to be imported for stone column construction and were therefore conserved. Vehicle movements to and from the site and associated environmental impact were significantly reduced.

Table 2. Recorded deformation moduli from 600 mm diameter plate load tests on stone columns constructed with recycled aggregate, based on Poisson Ratio of 0.3

SITE	RECYCLED AGGREGATE	STONE COLUMNS DEFORMATION Moduli	
		Range	Mean Value
Coatbridge (UK)	Crushed Concrete	35-55 MN/m²	45 MN/m²
Liverpool (UK)	Rail Ballast	31-40 MN/m²	35 MN/m²
Manchester (UK)	Rail Ballast	34-49 MN/m²	40 MN/m²

RECYCLED RAILWAY TRACK BALLAST

Within the railway network, there is a requirement to periodically remove and replace ballast primarily because of inter-particle attrition and aggregate breakdown. Following the privitisation of the rail industry within the UK in the mid 1990's, management of the supply chain for materials had become fragmented and it became evident that significant quantities of spent railway ballast were being landfilled unnecessarily. Having recognized this deficiency, the rail authority developed initiatives to address this, amongst which was the establishment of some 15 or so well-equipped regional supply and recycling depots.

The engineering properties of ballast in the UK, render the material particularly suitable for use in stone column construction. For example, the BR Ballast Specification – BR 1203 [18] required ballast aggregate to be finer than 63 mm, with not more than 0.8% finer than 1.18 mm, with most of the material lying between 50 mm and 28 mm. The material was otherwise required to be hard, durable stone, angular in shape with all dimensions almost equal and free from dust. The flakiness index nor the elongation index was to exceed 50%; and not more than 2% of the particles by weight were to have a dimension exceeding 75 mm. The new European Standard for Railway Ballast is BS EN 13450 [19]. Whilst the majority of spent railway ballast is supplied to the construction industry as a fill or sub-base material, it is also proving an appropriate resource for higher grade applications such as vibro stone columns.

Recycled spent railway ballast has been utilized to good effect on sites in Manchester, Liverpool and Birmingham (UK), for vibro stone column construction in weak fill materials, to provide an effective ground improvement solution, prior to construction of pad and strip foundations and ground bearing floor slabs, for low rise structures. Deformation moduli for the stone column material have been determined from 600 mm diameter plate load tests. The results for spent railway ballast are detailed in Table 2 and correspond well with the average value of 40 MN/m^2 normally adopted/reported for primary aggregates in stone column design.

Selig and Waters [20] have identified the main sources of rail ballast fouling (contamination), in service on open running track, and also defined a fouling index on the basis of the sum of the percentage of material passing 5mm and 0.075 mm sieve sizes. Dependent upon the extent of fouling, the recycled ballast can be processed and washed, as necessary, to ensure the material is "fit for purpose" in higher value applications such as vibro stone columns.

a) 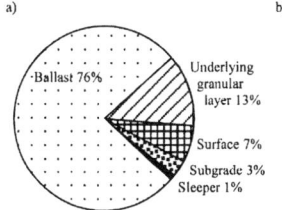   b)

Category	Fouling Index (F_I)
Clean	< 1
Moderately Clean	1 to < 10
Moderately Fouled	10 to < 20
Fouled	20 to < 40
Highly Fouled	> or = 40

$F_I = P_4 + P_{200}$
$P_4 = \%$ by weight of material < 5 mm
$P_{200} = \%$ by weight of material < 0.075 mm

Figure 5 a) Main sources of rail ballast fouling. b) Rail ballast fouling index, after Selig and Waters [20]

In recent years, higher train speeds and increased gross weights have necessitated the use of higher quality aggregates such as granite, which will be of future benefit for vibro stone columns.

OTHER POTENTIAL SOURCES OF RECYCLED AGGREGATE

Other potential sources of reclaimed aggregate for use in vibro stone columns include: general demolition materials and building debris (Recycled Aggregate, RA); waste rock and steel slag (Secondary aggregate, SA). However, the availability of secondary aggregates is dependent upon location. They are not as geographically widespread as recycled aggregates and there are various factors which may have some limitations on their use, which are discussed below.

General demolition materials and building debris tends to be weaker than crushed concrete, and has greater potential to further break down under the vibratory impact of the vibroflot during stone column construction, particularly below the water table. Before consideration could be given to use such materials in higher grade applications, such as stone columns, segregation at source, of hard inert materials and unsuitable weaker deleterious materials (wood, metal, plasterboard etc..), would be essential. Strict vigilence is required to ensure no significant older bricks are present, because of their extreme fragility. In view of this, a greater number of stone columns would probably be necessary, compared to column numbers utilized with primary aggregates, in order to provide similar/equivalent engineering performance, if the aggregate is deemed "fit for purpose". Lack of confidence in the product related to segregation issues, appear to have limited its use in stone columns to date.

The term "waste rock" (SA) is applicable to rocks such as hard limestone that have had certain chemicals removed to leave a hard rock "waste" which can be considered for use in stone columns, providing that excessive fines are not present. Geographical distribution and therefore availability of these materials are likely to be limited.

Metallurgical slags (SA) are essentially defined as non-metallic secondary products of the refining of metal from metallic ores. Slags associated with the iron and steel industries are the most common slag types produced and are utilized as secondary aggregates in construction within the UK. Information on Blast furnace slag and steel slag and their use as aggregates is provided in Building Research Information Paper IP18/01 [21].The author has no direct experience of use of these materials as secondary aggregates, for use in stone columns. However, some commentary is given by Slocombe [22]. It should be recognized, however, that appropriate risk assessment is necessary and caution should be exercised when giving any consideration to their potential use. For example, the material needs to be well weathered and swelling potential and particle disintegration risk would need to be adequately addressed.

CONCLUSIONS

The use of recycled aggregates in vibro stone columns makes an important contribution to sustainable construction. It reduces demand on primary aggregates and associated environmental disturbance.

Specifications and Quality control systems/protocols are available which will assist in ensuring that recycled aggregates selected for application in vibro stone columns are "fit for purpose". There is a need, however, to further disseminate information on their availability together with information services such as AggRegain, to increase awareness. Improved guidance for recycled and secondary aggregates is needed in the National Guidance Documents.

Based upon the author's experience, the best opportunities in respect of recycled aggregates, for use in vibro stone columns, have been with recycled railway ballast and crushed concrete. The recorded compressibility moduli for stone columns (as determined from plate load tests), constructed using these materials, is similar to values reported in the geotechnical literature and adopted in design, for primary aggregates. In the longer term there is a need to develop performance based specifications to optimize use of recycled and secondary aggregates for stone columns.

The fines content of recycled aggregate has a significant effect on the angle of internal friction of the stone column material and therefore the load carrying capacity and settlement potential o f t he t reated ground. S torage conditions for s tone c olumn aggregate s hould b e such that contamination and introduction of excessive fines (clay/silt) is avoided, which would otherwise lead to poor inter-granular contact within the completed column.

REFERENCES

1. MOSELEY, M.P. and PRIEBE, H.J.: Vibro Techniques, Ground Improvement, Edited by M.P. Moseley, Blackie Academic and Professional, pp 1-19, 1993.

2. BUILDING RESEARCH ESTABLISHMENT: Specifying Vibro Stone Columns, BRE Report 391,CRC, Garston, UK, 2000.

3. SERRIDGE, C.J.: A review of vibro stone columns in soft clay soils incorporating field trials at the Bothkennar soft clay research site, MSc Thesis (unpubl.) Bolton Institute of Higher Education, Bolton, UK, 2001.

4. HUGHES, J.M. and WITHERS, N.J.: Reinforcing of soft cohesive soils with stone columns, Ground Engineering, May, pp 42-49, 1974.

5. BAUMANN,V. and BAUER, G.E.A.: The performance of foundations on various soils stabilized by the vibrocompaction method. Canadian Geotech. Journal. Vol. 3, pp 509-530, 1974.

6. PRIEBE, H.J.: The design of vibro replacement, Ground Engineering, December, pp 31-37, 1995.

7. INSTITUTION OF CIVIL ENGINEERS: Specification for Ground treatment, I CE, London, 1987.

8. BRITISH STANDARDS INSTITUTION: Methods of sampling and testing of mineral aggregates and fillers. BS 812: 100 Series, Testing aggregates –Part 112: Method for determination of Aggregate Impact Value (AIV) – Part 110: Method for determination of Aggregate Crushing Value (AIV) – Part 111: Method for determination of Ten per cent Fines Value (TFV), 1990.

9. BRITISH STANDARDS INSTITUTION: BS EN 1097 Test Methods – Physical and Mechanical Part 2 – Methods for the determination of resistance to fragmentation.

10. BUILDING RESEARCH ESTABLISHMENT: Recycled Aggregates, BRE Digest 433, Garston, UK, 1998.

11. BUILDING RESEARCH ESTABLISHMENT: Quality Control - The Production of Recycled Aggregate. BRE Report BR 392, CRC, Garston, UK, 2001.

12. BRITISH STANDARDS INSTITUTION: BS EN 12620 Aggregates for Concrete, 2002.

13. AggRegain web site (www.aggregain.org.uk)

14. CIRIA web site (www.ciria.org/recycling)

15. BRE web site (www.bremap.co.uk)

16. QUARRY PRODUCTS ASSOCIATION web site (www.qpa.org)

17. McKELVEY, D., SIVAKUMAR, V., BELL, A. and McLAVERTY, G.: Shear strength of recycled construction materials intended for use in ground improvement, Ground Improvement, Vol 6, No.2 pp 59-68, Thomas Telford, London, 2002.

18. BRITISH RAIL: Ballast Specification, BR 1203, July 1985 (amended May 1988).

19. BRITISH STANDARDS INSTITUTION: BS EN 13450 Railway Ballast, 2002.

20. SELIG, E and WATERS, J.: Track Geotechnology and Substructure Management (Chapter 7 – Properties of Ballast Material), Thomas Telford, London, 1995.

21. BUILDING RESEARCH ESTABLISHMENT: Blast Furnace Slag and Steel Slag: Their Uses as Aggregates. BRE Information Paper IP18/01,CRC, Garston, UK, 2001.

22. SLOCOMBE, B.: Ground Improvement "Nature versus nurture", Ground Engineering, May, pp 20-22, 2003.

CHARACTERIZATION OF RECYCLED AGGREGATES PRODUCED FROM BUILDING RUBBLES IN GREECE

C-T Galbenis

S Tsimas

National Technical University of Athens

Greece

ABSTRACT. This paper focuses on the characterization of recycled concrete aggregates (RCA) and recycled masonry aggregates (RMA), obtained from demolished buildings in Attica region, Greece. Thermal and X-Ray Diffraction (XRD) analyses results from the above materials are presented. XRD data, in RCA samples, indicated the presence of calcite, portlandite and dolomite while thermal analysis revealed the amount of $CaCO_3$ and the hydration degree through the amount of portlandite contained. With regards to RMA sample, XRD data detected mainly the presence of quartz, diopside and gehlenite. Thermal analysis, in RMA sample, indicated no significant reactions up to 1100°C. According to the investigation process, there is a strong evidence that the examined materials could be suitable for reuse as raw materials in cement clinker manufacturing.

Keywords: Recycled concrete aggregates, Recycled masonry aggregates, X-ray dffraction analysis, Thermal analysis.

Mr C-T Galbenis is a Mining and Metallurgical Engineer and a Ph.D. candidate in the School of Chemical Engineering at the National University of Athens, Greece. His research interests are concerned with the recycling and reuse of construction and demolition wastes and cement and concrete technology.

Dr S Tsimas is a Chemical Engineer and Professor of Applied Inorganic Chemistry in the School of Chemical Engineering at the National University of Athens, Greece. His major interests are in chemistry and technology of aluminosilicates, specializing in cement technology and upgrade techniques of industrial minerals and byproducts. Prof. Tsimas has published over 50 papers in related Journals, Congresses and Conferences.

INTRODUCTION

Construction a nd D emolition (C&D) waste i s a n i mportant p art o f t he waste m anagement sector of most industrialized countries. Nowadays, large quantities of C&D wastes are produced in urban areas, raising the recycling and reuse of such types of wastes as an urgent and imperative environmental issue. Particularly, recycled concrete, masonry and tiles, which comprise the biggest amount of C&D wastes [1], offer many opportunities for their reutilization as aggregates, substituting quarried (primary) aggregates [2]. Recycled aggregates are already used in many countries in various applications as road pavement materials, sub-basements, soil stabilization, improvement of sub-ground, production of concrete of many categories, etc. [2, 3].

Greece is generally characterized by the absence of a network for the collection and utilization of the materials contained in the C&D waste generated [4]. From the 3 million tons of C&D wastes that are estimated to be annually produced in Greece, less than 5% are recycled and reused, whereas in the European Union Countries almost 30% of the relevant annual production (more than 300 million tons) are recycled [2,5]. Some quantities of C&D wastes end up to landfills, while the major quantities end up in uncontrolled disposal areas or in other inappropriate sites, polluting the environment, the soil, as well as the underground water [4]. With regards t o Attica region, it should be remarked that the C &D activity has been developing rapidly over the last years due to a great number of running public projects for the 2004 Athens Olympics [4,6]. Considering the increasing rate of C&D works in Greece, the implementation of alternative management methods of C&D wastes with special emphasis in their recycling and reuse is of vital importance. In 2001 a new Law (Law 2939) was published in Greece regarding the alternative management of packaging and other waste [7]. The implementation of Law 2939 sets the legislation framework for the collection, recycling and reuse of wastes that are produced from construction activities, technical infrastructure activities, excavations and physical and technological disasters.

The current paper deals with the characterization of RCA and RMA obtained from demolished buildings in Attica region (mainly in Athens) Greece by means of XRD and thermal analyses. This is the first step of an extended study aiming to investigate the possibility of utilizing recycled aggregates as raw materials in cement clinker manufacturing.

MATERIALS AND METHODS

Materials

RCA and RMA samples were collected from a recycling unit, which is established in Attica region, Greece. Currently, this unit is the only one, which specializes on the collection and recycling of C&D wastes in Attica region. It should be noted that the accurate source of the building rubbles was unknown.

RCA samples were taken in the fractions 0-8mm (code RCA 1), 8-16mm (code RCA 2), 16-32mm (code RCA 3) and >32mm (code RCA 4). RMA sample was taken in the fraction >32mm (code RMA). For the mineralogical and thermal analysis experiments, the samples were prepared as a fine powder (<90μm) by grinding. With regards to the origin of the examined samples, RCA samples were found to be mostly calcareous contrary to RMA sample where its siliceous origin was confirmed. Table 1 presents the chemical composition of the examined samples. The chemical analysis was carried out by XRF using an Oxford MDX 1000 spectrometer.

Table 1. Chemical composition of RCA and RMA samples

COMPONENT	SAMPLES				
%	RCA 1	RCA 2	RCA 3	RCA 4	RMA
SiO_2	2,70	1,25	1,23	2,84	44,59
Al_2O_3	0,68	0,34	0,37	0,68	9,70
Fe_2O_3	0,43	0,27	0,28	0,36	4,58
CaO	52,75	54,36	54,79	53,10	26,77
MgO	0,96	0,62	0,57	0,86	2,31
K_2O	0,10	0,04	0,04	0,10	1,49
Na_2O	0,08	0,04	0,04	0,10	0,83
SO_3	0,07	0,01	0,01	0,01	0,14
LOI	42,25	42,94	42,85	41,94	9,14

Methods

Examined samples were evaluated for their mineralogical composition by XRD using a Siemens D-5000 diffractometer with nickel-filtered CuKα radiation (λ=1.5405 Å). All patterns were obtained in a scanning range from 5° to 65° in 2θ scale. The testing rate that was applied was 0,02°/min for all samples. Thermal analysis was carried out using a Mettler Toledo TGA/SDTA 851 instrument. Powdered samples were heated from ambient temperature to 1100°C at a rate of 10°C/min. All measurements were accomplished in nitrogen atmosphere with a flow rate of 50 ml/min. Platinum crucible was used in each measurement.

RESULTS AND DISCUSSION

XRD Analysis

XRD analysis results of RCA samples are presented in Figure 1. As it is shown in this figure, the most predominant mineral present in all RCA samples was calcite. This observation was expected considering that Hellenic aggregates are mostly calcareous [8]. XRD data also indicated the presence of portlandite and dolomite. The detection of portlandite confirmed the cement paste presence in RCA samples. It should be remarked that portlandite constitutes one of the main mineralogical phases in Portland cement hydration [8]. Relative to portlandite, it is also observed that samples deriving from coarse fractions of the examined materials (RCA 3, RCA 4) revealed stronger XRD peaks in comparison with those originating from fine fractions (RCA 1, RCA 2). An explanation for this is probably that in fine fractions the separation of cement paste is more effective than in coarse fractions during the crushing process.

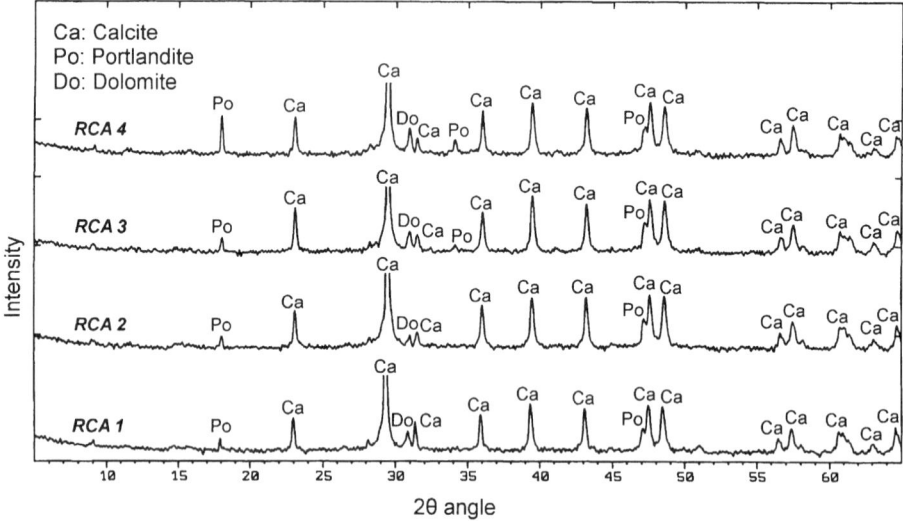

Figure 1. XRD profile of RCA samples

Figure 2 presents the XRD profile of RMA sample. As it is observed, the presence of quartz was predominant in RMA sample. This fact is an indication concerning the siliceous origin of RMA sample. Strong XRD peaks of diopside and gehlenite were also observed. The presence of diopside and gehlenite confirms that RMA sample is originated from bricks which have been fired above 900°C [9]. Moreover, the presence of diopside and gehlenite indicates that a calcareous paste was used in RMA sample [9]. Additionally, the sample contained hematite as well as little amount of feldspars was also detected.

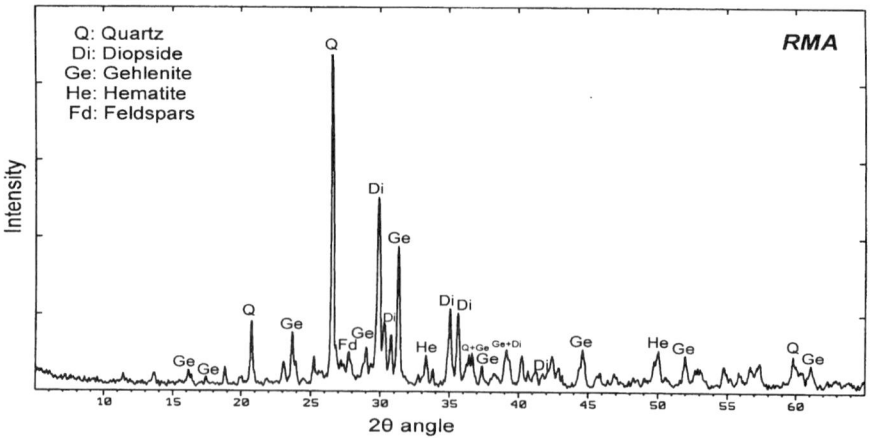

Figure 2. XRD profile of RMA sample

Thermal Analysis

Figures 3-6 show the thermogravimetric (TG) and differential thermogravimetric analysis (DTG) curves of RCA samples. It is observed from the curves that all RCA samples revealed loss of weight in the same temperature regions. The first weight loss, which is related to the humidity of the samples, was occurred in the region 60-140°C. A continuous loss of weight was observed in the region from 140°C to 400°C at a very slow rate. This event is associated with the loss from hydrated calcium silicates present in the samples [10]. The next loss of weight was between 400°C and 460°C corresponding to the decomposition of portlandite - $Ca(OH)_2$. The reaction concerning the decomposition of portlandite is given below.

$$Ca(OH)_2 \text{ (s)} \rightarrow CaO \text{ (s)} + H_2O \text{ (g)}$$

The final loss of weight started at 660°C and finished at 920°C and was due to the decomposition of carbonates present in RCA samples. According to TG/DTG data, the amounts of portlandite and calcite- $CaCO_3$ can be calculated. These amounts are shown in Table 2.

Table 2. TG/DTG data on the amounts of portlandite and calcite present in RCA samples

SAMPLES	FROM TG/DTG DATA	
	% Portlandite	% Calcite
RCA 1	0,68	82,59
RCA 2	0,62	90,64
RCA 3	0,86	91,84
RCA 4	1,42	89,05

The amount of portlandite was very low in all RCA samples and this fact must be attributed to the carbonation of portlandite that is taken place over many years [10, 11]. It should be noted that the amount of portlandite is expected to be 10-20% for uncarbonated cement paste [11]. Moreover, the amount of portlandite was higher in RCA samples that have been derived from coarse fractions of the examined materials (RCA 3 and RCA 4) than those coming from fine fractions (RCA 1 and RCA 2). The above ascertainment is in aggreement with the XRD data, verifying that the presence of portlandite is more obvious in RCA 3 and RCA 4 samples.

In terms of calcite, all RCA samples revealed high amount of $CaCO_3$, confirming its strong presence in these samples. The predominant presence of calcite was also confirmed by XRD analysis. The high content of calcite can be explained considering that the total amount of calcite is originated from: i) the amount of calcite present as aggregate as well as ii) that coming from the carbonation of portlandite over many years [10,11].

The TG and DTG curves of RMA sample are shown in Figure 7. It is observed that there was no significant loss of weight in RMA sample. The total loss of weight represents 1,68% of the initial weight. Loss of weight in RMA sample is associated with the loss of absorbed water in the region 80-140°C as well as the loss of CO_2 due to the decomposition of carbonates in the temperature region from 600°C to 800°C. As it is mentioned above, RMA sample was derived from bricks that have already been fired above 900°C and this was the basic reason that no significant reactions took place during the thermal analysis experiments of the sample.

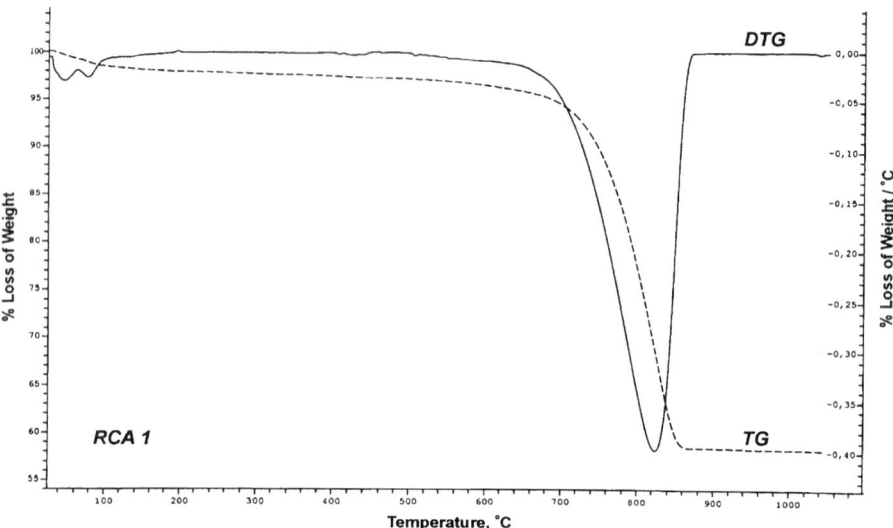

Figure 3. TG and DTG curves of RCA 1 sample

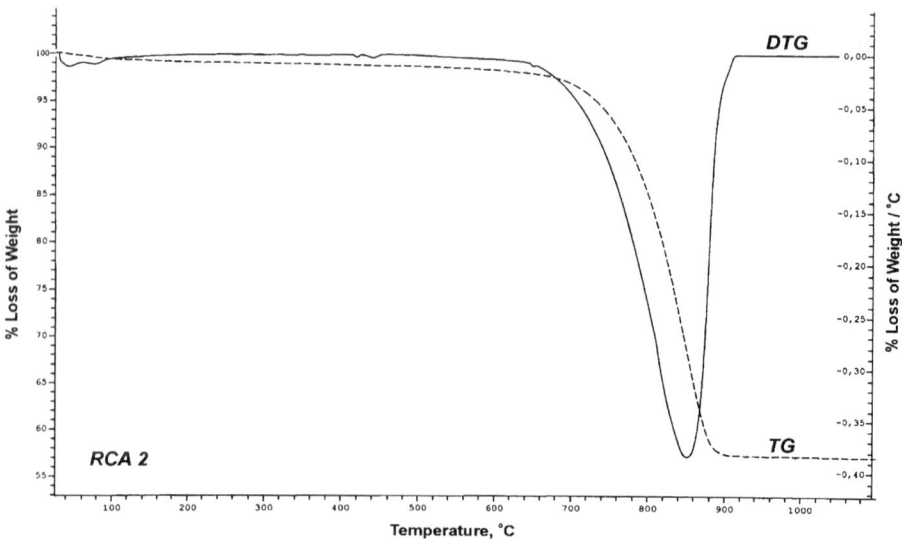

Figure 4. TG and DTG curves of RCA 2 sample

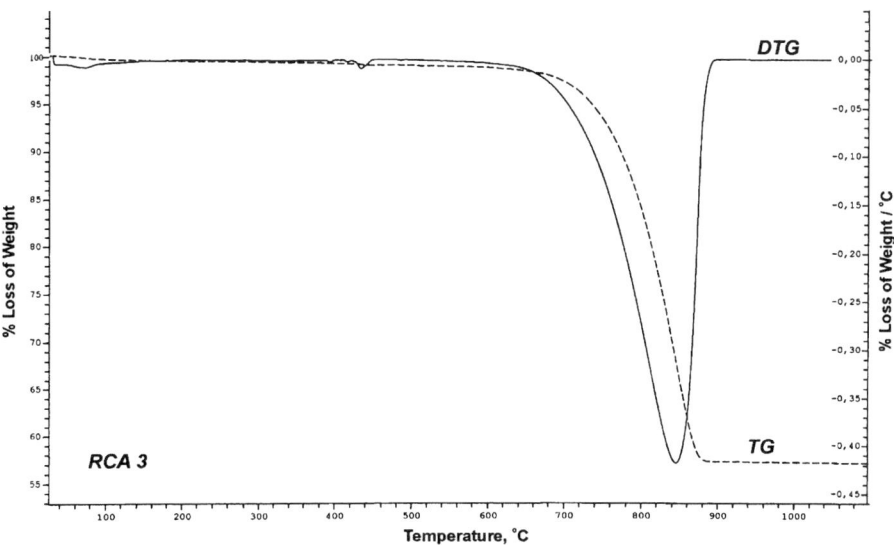

Figure 5. TG and DTG curves of RCA 3 sample

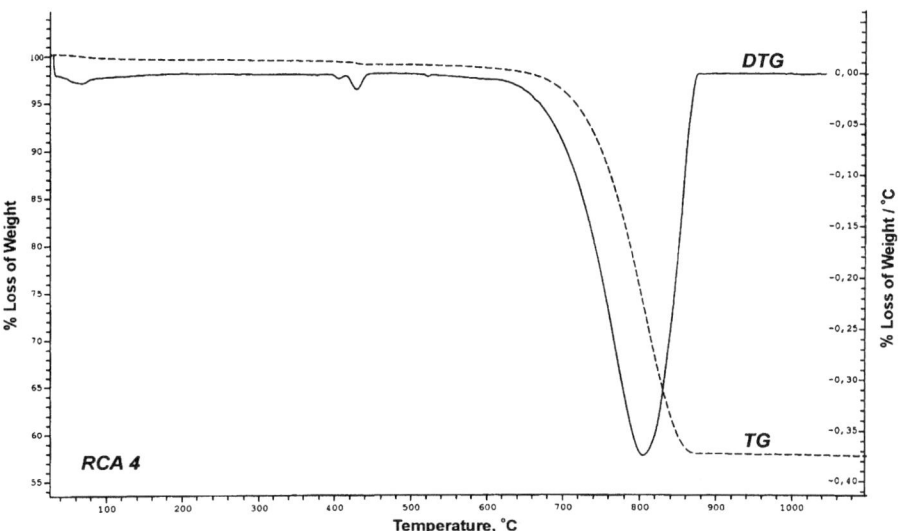

Figure 6. TG and DTG curves of RCA 4 sample

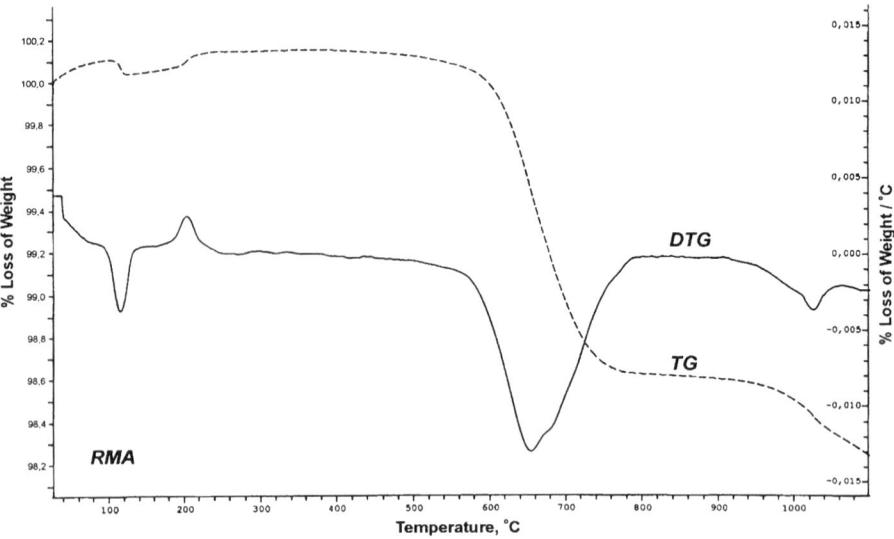

Figure 7. TG and DTG curves of RMA sample

CONCLUSIONS

XRD data in combination with thermal analysis data permit the composition of the examined materials to be estimated. With regards to RCA samples, the predominant presence of calcite was confirmed. High amounts of calcite can be ascribed from the amount of calcite present as aggregate as well as that of the carbonation of portlandite that is taken place over many years. Moreover, the low cement paste presence was detected in all RCA samples through the amount of portlandite. The cement paste presence was more obvious in RCA samples derived from coarse fractions than those coming from fine fractions. In terms of RMA sample, the predominant presence of quartz indicated its siliceous origin. The presence of gehlenite and diopside revealed that a calcareous paste was used in RMA sample.

Taking into account the calcareous and siliceous origin of RCA and RMA samples respectively, there is an indication that these materials could partially replace the raw materials used in cement industries in order to produce cement clinker. However, the potential demand for recycled aggregates as cement raw materials depends not only on their composition but also on their steady supply and reliable logistics and this is an important key-point for the specific application.

REFERENCES

1. OIKONOMOU, N.: Recycled concrete aggregates, Cement and Concrete Composites, 2004, Article in Press, Corrected Proof.

2. SYMONDS GROUP LTD: Construction and Demolition management practices and their economic impacts, Brussels, February 1999, pp1-6.

3.   CUPERUS, J.G. AND BOONE, J.: International experiences in the use of recycled aggregates, Recycling and Reuse of Waste Materials, Proceedings of International Symposium, Dundee, September 2003, pp383-387.

4.   FATTA, D., PAPADOPOULOS, A., AVRAMIKOS, E., SGOUROU, E., MOUSTAKAS, K., KOURMOUSSIS, F., MENTZIS, A. AND LOIZIDOU, M.: Generation and management of Construction and Demolition waste in Greece - An existing challenge, Resources, Conservation and Recycling, 2003, Volume 40, pp81-91.

5.   KALDELLIS, J.K. AND KONSTANTINIDIS, P.: Recent progress concerning the Construction & Demolition debris in Europe, HELECO '03, Proceedings of 4th International Exhibition and Conference on Environmental Technology, Athens, February 2003, pp. 255-263 (in Greek).

6.   LIFE 00 ENV/GR/000739: Construction and Demolition waste management – Recycling unit for inert materials, Life-Environment, EU funded project, Athens, 2003.

7.   LAW NO. 2939: On the alternative management of packaging and other waste, Hellenic Government, Greece, 2001 (in Greek).

8.   TSIMAS, S. AND TSIVILIS, S.: Science and Technology of Cement, Laboratory of Inorganic and Analytical Chemistry, School of Chemical Engineering, National Technical University of Athens, Athens, 2004, pp52-56, 224-241 (in Greek).

9.   LOPEZ-ARCE, P., GARCIA-GUINEA, J., GARCIA, M. AND OBIS, J.: Bricks in Historical Buildings of Toledo City: Characterization and Restoration, Materials Characterization, 2003, Volume 50, pp. 59-68.

10. DOLLIMORE, D., LERDKANCHANAPORN, S., GUPTA, J.D. AND NIPPANI, S: An examination o f r ecycled P ortland c ement c oncrete r ich i n d olomite a nd l ow i n c alcite obtained from various locations in Ohio, Thermochimica Acta, 2001, Volumes 367-368, pp.311-319.

11. DOLLIMORE, D., LERDKANCHANAPORN, S., GUPTA, J.D. AND NIPPANI, S: A thermal analysis study of recycled Portland cement concrete (RPCC) aggregates, Thermochimica Acta, 2000, Volumes 357-358, pp.31-40.

THE PERFORMANCE OF CONCRETE MADE WITH COMMERCIALLY PRODUCED RECYCLED COARSE AND FINE AGGREGATES IN THE WESTERN CAPE

B Kutegeza

M G Alexander

University of Cape Town

South Africa

ABSTRACT. The use of recycled coarse and fine aggregate in concrete production is the most desirable form of achieving a closed life cycle for concrete as a construction material. This paper describes the results of a study undertaken to examine the properties of recycled aggregate as well as the properties of concrete made with recycled aggregate from the Cape Peninsula, South Africa. Sieve analysis showed that both recycled fine and coarse aggregates are continuously graded. Absorption of recycled coarse and fine aggregate was 7 and 10 times higher than natural coarse and fine aggregate respectively. Using saturated recycled aggregates in the concrete mixes reduced the possible effect of absorption on fresh concrete properties. Reduction of compressive strength up to 40% was observed in concrete containing both recycled coarse and fine aggregates, while concrete with only recycled coarse aggregate had 2 to 10% lower compressive strength in comparison with concrete made with natural aggregates. Thus, recycled coarse aggregates can produce a range of concretes with acceptable compressive strength.

Keywords: Recycled fine aggregates, Recycled coarse aggregates, Physical properties, Concrete mix proportions, Bleeding, Compressive strength

Mr B Kutegeza is a Post Graduate student in Civil Engineering Department at the University of Cape Town. For his MSc (Eng.), he is investigating technical aspects of the recycling and re-use of construction and demolition waste as a source of aggregates for concrete production in the Western Cape. His research interests include various aspects of structures and structural materials, durability of concrete structures and waste utilization.

Professor M G Alexander is Professor of Civil Engineering in the University of Cape Town. He holds BSc (Eng), MSc (Eng), and PhD degrees from the University of the Witwatersrand. His teaching and research interests are in cement and concrete technology, with experience in materials and application to design and construction. He heads the Concrete Materials Research Group at the University of Cape Town, where extensive work is being done on problems of marine concrete durability. He has published papers both in South Africa and abroad. He also acts as a specialist consultant to industry and the profession on concrete materials problems.

INTRODUCTION

It is estimated that the South African construction industry generates between 5 and 8 millions tons of construction and demolition wastes (C&DW) per annum, most of it being concrete rubble. If this huge volume of rubble could be recycled and re-used as aggregates, the amount of land needed for disposal would be reduced and existing aggregate resources would not be depleted as quickly. Recycling and use of aggregate from demolition and construction wastes should therefore be dramatically increased, not only because this will conserve and extend natural resources, but also because such materials are likely to provide both environmental and economic advantages.

The increasing charges for landfill, on the one hand, and the scarcity of natural resources for virgin aggregates on the other, encourages the use of waste from construction and demolition as a source for aggregates. The usage of inferior materials for construction purposes is not acceptable, but changing the perception of what is an acceptable material for construction is important. Barriers to the use of recycled aggregate include those created by the use of specifications which were originally based on the characteristics of natural aggregates. The changing of standards and specifications from being based on primary materials to being more performance-based needs to be considered. Many countries throughout the world have now introduced measures aimed at reducing the use of primary aggregates and increasing recycling of C&DW, and using such aggregates where it is technically, economically or environmentally acceptable [1].

The production of concrete with recycled aggregate is the most desirable form of achieving a closed life cycle of concrete as a construction material. While the properties and performance of concrete made with laboratory-crushed recycled aggregate have been extensively investigated and reported by various researchers [2-7], only limited data are available on properties and performance of commercially produced recycled fine and coarse aggregates.

Performance tests [2-7] on fresh and hardened concrete indicate that the differences between the concrete characteristics made with recycled concrete aggregates and natural aggregate concrete are within manageable limits. Very little is known in South Africa about the use of recycled aggregates in concrete production, possibly because natural aggregates are still readily available in urban areas, or engineers are not confident with the use of recycled aggregate materials due to lack of information on availability and suitability of recycled aggregates. In this paper, aggregates from construction and demolition waste (C&DW) will be referred as recycled fine aggregates (RFA) and recycled coarse aggregate (RCA), and concrete made from these aggregates will be referred as recycled concrete (RC). Natural coarse (NCA) and fine aggregates (NFA) will refer to primary aggregates (greywacke stone and Klipheuwel sand).

STATISTICS OF C&DW PRODUCTION IN THE CAPE PENINSULA

Construction and demolition waste

A report by Macozoma [8] shows that the South African construction industry generates between 5 and 8 millions tons of C&DW per annum. Over 1 million tons, mostly concrete rubble reaches landfill sites every year and the remainder is recycled and reused or dumped illegally. More than 200 000 tons of C&DW reach the Cape Peninsula landfills. The sources of construction industry waste include waste from over production, rejects, repair or demolition to mention a few.

Table 1 Landfill sites and estimated quantity of C&DW in the Cape Peninsular [9]

REGIONAL AUTHORITY	LANDFILL SITE	TOTAL SOLID WASTE	QUANTITY OF C&DW
		received/year	@15%/year
		(Tons)	*(Tons)*
	Coastal Park	221 796	33 269
	Swartklip	185 316	27 797
Cape Metro	Vissershok (WMF*)	327 687	49 153
	Bellville Park	329 344	49 402
	Vissershok (CMC**)	295 440	44 316
	Faure	165 568	24 835
	Brackenfell	78 743	11 811
TOTAL		**1 603 894**	**240 584**

Waste management facility   ** *Cape metro council*

Macozoma also reported that illegal dumping is becoming a serious problem in South African open space, as perpetrators try to avoid transport and disposal costs. Details of the quantities of Municipal solid waste (MSW) and C&DW rubble produced in the Cape Peninsula are shown in Table 1. The quantities of C&DW disposed at the landfill sites within the Cape Peninsula are not well known, owing to a lack of record keeping. The percentage of C&DW making up total MSW is estimated at about 15% of the total solid waste.

Application of recycled aggregate

Data in Table 2 from unpublished results of questionnaires show the amount of recycled aggregate produced in the Cape Peninsula by recycling companies, and their applications. Generally in South Africa, a large percentage of recycled aggregates finds low-level applications in backfilling, landscaping and site leveling and landfill to mention a few. Aggregate from recycled C&DW rubble finds application in road works as base and sub-base course, foundations in building construction and brick/block manufacturing. The crushing of C&DW rubble to produce aggregates has been occurring in the Cape Peninsula from about 1996 [9].

There are three types of companies involved in recycling and using of recycled C&DW aggregates in the Cape Peninsula: - commercial crushing companies; brick manufacturers and civil engineering contractors. Although recycling and use of recycled C&DW aggregates has been practised in the Cape Peninsula for more than seven years, there are few data showing how much C&DW materials are generated, recycled and reused annually.

Table 2 Companies, production and application of recycled aggregates in the Cape Peninsular

COMPANY	PRODUCTION/YEAR (Tons)	DEMAND/YEAR (Tons)	APPLICATION
Malans Quarries	> 150 000	> 200 000	Road works and site fill (Sub-base & base course)
Bradis (Pty) Ltd	12 000-14 000	> 10 000	Road works, concrete works and brick manufacturing
Ross & Sons Demolition	> 15 000	> 20 000	Road works and site fill
Cape Brick (Pty)	> 2 000	> 2 000	Brick manufacturing

EXPERIMENTAL PROGRAMME

Materials

Natural coarse aggregate was a 19 mm crushed greywacke, a fined-grained stone consisting of quartz, feldspar, mica and iron oxides developed by thermal metamorphism of argillaceous rocks of the Malmesbury group. The rock does not crush to a good cubical particle shape and tends to be elongated and flaky, thus negatively affecting the workability of concrete. The natural fine aggregate was a Klipheuwel sand, which is a siliceous pit sand having rounded particle shape. Both of these aggregates are widely used in concrete production in the Cape Peninsula.

The recycled aggregates, both coarse and fine, were obtained from Bradis Crushing and Recycling (Pty) Ltd, which produces recycled coarse and fine aggregates of variable sizes. The recycled aggregates were used as taken from the supplier without further treatment. The size ranged from 19 mm down to fine material as shown in figures 1 and 2. Recycled fine aggregate had 30% of its particles greater than 4.5 mm, which was taken into consideration during mix proportioning. Recycled aggregates from one batch and the same source were used to minimize variations.

The binder used was Ordinary Portland cement (CEM I 42.5) and corex slag from Saldanha. Corex slag is a by-product of the reduction of iron ore to metallic iron. It is a new product, but has been widely used as binder material in the Western Cape. The mineralogy of corex slag is similar to that of ordinary Portland cement [10], containing the same cementitious minerals but in different proportions.

Performed Tests

The tested properties of recycled and natural aggregates included grading, fineness modulus, composition, particle shape and texture, bulk and relative density, absorption and 10% FACT. The procedures and test methods were in accordance with South African standards (SABS) and specifications. Fresh concrete was tested for workability/cohesiveness and bleeding. The slump of the fresh concrete was measured to determine the consistence and workability, and 75±25 mm slump was aimed for.

A modified version of the ASTM standard C 232-92 [11] was used to determine the bleeding potential of the concrete. A rigid steel cylinder with inside diameter of 155 mm and height of 150 mm was used instead of the prescribed standard one. Bleeding measurements were taken at 30-minute intervals for a period of two hours and a syringe was used to draw out the bleeding water. The bleed cylinder was kept at a constant temperature of 23±1 ^{0}C and covered to prevent evaporation. Compressive strength of hardened concrete was determined on 100 mm concrete cubes which were cast and water-cured before being tested at ages of 3, 7, 28, and 56 days.

Concrete Mix Design

Since there is no existing method in South Africa for designing concrete mixes containing recycled aggregates, the Cement and Concrete Institute (C&CI) method [12] was used as a guide to design the concrete mixes. This method is normally used for single sized natural stone while the recycled aggregates were graded. To overcome high absorption properties of recycled aggregates, both coarse and fine recycled aggregate were pre-saturated prior to mixing. To maintain the designed mix proportions, the amount of water and aggregate used in the mixing were adjusted according to the actual saturated moisture contents of the aggregates. No other admixtures were used.

Various trial mixes were performed to obtain concrete of good workability and consistency. Six mixes (Mix A, B, C, D, E, and F) were prepared. Mixes A and E were control mixes containing natural coarse and fine aggregate, 50% of CEM I 42.5 and 50% of corex slag for mix A and 100% CEM I 42.5 for mix E. the 30% fraction of coarse particles greater than 4.5 mm in recycled fine aggregate was taken into account in all mixes containing RFA and the stone content adjusted accordingly. Four water-binder ratios, 0.45, 0.60, 0.75 and 0.90, were used. Table 3 shows the constituents and quantities of materials used for all mixes.

RESULTS AND DISCUSSION

Properties of Aggregates

The sieve analyses for RCA and RFA are shown in figure 1 and 2 respectively. Three samples were taken randomly from the same batch, using sampling methods in accordance with SABS 827:1994. The sieve analyses show that both RCA and RFA are continuously graded. The grading of the aggregates has an influence on the workability, cohesiveness and bleeding properties of concrete.

The fineness modulus of the RFA fraction less than 4.5 mm was found to be 3.0, which can be regarded as coarse sand. RCA fineness modulus was 6.9. Based on visual inspection, the contaminants in recycled aggregates were mainly timber, chipboard and plastic particles. Contrary to other literature [1, 3, 5], the composition of the RCA was such that the stone content was higher than stone/mortar conglomerate, Figure 3.

Table 3. Mix proportions (kg/m^3)

MIX TYPE	W/C RATIO	CEMENT	SLAG	COARSE AGGREGATE		FINE AGGREGATE		WATER
		CEM I	Corex	NCA	RCA	NFA	RFA*	
A	0.45	200	200	1100	-	725	-	180
	0.60	150	150	1100	-	810	-	180
	0.75	120	120	1000	-	965	-	180
	0.90	100	100	1000	-	1000	-	180
B	0.45	200	200	610	345	-	800	180
	0.60	150	150	610	370	-	855	180
	0.75	120	120	610	380	-	895	180
	0.90	100	100	610	395	-	915	180
C	0.45	200	200	-	1060	700	-	180
	0.60	150	150	-	1060	785	-	180
	0.75	120	120	-	1060	840	-	180
	0.90	100	100	-	1060	875	-	180
D	0.45	200	200	-	1055	-	665	180
	0.60	150	150	-	1080	-	720	180
	0.75	120	120	-	1095	-	755	180
	0.90	100	100	-	1105	-	785	180
E	0.45	400	-	1100	-	736	-	180
	0.60	300	-	1100	-	820	-	180
	0.75	240	-	1000	-	970	-	180
	0.90	200	-	1000	-	1000	-	180
F	0.45	400	-	-	1060	-	670	180
	0.60	300	-	-	1080	-	730	180
	0.75	240	-	-	1095	-	765	180
	0.90	200	-	-	1105	-	785	180

* Portion of cycled fine aggregates less than 4.5 mm

The stone fraction comprised mainly greywacke natural coarse aggregates, and the brick fractions were remnants of burnt clay bricks that may have been used in the original structures. Generally the surface texture was rough and particle shape was angular classified in accordance with BS 812: Part 102:1989. Other properties were as shown summarized in Table 4.

Table 4. Aggregate properties

PROPERTY	RCA	RFA*	NCA	NFA
Compacted bulk density kg/m^3	1510	1533	1585	1778
Loose bulk density kg/m^3	1330	1430	1388	1608
Relative density	2.50	2.50	2.65	2.65
Fineness modulus	6.9	3.0	7.0	2.4
Dust content % (by mass)	1.0	3.2	0.4	2.2
Absorption % (by mass)	3.0	7.4	0.4	0.7
10 % FACT (kN)	195	-	285	-

* Less than 4.5 mm fraction

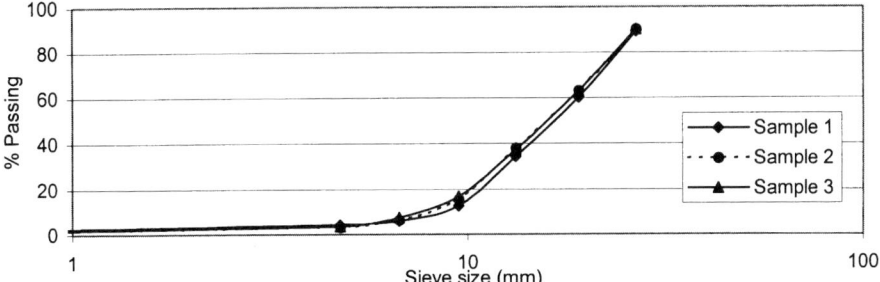

Figure 1: Grading curves for coarse recycled aggregates.

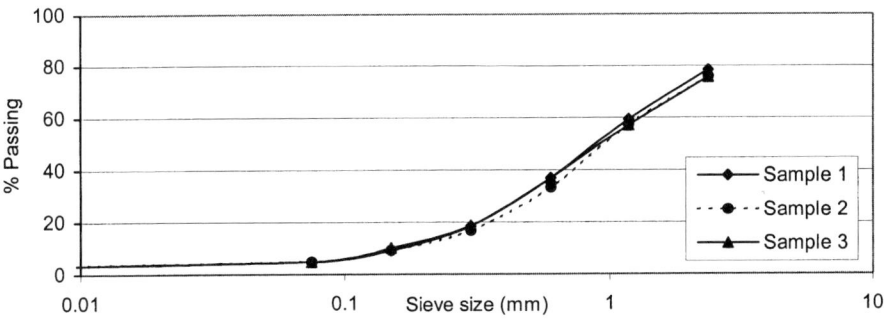

Figure 2: Grading curves for recycled fine aggregates (<4.5 mm fractions).

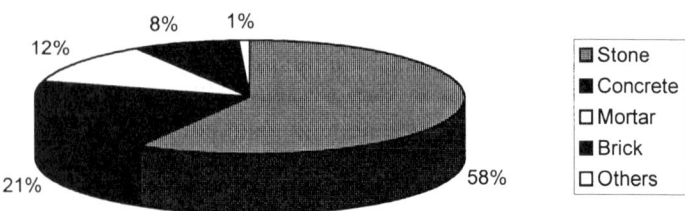

Figure 3: The composition of recycled coarse aggregate

Workability, Cohesiveness and Bleeding

Since both coarse and fine recycled aggregates were saturated before mixing, there was little absorption effect on the fresh concrete. Generally the workability and cohesiveness of all mixes were acceptable, although the mixes containing recycled aggregates were somewhat harsher and stiffer than mixes containing natural aggregates. Slumps varied between 70 mm and 100 mm for mixes containing natural aggregates and between 50 mm and 75 mm for mixes containing recycled aggregates. Bleeding was lower for all concrete mixes containing recycled aggregate for all water-binder ratios in comparison with control mixes.

Corex slag had an influence on the reduction of the bleeding as well. Mix A which had 50% Corex slag had lower bleeding volume than mix E which had only CEM I 42.5. The reduction of bleeding by the recycled aggregate may be attributed to the shape and texture as well as dust content of the recycled aggregate. Corex slag has finer particles than Portland cement, which hold water in the mix and prevents excessive bleeding [10].

Compressive Strength

The compressive strength of the concrete mixes was determined on concrete cubes continuously cured in water for up to 56 days. As shown in Figure 4, compressive strength of mix C containing RCA was marginally lower than control mix A, but higher than control mix E. In this case the compressive strength depended more on binder type than type of aggregate. Binder type had an effect on mix D and F as well.

As expected, the mixes containing corex slag achieved higher compressive strength at later ages owing to the excellent hydraulic properties of the slag. The increase in strength of mixes containing slag between 7 and 56 days is significantly higher than mixes containing only ordinary Portland cement. All mixes that contained RFA had very low compressive strength. This may have been due to contaminants in the RFA. At 56 days of age strength loss of 2% to 40% was observed in concrete containing recycled aggregates for all water-binder ratios.

Maximum reduction of 40% compressive strength was observed in concrete mix B and D, which contained RCA, RFA and 50:50 CEM, I 42.5: Corex slag at water binder ratios of 0.75 and 0.90, Figure 5. Concrete mix C containing only RCA and 50:50 CEM I 42.5: Corex slag had between 2% and 10% strength lower than control mix over all water binder ratios.

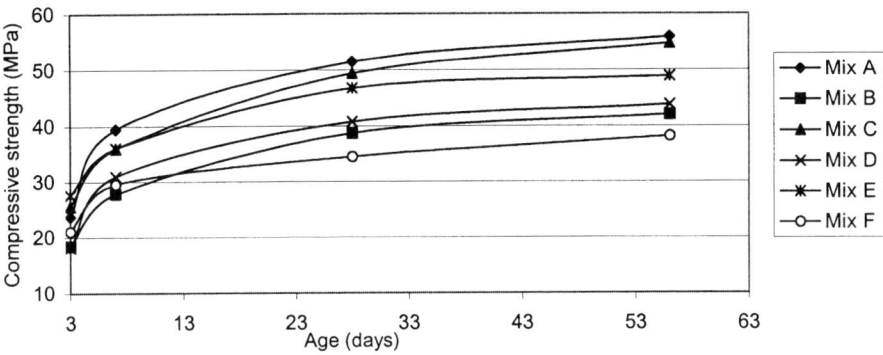

Figure 4: Concrete compressive strength at age of 3,7,28 and 56 days

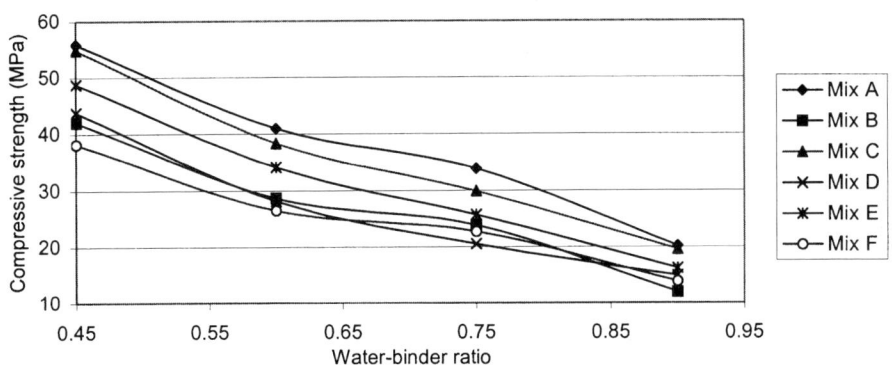

Figure 5: Variation of compressive strength with water-binder ratio after 56 days

CONCLUSIONS

- Sieve analysis showed that both recycled fine and coarse aggregates are continuously graded. The observed contaminants based on visual inspection were wood, chipboard and plastic particles.
- Recycled coarse aggregate contained about 58% of natural aggregate; surface texture was rough with angular particle shape.
- Absorption of both recycled coarse and fine aggregate was 7 and 10 times higher than natural coarse and fine aggregate respectively. Pre-saturation of both recycled coarse and fine aggregates minimized the absorption effect on fresh concrete. Bleeding volume was small for all mixes with recycled aggregates.
- Strength loss of 2 to 40% was observed for concrete containing recycled aggregate. Maximum reduction of compressive strength occurred in concrete containing both recycled coarse and fine aggregates. For all water-binder ratios, the compressive strength difference between concrete mix with only recycled coarse aggregate and natural aggregate was marginal. This shows that recycled coarse aggregates can produce a range of concrete with acceptable compressive strength.

REFERENCES:

1. HANSEN, T. C.: Recycling of demolished concrete and masonry, RILEM Report No. 6, London, E & FN Spon, 1992

2. LIMBACHIYA, M. C. Use of Recycled aggregate: Realizing application and overcoming reluctance. Paper presented at the conference: Concrete for the 21st Century 'Modern Concrete progress through innovation'. Midrand, South Africa, 2002

3. ALEXANDER, M. G, HEIYANTUDUWA, R. Recycled concrete coarse aggregates and its potential for use in semi – structural concrete mixes. Research report. Concrete Materials Research Group, University of Cape Town. 2002

4. SAGOE-CRENTSIL, K. K., BROWN, T., TAYLOR; A. H. (2001), Performance of concrete made with commercially produced coarse recycled concrete aggregate. Cement and Concrete Research Vol. 31, Issue 5, pp. 707 – 712

5. DHIR, R. K at el. (editors): Sustainable construction: Use of recycled concrete aggregate. Proceedings of the International Symposium organized by Concrete Technology Unity, University of Dundee and held at the Department of Trade and Industry Conference center, London UK on 11-12 November 1998. Thomas Telford ltd, 1 Heron Quay, London E14 4JD UK.

6. HENDRIKS, C. F AND JANSSEN, G. M. T.: Use of recycled materials in constructions. RILEM TC URM 'Use of Recycled Materials'. Materials and structures, Vol.36, 2003- pp 604-608, 2003.

7. CHEN, H., YEN, T., CHEN, K.: Use of building rubbles as recycled aggregates. Cement and Concrete Research 33, Issue 1, pp.125 – 132, 2003.

8. MACOZOMA, D. S.: Towards an established secondary construction materials market in SA: Some bottlenecks and solution. CIB World Building Congress. Wellington New Zealand, 2001.

9. AYERS, M. J. Recycled aggregate: Production, usage and barriers within the Western Cape. Thesis No. 6. Department of Civil Engineering, University of Cape Town, 2002

10. JAUFEERALLY H, Performance and properties of structural concrete made with corex slag. A MSc (Eng.) thesis submitted to the Faculty of Engineering and Built Environment University of Cape Town, 2001

11. ASTM: American Society for Testing and Materials; Standard methods for bleeding of concrete; Annual book, pp 148-151, 1992. Philadelphia, USA.

12. ADDIS, B, Concrete mix design. Fulton's Concrete Technology, 8th edition. Cement and Concrete Institute, Midrand, South Africa, 2001

USE OF RECYCLED AGGREGATE IN CONCRETE APPLICATION: CASE STUDIES

A Koulouris

M C Limbachiya

A N Fried

J J Roberts

Kingston University

United Kingdom

ABSTRACT. This p aper d escribes i nitial f indings o f w ork u ndertaken t o d emonstrate t he suitability of coarse recycled aggregate for use in concrete paving applications. During this work, s cientific a nd a pplied r esearch w as c ombined. R ecycled a ggregates o btained f rom a local aggregate supplier were used. Concrete was produced using blend of natural gravel and up to 100% recycled aggregate contents. All mixes were designed for equivalent 28-day compressive cube strength of 30 and 35 N/mm^2 for specific paving applications- PAV1 and PAV2 mixes respectively. In the first instance fresh, engineering and durability performance of c oncrete w as assessed. F resh p roperty r esults s howed n o e ffect d ue t o t he c oarse R CA inclusion with results remaining within the required tolerance of ±20 mm. The study of engineering properties showed similar performance (strength development, flexural strength, static modulus of elasticity, drying shrinkage and swelling) between two types of concrete. Durability performance (near surface absorption, carbonation (CO_2) penetration, freeze thaw scaling, resistance to chloride ingress and abrasion) of RCA concrete was assessed and results for RCA equivalent strength concrete were very similar to NA concrete.

Keywords: Construction demolition waste, Recycled concrete aggregates, Concrete performance, Case studies.

Angelos Koulouris is a lecturer in the School of Engineering and member of the Sustainable Technology Research Centre within the Faculty of Technology at Kingston University UK. His research interests include the performance of by-products as materials in construction and concrete technology and construction.

Dr M C L imbachiya i s a Chartered Engineer and Reader in the School of Engineering, Faculty of Technology at the Kingston University UK. His research interests include the recycling and reuse of waste materials, concrete technology and construction, and repair and maintenance.

Dr A N Fried is a Chartered Civil Engineer and Reader in the School of Engineering, Faculty of Technology at Kingston University UK. He has carried out concrete and masonry research for 20 years, gaining experience in running research programmes including investigations into Concrete durability and masonry performance.

Professor J J Roberts is a Chartered Civil and Structural Engineer and Dean of Faculty of Technology at Kingston University UK. He is also Director of the Sustainable Technology Research Centre. His research interests include various aspects of the design and use of concrete and masonry and sustainable technology.

BACKGROUND

The construction industry has key role in any country's infrastructure development. With ever growing demand for infrastructure development, in recent years this industry, in many parts of world, has been identified as one of the largest waste producer with a potential for recycling. Industry recognised this situation and undertaking various initiatives to minimise the waste generation and promote recycling and reuse of construction demolition debris.

On other hand, construction industry world wide is promoting the use of recycled aggregates for concrete production, mainly to respond to the problem of the depletion of natural aggregates. The use of recycled concrete aggregate (RCA) is economically viable and at the same time environmentally sustainable practice. Recently published figures estimates construction demolition waste within the EU is around 180 million tonnes per year [1]. It has now w idely r ecognised and a cknowledged t hat t here i s a great p otential f or r ecycling t his waste for reuse as aggregate in new concrete production. Currently, in the UK approximately 78 m illion t onnes o f C &D w aste i s p roduced annually [2] a nd t his a mount i s e xpected t o increase considerably in the future with rapid growth in infrastructure development.

At the same time, construction industry constantly looking at value-added outlets for recycled aggregates, including use of RCA as direct replacement to primary aggregates obtained from natural resources, where technically and economically possible [2,3]. Against this background, combined scientific and applied research was undertaken by the Concrete and Masonry Research Group at Kingston University. During this work samples of coarse RCA obtained from various sources were used as a direct replacement of Thames valley gravel to examine associated practical and technical aspects. The performance of RCA concrete was examined and compared with NA concrete in PAV1 (30 N/mm^2) and PAV2 (35N/mm^2) designated mixes- meeting requirements of EN 206 8 and BS 8500 [4].

KINGSTON UNIVERSITY RESEARCH

The principle objective of research was to demonstrate through full-scale site trials, that coarse recycled aggregates can be used successfully in a range of concrete applications and establish case studies for monitoring of in-situ performance of recycled aggregate concrete elements. A laboratory support work was undertaken first, followed by full-scale demonstrations on real construction sites.

Aggregate Production and Characterisation

Construction and demolition waste from a range of demolished structures was delivered at the recycled aggregate production plant in Greenwich and processed to produce coarse RCA. The RCA was produced in accordance to EN 12620 [5] as graded aggregate (16-4 mm) and BS 882 [6, 7] as graded aggregate (20-5 mm). Table 1 gives typical composition of recycled aggregate sample use in the study and this meet requirements set in BS 8500 [4] for RCA.

An extensive test programme was devised to assess and compare key characteristics of RCA and NA. Results obtained, together with standard test procedures used are given in Table 2. Coarse RCA found to have comparable characteristics with coarse NA. The major difference was found to be in the water absorption where RCA present values 4 times higher than NA. This is mainly due to higher porosity of the RCA, resulting from the original cement paste attached on the particles. RCA was also found to have lower density but well within the limits for normal weight aggregates. RCA was found to be rounder and especially source III less flaky and elongated than NA. Figures 1 and 2 show Particle size distribution test results.

Table 1. Constituent materials of RCA used the study.

	PROPORTIONS % (m/m)			
	Test Sample 20 – 5 mm			
CONSTITUENTS PRESENT	Mean (%)	S.D.	Variability	Maximum According to BS 8500:2 4
Concrete	93.6	0.3	0.1	-
Masonry	1.7	0.4	0.0	5.0
Foreign materials	0.5	0	0.1	1.0
Asphalt	4.1	0.8	0.0	5.0
Fines	0.1	0.3	0.6	5.0
Lightweight material (ρ<1000 kg/m^3)	0.0	0.0	0.0	0.5

Table 2. Aggregate characterisation for recycled and natural aggregates

Physical and Mechanical Characteristics	Aggregate Fraction (mm)	Natural Aggregate	RCA			Test procedure
			Source 1	Source 2	Source 3	
Shape, Visual	All	Irregular	Round	Round	Round	BS 812-Part 105
Flakiness index (m/m)	28 – 6.3	13 - 16	9 - 10	8.25-9.2	5.8-6.2	EN 933-Part 3
Shape index (m/m)	28 – 6.3	20 - 22	18 - 19	14.73-15.5	5.5-6	EN 933- Part 4
Loose bulk density (Mg/m^3)	20-5 <5	1.483 1.667	1.271 -	1.36 -	1.251 -	EN 1097- Part 3
Voids (%)	20-5 <5	40.9 36.9	43.7 -	39.9 -	45.4 -	EN 1097- Part 3
Apparent particle density (Mg/m^3)	20-5 <5	2.603 2.701	2.561 -	2.585 -	2.548 -	EN 1097- Part 6
Particle density on an oven dried basis (Mg/m^3)	20-5 <5	2.511 2.644	2.260 -	2.263 -	2.291 -	EN 1097- Part 6
Particle density on a surface saturated dry basis (Mg/m^3)	20-5 <5	2.546 2.665	2.377 -	2.387 -	2.392 -	EN 1097- Part 6
Water absorption (%)	20-5 <5	1.2 0.8	5.2 -	5.5 -	4.8 -	EN 1097- Part 6
Aggregate crushing value (%)	14 - 10	12.4	22.0	17.5		BS 812- Part 110
Aggregate impact value (%)	14 - 10	6.3 – 7.3	13.0 – 15.4	11.82-12.2		BS 812- Part 112
10% fines value (%)	14 - 10	155	270	190		BS 812- Part 111

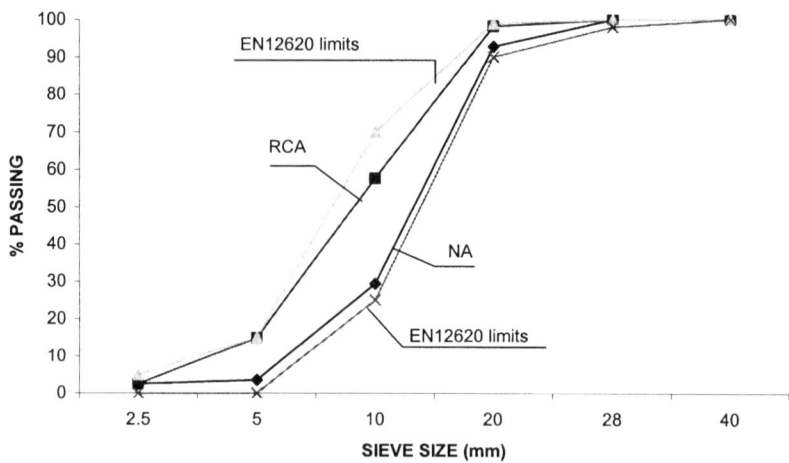

Figure 1: Particle size distribution of natural and recycled concrete aggregate used in the study

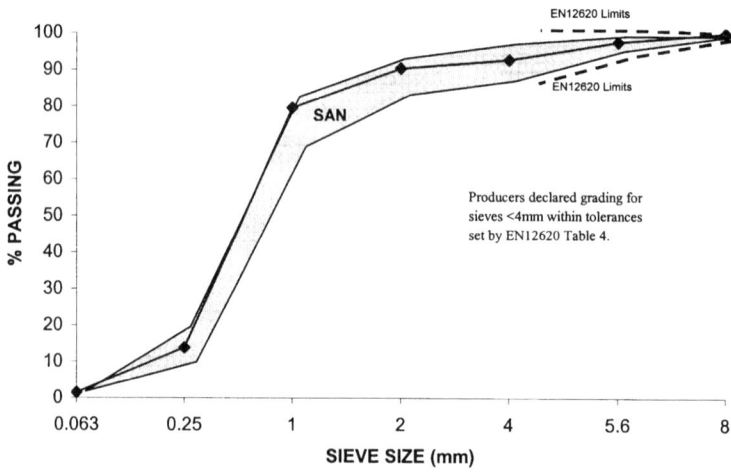

Figure 2: Particle size distribution of natural sand used in the study

Laboratory Support Work

Prior to full-scale on-site demonstrations laboratory based investigation was undertaken. Concrete mixes with Thames Valley gravel and natural sand were produced as reference concrete and, compared its performance with RCA concrete containing 30, 50, 75 and 100%

coarse RCA. A c onventional mix design method [9] was used for mix proportioning. The final selected proportions for paving applications (PAV1 and PAV 2- as per BS 8500-2 [4]) are given in Tables 4 and 5.

Table 4. Reference concrete mix proportions

| MIX | MIX PROPORTIONS, kg/m^3 | | | | W/C RATIO | AEA^2 ml/100kg CEMENT |
| | PC^1 | Water | Aggregates | | | |
			Coarse	Fine		
PAV1	345	165	1285	475	0.47	360
PAV2	385	165	1250	480	0.43	360

1 EN197 42.5N Portland cement
2 The concrete contains an air-entraining admixture to give a minimum air-content by volume of 3.5 % at delivery.

Table 5. RCA concrete mix proportions

RCA CONTENT %	AGGREGATE PROPORTIONS kg/m^3					
	Designated Mixes					
	PAV1			PAV2		
	Coarse		Fine	Coarse		Fine
	RCA	NA	NA	RCA	NA	NA
30	385	900	505	375	875	480
50	635	635	495	625	630	475
75	950	315	490	930	310	470
100	1260	0	485	1235	0	465

Fresh Properties

The fresh properties of NA and RCA concrete were examined by means of workability and stability tests. Slump and stability (bleeding and segregation) was measured immediately on completion of mixing. Table6 gives results obtained. It is clear that slump results remained well within the tolerances specified by EN 206-1 [8] 8 of ±20 mm.

Generally stability did not appear to be affected by the inclusion of coarse RCA regardless of % inclusion. All concrete mixes were found to be very cohesive with little or no further slump at all. There was no observation of any bleeding or segregation for any of the concrete mixes tested.

Engineering Properties

It is well accepted now that the inclusion of RCA into concrete in excess of 30% will most likely cause a reduction if not only a delay in strength development. This has as a result that the 28-day strength (characteristic strength) may not be achieved. For this reason this study took on board findings from previous research carried out by one of the authors [3, 12]. In order to achieve equivalent strength concrete, the water-cement (w/c) ratio was adjusted for mixes containing more than 30% coarse RCA. The adjustment was applied either by increasing the cement content or by reducing the water content of the mix. In the later case superplasticiser can be used to improve the reduced workability.

Table 6. Workability results

MIX, % COARSE RCA	SLUMP (mm)	TARGET Slump (mm)	CLASS	COHESION PROPERTY	
				Description	Observation
PAV 1 0%	50		2a	Very cohesive	Gradually slumps further, no shearing
PAV 1 30%	60	60 S2*	2a	Very cohesive	Gradually slumps further, no shearing
PAV 1 50%	65		2a	Very cohesive	Gradually slumps further, no shearing
PAV 1 100%	45		2b	Cohesive	Gradually slumps further, some shearing
PAV 2 0%	55		2a	Very cohesive	Gradually slumps further, no shearing
PAV 2 30%	60	55 S2*	2a	Very cohesive	Gradually slumps further, no shearing
PAV 2 50%	55		2a	Very cohesive	Gradually slumps further, no shearing
PAV 2 100%	70		1	Over cohesive	Little further slump

* EN206 Part1 Table 3 Slump classes 8

The strength development of NA and RCA concrete was examined according to EN12390-3 [10] using 100 mm cubes and cylinders with diameter D=150 mm and height h=300 mm according to EN12390-1 [11]. The results for 28-days cube and cylinder compressive strengths are presented in Table 7. Clearly equivalent strength concrete is achieved and the adjustment method used verified.

Table 7. Results for engineering properties for different coarse RCA contents at 28 days

ENGINEERING PROPERTIES	PAV I					PAV II				
	RCA Content %									
	0	30	50	75	100	0	30	50	75	100
Compressive strength f_{cube}, N/mm^2	34	35	37	35	33	39	36	37	37	37
Compressive strength $f_{cylinder}$ N/mm^2	27	26	27	27	27	28	28	28	28	29
Flexural strength, N/mm^2	5.2	5.2	5.2	5.1	5.1	5.6	5.6	5.7	5.7	5.7
E-Value, kN/mm^2	30	34	41	31	31	32	28	30	29	30
Shrinkage max, $\mu\varepsilon$	277	435	543	415	313	220	298	397	395	380
Swelling max, $\mu\varepsilon$	175	343	237	240	280	185	228	360	310	285

A typical strength development of these mixes is graphically presented in Figure 3 for PAV1. The concrete was water cured at 20 ±2° C up to the day before testing. Before testing the density of each specimen was examined to identify any direct relations with strength results.

Differences in densities found to be negligible between concretes with various RCA contents and most importantly, no indication whatsoever was observed that compressive strength was affected by these differences.

The strength development was found to be unaffected by the RCA inclusion at any content. Concrete strength developed as expected and differences in strengths were of small magnitude mainly because of variability. The adjustment of the w/c ratio resulted to equivalent strength concrete and in some cases there is an indication that the adjustment may have been somewhat generous. The flexural strength and modulus of elasticity results are shown in Table 7. Flexural strength was examined according to EN 12390-5 [13]. The inclusion of coarse RCA proved to have no effect at all on flexural strength of concrete. In fact concrete with 50, 75, and 100% coarse RCA found to have a tendency towards higher flexural strength values. Having said that it will be insatiable to say that RCA improves flexural strength, however this tendency to higher values provides some confidence to the suitability of coarse RCA.

Modulus of elasticity was examined according to BS 1881 Part 121 [14]. The results are presented in Table 7. The results are showing a similar behaviour as for flexural strength. There is no indication of any effect due to the inclusion of RCA except of the fact that as in flexural strength there is a tendency of higher E-value for 75 and 100% RCA concrete. The values ranged between 29 and 34 kN/mm^2 in agreement with the BS 8110-2 (Table 7.2) [16] typical values, with the exception of PAV 1 with 50% RCA where the value was 41 kN/mm^2.

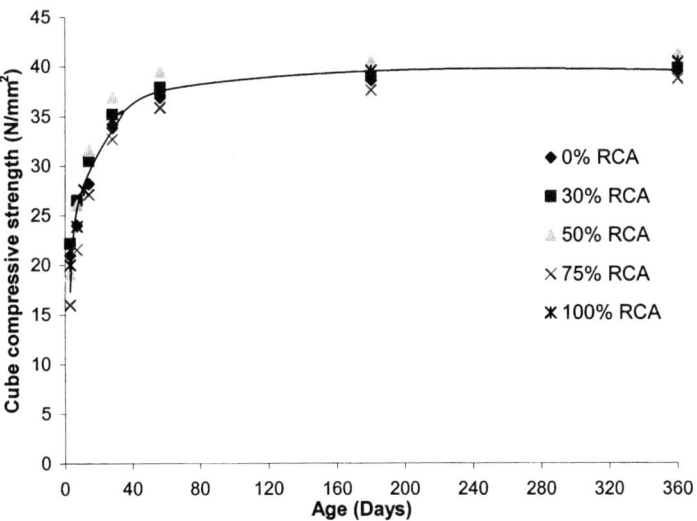

Figure 3: Strength development of RCA concrete (PAV1)

Durability Related Performance
The aim of this part of the study was to assess the effect of RCA content on various durability aspects of concrete relevant to particular applications. Properties assessed include near surface absorption, carbonation (CO_2) penetration, freeze thaw scaling, resistance to chloride ingress and abrasion. Test results are given in Table 8 and test procedures and testing ages briefly outline below.

Initial surface absorption tests were carried out on 150 mm oven-dried cube specimens according to BS 1881 Part 208 [17]. The ISAT-10 results for PAV1 were found to be slightly higher than 0.5 ml/m^2/sec which is considered to by a high initial absorption value. PAV2 initial absorption results ranged between 0.48 and 0.65 ml/m^2/sec. On comparison, it is clear that up to 50% RCA inclusion the initial surface absorption is very similar. However a noticeable increase occurs for concrete containing 100% RCA. Carbonation resistance was measured at 2, 4, 8, 12 and 20 weeks for PAV1. The assessment was carried out by means of the phenolphthalein solution using 100 mm cubes. From the results it is clear that RCA had very little effect in CO$_2$ penetration with the exception of 100% RCA concrete which presents a slightly higher penetration.

Limited testing was carried out for abrasion resistance in order to assess the effect of coarse RCA on the resistance to wear. The method used is described in BS 8204 Part 2 [18]. Concrete slabs (1 m x 1 m by 0.1 m) were prepared in the lab and three tests were carried out on each one of them. From the result there is an indication that the inclusion of 30% coarse RCA provided concrete with better wear resistance than NA concrete.

Table 8. Summary results of durability performance of RCA and NA concrete mixes

PROPERTY	DESIGN STRENGTH (N/mm^2)	NA	COARSE RCA, %		
			30	50	100
ISAT-10, ml/m^2/s	30	0.60	0.60	0.63	0.73
	35	0.48	0.49	0.53	0.64
Carbonation depth mm	30	20.5	22.5	19.5	23.00
	35	-	-	-	
Abrasion depth: mm	30	1.25	0.85	-	-
	35	1.13	0.77	-	-
Freeze-thaw resistance	30	0.023	0.056	0.046	0.065
Surface scale: kg/m^2	35	0.013	0.023	0.023	0.058

CASE STUDIES

Lessons learned and information obtained during laboratory support work was then taken on real construction sites with a view to undertake full-scale site demonstrations, building heavy duty industrial pavements. Brief description on these demonstrations is given in this section.

CASES: Construction of Heavy Duty Industrial Pavements, using coarse and fine RCA

Location: Reported two cases were undertaken at Day Group Ltd., Greenwich – London, UK

Main Activity: Demonstration of quality recycled aggregate production. Followed by, Use of RCA (produced using C&D waste originated from various concrete structures around London) in pavement concrete production and construction Concrete mixes meeting requirements for heavy duty industrial pavements, as per BS 8500 designated concrete PAV2 (C28/35) were designed. Final mix proportions were decided following initial trial mixes in the laboratory. In this case, both coarse and fine natural aggregates were replaced with RCA of same sizes.

Coarse RCA Production

Construction demolition debris obtained from reinforced c oncrete structures in the Greater London Area were used for RCA production. First, C&D waste was transported and processed at the Day Group's recycling plant at Greenwich. C&D debris were fed through primary jaw and secondary impact crushers, and foreign materials were removed using various techniques such as electromagnets for reinforcement removal, manual separation, blowers for lightweight material separation and screens for size distribution.

Trial Concrete Mixes: Production and Performance

Series of trial mixes were casted to assess key fresh and strength properties. The performance of concrete produced using coarse RCA (of 20-5mm size- meeting grading requirements) proved to be c omparable to concrete produced u sing c oarse natural a ggregates. The fresh, engineering and durability properties were assessed in depth through an exhaustive parallel research programme for both NA and RCA concretes. The laboratory results were positive and compatible with those collected from the sites during the full-scale demonstrations.

Full-scale RCA Concrete Construction

Following trial mixes, series of full-scale demonstrations were undertaken to demonstrate production of quality RCA, RCA concrete, and it's handling and placing for heavy duty industrial paving construction. Concrete mix proportions used in two cases are given in Tables 9 and 10. Practical aspects considered and some technical aspects of full-scale demonstrations are schematically presented in Figure 4.

Table 9. P330 Concrete Mix Proportions (in kg/m^3) Used for the Pavement Concrete- Case 1.

	CEMENT	WATER	RECYCLED CONCRETE AGGREGATES	
			Sand	Coarse
			≤ 5 mm	5 – 20 mm
100% RCA -Batch 1	345	158	517	1353
100% RCA - Batch 2	350	153	512	1365

Table 10. C35 - Concrete Mix Proportions (kg/m^3) used for Demonstration 2

RCA CONTENT	CEMENT	WATER	SAND < 5 mm	COARSE AGGREGATES NA	COARSE AGGREGATES RCA	ADMIXTURE WRA (litres)	REINFORCEM-ENT, XT Fibres
0%	340	167	791	770	330	1.02	0.90
30%	340	167	791	770	330	1.02	0.90
100%	360	180	750	0	1050	1.08	0.90

Figure 4. On Site Demonstration of Industrial Pavement RCA Concrete Construction.

KEY OBSERVATIONS AND PRACTICAL IMPLICATIONS

Results obtained in this research and other studies have shown that good quality recycled aggregates can be produced with the commercial plants that are used for the production of crushed-rock aggregates. Clearly, this information could encourage clients and demolition contractors to direct C & D waste for production of RCA, while reducing disposal to landfill. Work undertaken at Kingston University, in collaboration with Day Group Ltd., and London

Remade, has demonstrated the suitability of coarse RCA for use in a range of normal grade concrete applications. Indeed, the results have shown that up to 30% coarse RCA can be used, without any modification in the mix design, in concrete construction with performance similar to natural aggregate concrete. For concrete producers, the use of coarse RCA is unlikely to pose any problem in the production of concrete that is stable in the fresh state and able to develop properties comparable to the corresponding NA concrete in hardened state. This is of great importance to reduce inhibition of concrete specifiers and producers towards using RCA. Final finished surfaces with varying level of coarse RCA contents of case 2 are shown in Figure 5.

A m ethod o f m ix p roportioning t hat t akes a ccount o f t he e ffect o f R CA c ontent (greater t han 30%) on concrete strength has been devised. It is simple a nd can be integrated into existing concrete mix design procedures and production techniques. However it should be recognised, as is the case of existing design methods, that the proposed methodology will give initial proportions for trial mix purposes only. Particular adjustments for individual material characteristics and proportions will have to be established thereafter. The key engineering and durability properties of RCA concrete are found to be similar to corresponding NA concrete, providing the mixes are of equivalent strength achieved through adjustment in the w/c ratio. However, given the influence of the RCA content on the shrinkage and creep strains, their use in structural elements prone to such deformations may require some special considerations. The work s uggest t hat t he u se o f c oarse R CA m ay b e e xtended t o h igh s trength c oncrete (C50 o r greater), thus offering further added value outlets for the material- as reported previously [3, 12].

Figure 5. Finished Surfaces of RCA Concrete Pavements

ACKNOWLEDGMENTS

The e xperimental r esults a nd i nformation g iven this p aper a re b ased o n research w ork funded by the UK's London ReMade and Day Group Ltd. The authors acknowledge the financial support and technical advice received during the course of reported work.

REFERENCES

1. European Environment Agency. Indicator: Total waste generation- 2003.

2. www.environmentagency.gov.uk ,Construction Waste, 2002.

3. LIMBACHIYA M. C. LEELAWAT T, DHIR R K. Suitability of recycled concrete aggregate for use in BS 5328 designated mixes. Proceedings of the Institution of Civil Engineers, Structures and Buildings, Vol.134, No.3, August, 1999.

4. BRITISH S TANDARDS INSTITUTION: B S 8 500-2:2002 C oncrete – C omplementary British Standard to BS EN 206-1:2000, Concrete - Part 1: Specification, performance, production and conformity.

5. BRITISH STANDARDS INSTITUTION: BS EN 12620:2002 Aggregates for concrete. CEN, 2002.

6. BRITISH STANDARDS INSTITUTION: BS 882:1992 Specification for aggregates from natural sources for concrete.

7. BRITISH STANDARDS INSTITUTION: BS 882:1992 Incorporating Amendment No.1 Specifications for aggregates from natural sources for concrete. BSI, 1992.

8. BRITISH STANDARDS INSTITUTION: BS EN 206-1:2000, Concrete - Part 1: Specification, performance, production and conformity.

9. TEYCHENNE, D.C. FRANKLIN, R.E. and ERNTROY, H.C. eds, Design of normal concrete mixes, Building Research Establishment: 1988, pp.13-39

10. BRITISH STANDARDS INSTITUTION: BS EN 12390-3:2002 Testing hardened concrete - Compressive strength of test specimens.

11. BRITISH STANDARDS INSTITUTION BS EN 12390-1:2002 Testing hardened concrete - Part 1: Shape, dimensions and other requirements for specimens and moulds.

12. LIMBACHIYA M C, LEELAWAT T, DHIR R K, RCA Concrete: A study of properties in the fresh state, strength development and durability, Proceedings of International Symposium, Thomas Telford, London, 1998, pp 227-237.

13. BRITISH STANDARDS INSTITUTION: BS EN 12390-5:2000 Testing hardened concrete – Part 5: Flexural strength of test specimens.

14. BRITISH STANDARDS INSTITUTION: BS 1881-121:1983 Testing concrete - Part 121: Method for determination of static modulus of elasticity in compression.

15. BRITISH STANDARDS INSTITUTION: BS 1881Testing concrete - Part 208: Recommendations for the determination of the initial surface absorption of concrete.

16. BRITISH STANDARDS INSTITUTION: BS 8204-2:2003 Screeds, bases and in-situ floorings. Concrete wearing surfaces. Code of practice.

17. BRITISH STANDARDS INSTITUTION: BS 8110-2:1985 Structural use of concrete - Part 2: Code of practice for special circumstances.

18. BRITISH STANDARDS INSTITUTION: BS 1881-208:1996 Testing concrete - Part 208: Recommendations for the determination of the initial surface absorption of concrete.

19. BRITISH STANDARDS INSTITUTION: BS 8204-2:1999 Screeds, bases and in-situ floorings – Part 2: Concrete wearing surfaces – Code of practice.

AN INVESTIGATION INTO THE EFFECTS OF THE ADDITION OF WASTE RECYCLED PLASTIC AGGREGATES ON THE PROPERTIES OF LIGHT WEIGHT AGGREGATE CONCRETE PAVEMENT BLOCKS

H K Al Nageim

Liverpool John Moores University

H Robinson

N Ghazireh

Tarmac Group

United Kingdom

ABSTRACT. Approximately 214 million tonnes of aggregates are used each year in England as raw materials. Of this supply, 50 million tonnes are already derived from recycled or secondary sources. England is a leading user of these materials in Europe and has already established large and successful markets for these products. Scope for obtaining additional supplies already exists from construction and demolition wastes that are currently sent to landfill and through better utilisation of secondary resources. Another potential source of alternative aggregate is from waste plastic produced from a variety of sources including; automotive, agriculture and domestic. This Paper highlights the need to develop higher value applications for secondary aggregates, which can be achieved through research to broaden the range of materials using recycled and secondary aggregates. This paper looks at using recycled plastics in lightweight concrete blocks. In our preliminary investigation, physical and mechanical properties of the concrete were evaluated. The tests included properties of the concrete's fresh state, compressive strength, splitting tensile strength, flexural strength, and water permeability. In this study mixed polymer waste plastic as received direct from the suppliers was used. This consisted of particles 5-10mm in diameter and 1-2mm in thickness. It was found that the addition of plastic particles decreased the overall concrete strength properties. The permeability of the blocks was also adversely affected, though the workability of each mix remained relatively unchanged.

Keywords: Lightweight concrete, Recycled plastic aggregates, Light weight concrete blocks

Professor Hassan K Al Nageim is a Chartered Engineer, Professor of Structural Engineering at Liverpool John Moores University and Head of JMU Liverpool Centre for Material Technology. His research interests include various aspects of the design and evaluations of steel, concrete and pavement structures including the recycling and use of waste materials.

Dr Howard Robinson has been actively involved with industry level research related to; aggregates, asphalt, bitumen and concrete for almost 20 years. He is currently Head of Product Development for the Tarmac Group based in Wolverhampton.

Dr Nizar Ghazireh is a Civil Engineer by profession specialising in geotechnical engineering and foundations and has worked at WS Atkins for 9 years as a geotechnical engineer responsible for the design and construction of shallow and deep foundations. He is currently Product Development Manager for the Tarmac Group based in Wolverhampton.

INTRODUCTION

The Case For Using Recycled and Secondary aggregates

As reported by the Waste and Resource Action Programme (WRAP) annual aggregates progress report [1] approximately 214 million tonnes of aggregates are used each year in England as raw materials. Of this supply, 50 million tonnes are already derived from recycled or secondary sources. England is a leading user of these materials in Europe and has already established large and successful markets for these products.

By 2012, if UK demand for aggregates increases by an expected 1% per cent per annum, an extra 20 million tonnes of aggregates will be needed each year. The report states; we can either satisfy this additional demand by extracting further primary aggregates, or we can follow a more sustainable route and continue to increase our use of recycled and secondary aggregates.
This more environmentally friendly option is achievable. Scope for obtaining additional supplies already exists in the construction / demolition and excavation wastes that are currently sent to landfill and through better utilisation of secondary resources. This report clearly highlights the need for more uses of secondary aggregates to be found. With this in mind the report concludes that one of WRAP's targets for future development is through research to broaden the range of materials and applications using recycled and secondary aggregates.

Background Studies

Although the use of recycled plastic in asphalt pavements has been emphasised and well documented in several publications, very few papers have been published regarding the use of recycled plastics in concrete. In the cases where the use of plastic in concrete has been reported, the plastic has been used to modify the aggregates used in the mix designs. This has been highlighted in Rossignolo and Agnesini's [2] work into the mechanical properties of polymer-modified lightweight aggregate concrete which investigates the influence of various percentages of styrene–butadiene rubber (SBR)-modified lightweight aggregate and the affect they have on the properties of the modified concrete. The tests included properties of the concrete's fresh state, compressive strength, splitting tensile strength, flexural strength, and water absorption.

They report that the workability of both the modified and unmodified concretes remain fairly unchanged and both exhibit good workability even an hour after completion of mixing and also report that the air content of the modified aggregate concrete is on average higher than the unmodified concrete, which they report plays a significant factor in the lower compressive strengths gained with the increase in use of the modified aggregates. Decreases in strength of up to 4.4% were found at the 10% dosage level.

Rossignolo and Agnesini also report that results gained for both tensile and flexural strengths indicate an increase for these strengths with increased levels of the modified aggregate, this they attribute to an overall improvement in cement hydrate–aggregate bond because of a decrease in the water – (cement + silica fume) ratio and the high tensile strength of SBR films present in SBR-modified lightweight aggregate concretes. The report also indicates a decrease in the water absorption of SBR-modified LWAC's is attributed mainly to a reduction in permeability, caused by a reduction in W/(C + S). Such W/(C + S) reduction ultimately affects the gel–space ratio and causes a reduction in the capillary porosity of the system.

Nehdi and Khan's [3] investigation into the use of cementitious composites containing recycled tyre rubber provides an interesting insight to some of the physical properties of the composite concrete investigated. This report is similar in many ways to this investigation, as Nehdi and Khan discuss their findings into the use of recycled crumb rubber as a replacement aggregate.

The report is in the form of a critical overview of the findings of others on the subject of using rubber as a replacement aggregate. Fedroff et al. [4], and Khatib and Bayomy [5] all concur that increasing the rubber content in the concrete increases the air content, this, they suggest may be due to the non-polar nature of the rubber particles, in that they may attract air as they repel water. They also conclude that air may be entrapped by the particles' jagged surface texture.

The mechanical strength properties gained from results of various investigations indicate that the size, proportion, and surface texture of the rubber particles greatly affect the strength of the various concrete mixes. Eldin and Senouci [6] indicate that there was approximately an 85% reduction in compressive strength and a 50% reduction in tensile strength when the coarse aggregate was fully replaced by coarse rubber chips. Specimens containing fine crumb rubber lost 65% of their compressive strength and 50% of their tensile strength. However, results of tests carried out by Fatuhi and Clark [7] indicate the opposite trend.

Various studies show that the rougher the rubber particles used in concrete mixtures, the better the bonding they develop and hence a higher compressive strength achieved, as argued by Tantala et al. [8] who suggest that pre-treating the particles with an acid solution increases the microscopic surface roughness and therefore an increase in compressive strength. The same report also suggests that rubber inclusion in concrete also makes the material a better thermal insulator.

A further study by Huynh, Raghavan and Ferrais [9] adds further evidence to the argument that weak interfacial bonding between rubber particles and concrete matrix exists. They also found that an increase in rubber content decreases both the compressive and flexural strength. This report however, suggests that an increase in rubber does have its advantages; this is indicated in their findings of improved plastic shrinkage.

The work of Naik et al. [10] provides an interesting insight into the use of post-consumer waste plastics in cement-based composites. During their investigation, they focussed on the pre-treatment of the plastic using three substances: water, bleach and bleach + NaOH, with the proportions of plastic added to the mix being 0-5% by weight. The results of which showed that the chemical treatment had a significant effect on the performance of the plastic as an aggregate filler. It was also noted by the authors that; In general, plastics do not achieve chemical bonds with cementisious materials. Therefore, chemical treatments of plastics are needed to enhance bonding characteristics with the cementitious matrix.

Mechanisms of Strength Reduction

Khatib and Bayomy [5] hypothesised that there are three major causes for the strength reduction displayed in concrete samples containing crumb rubber. Firstly, because rubber is much softer than the surrounding cement paste, when loaded, cracks are initiated quickly around the rubber particles due to difference in elastic behaviour. These cracks then propagate to bring about failure of the rubber-cement matrix. Secondly, they reported that due to the weak bonding between the rubber particles and the cement paste, the softer rubber particles may be viewed as voids in the concrete mix, and this assumed increase in void content causes a reduction in strength. The third possible reason is due to the density, size and hardness of the aggregates. Khatib and Bayomy also found that the flexural strength also decreased similarly to that of compressive strength, again due to the aforementioned mechanisms.

Having undertaken a comprehensive literature review the following hypothesis has been identified: "Can recycled waste plastic be used as a replacement aggregate in the production of lightweight aggregate concrete blocks? If so, how does the replacement plastic improve the blocks' properties?"

THE TEST SPECIMENS

Materials

The materials used in the concrete mixes were:

- **Cement:** Portland cement.
- **PFA:** Pulverised Fuel Ash from Drax Power Station, Selby, North Yorkshire, UK.

The problem of variation in cement and PFA fineness was eliminated by using cement or PFA from one batch only.

Bottom Ash Fines: Supplied by Tarmac Group, UK - Figure 1.

Figure 1. Bottom Ash Fines

The bottom ash collected from the base of furnaces in the power station is usually water cooled and transferred to stockpiles. The material is then crushed and screened in different size fractions either at the station or at the block manufacturer's site. Typical grades in block production are 14-0, 14-5 and 5-0mm, depending on the type of block being manufactured and has a density in the range of 800-1100 kg/m^3. Bottom ash and boiler slag are composed principally of silica, alumina, and iron, with smaller percentages of calcium, magnesium, sulfates, and other compounds. The composition of the bottom ash or boiler slag particles is controlled primarily by the source of the coal and not by the type of furnace.

Ground Granulated Blastfurnace Slag: Supplied by Frodingham Cement located in Scunthorpe, UK, see Figure 2.

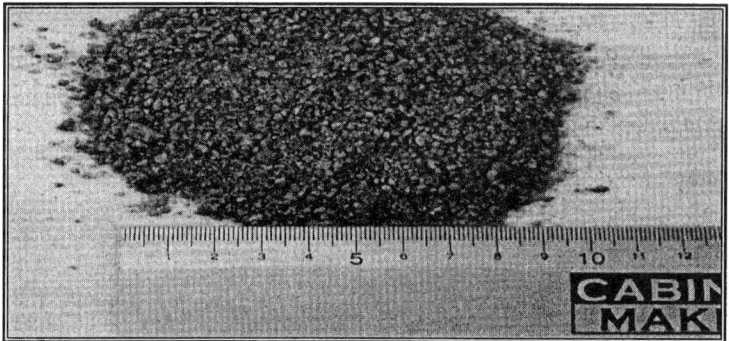

Figure 2. Ground Granulated Blastfurnace Slag.

Blast furnace slag is a non-metallic co-product produced during the manufacture of iron from iron ore and limestone fluxes which have been charged into a blast furnace along with coke for fuel. It consists primarily of silicates, aluminosilicates, and calcium-alumina-silicates. The molten slag, which absorbs much of the sulfur from the charge, comprises about 20 percent by mass of iron production. When ground to very fine cement-sized particles, ground granulated b lastfurnace slag o r ggbs, h as c ementitious p roperties w hich make i t a s uitable partial replacement for Portland cement.

Water: Clean, potable 'tap' water at room temperature was used in the concrete mixtures. Water content taken as 7.5% of total weight of aggregates of control mixture.

Plastic: In this study recycled mixed polymer waste plastic particles approximately 5-10 mm diameter and approximately 1-2mm thickness, were used supplied by G.W. Webb Plastics of Birmingham, UK, (see figure 3).

Figure 3 – Recycled plastic chips used in the design mixes.

The Design Mixes

Plastic replacement proportions of 5%, 10%, 15% and 20% were used along with a control sample of 0% plastic to give five design mixtures as follows:

1) 52% Furnace bottom ash fines, 31% granulated blastfurnace slag, 9% PC, 8% PFA
2) 52% Furnace bottom ash fines, 26% granulated blastfurnace slag, 9% PC, 8% PFA, 5% plastic (by volume)
3) 52% Furnace bottom ash fines, 21% granulated blastfurnace slag, 9% PC, 8% PFA, 10% plastic (by volume)
4) 52% Furnace bottom ash fines, 16% granulated blastfurnace slag, 9% PC, 8% PFA, 15% plastic (by volume)
5) 52% Furnace bottom ash fines, 11% granulated blastfurnace slag, 9% PC, 8% PFA, 20% plastic (by volume)

Casting cubes, cylinders and beams

Three, 100mm cubes, 150mm diameter by 300mm long cylinders and (100 x 100 x 500) mm long beams were used for testing each design mix and for each property to be calculated/ measured. Specimens were cured in a half-full water bath, pre-heated to 50°C for twenty-four hours. The shelves of the water bath were positioned so that the specimens, when placed on the shelves were above the water. The water bath was then covered with polythene sheeting, weighted down to ensure no escape of the steam produced to ensure 100% relative humidity employed. After twenty-four hours curing in this condition, the samples were then placed into a water bath at 20°C until the required age of testing.

PROPERTIES OF THE CONCRETE MIXES

The compressive [11], indirect tensile [12], flexural [14] and shear strength of each design mix was tested at 3, 7 and 28 days. The load was applied so that the load increased continuously at a rate of $0.2N/(mm^2 s)$ until the cube failed and the maximum load recorded. Five cubes were crushed at each age for each mix and the mean value taken as the compressive strength. Falling head permeability test was also performed on samples at 28 days age.

ANALYSIS OF RESULTS AND DISCUSSIONS

Mix Workability Results

As can be seen in figure 4, the workability of each mix remained largely unchanged. During the test, when the cone was removed, all the samples tested exhibited no slump, this can be attributed to the dry nature of the mix, therefore Vebe Time is measured instead of the slump[13]. The small changes in vebe time indicate that workability is increased with an increase in plastic; this is because plastic is impermeable and does not absorb the water, therefore there is more water dispersed throughout the rest of the mix matrix and acts as a lubricant between the particles in the mix. The surface texture of the plastic also plays a part in increasing workability. The plastic, when compared with the other aggregates has a much smoother surface texture and hence less friction between the particles in the concrete matrix and can move about more easily and hence increase workability.

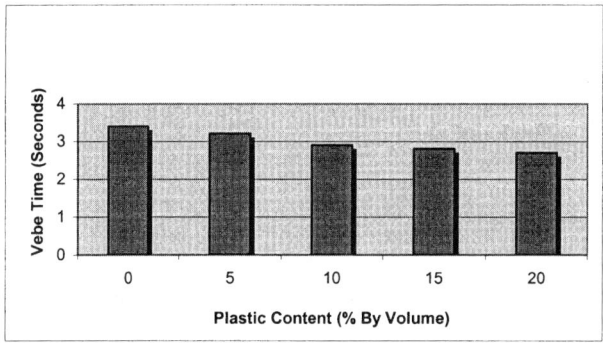

Figure 4. Bar Chart to Show Vebe Time of Concrete Mix at Varying Plastic Contents

Compressive Strength Results

Figure 5 shows the mean compressive strength of the tested samples at each specified age. It can clearly be seen that as the plastic content increases the compressive strength decreases. Figure 5 also shows that the samples' rate of increase of strength are not similar for each mix design and the initial rate of gain in strength is greatest between 0-7 days, by which time each mix has gained approximately 80-95% of the 28 day strength. The early strength gain is attributable to the nature of hydration of cement in the matrix, where the reaction between free water and cement is greatest in this early period, which has also been exaggerated by the method of curing, where an increased temperature of 50°C increases the hydration of cement and hence, greater relative early strengths.

The main influencing factor in the decrease of compressive strength is that of the relatively smoother surface texture of the plastic particles. The smoother surface means that the plastic particles are unable to achieve a good bond with the surrounding matrix, compared to conventional aggregates which have rougher surfaces, resulting in the 0% plastic cubes having a considerably higher compressive strength than that of the 5%, 10%, 15%, and 20% plastic cubes.

It should be noted that during mixing it was noticeable that whereas the conventional aggregates physically bonded together, the plastic was unable to form a physical bond. It was also observed that some of the plastics were more weaker than others, some being almost paper-like, therefore it can be seen that the plastic aggregates are much softer, weaker and are also less dense, which decreases the particle strength, supporting the findings reported in the literature. [3-9].

During testing, no abnormal modes of failure were encountered. This indicates to some degree that bonding between the plastic and the concrete matrix does exist as the mode of failure is at 45°, representing some shear resistance.

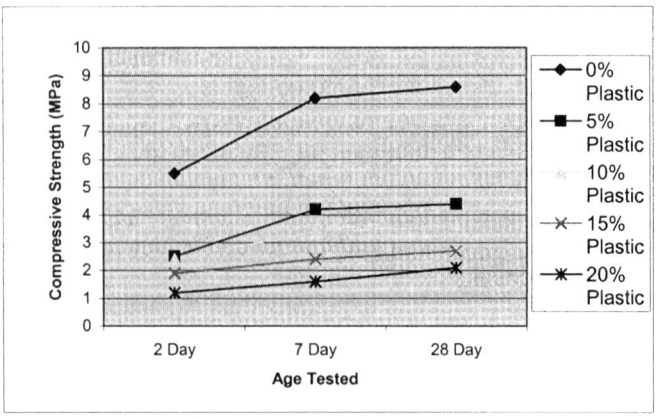

Figure 5. Graph to Show Compressive Strengths of Cubes Tested at Various Ages and at Varying Plastic Contents

Indirect Tensile Strength Results

As can be seen in figure 6, the results gained show the indirect tensile strength decreases with an increase in plastic content.   It has been reported in the FIP Manual of Lightweight Aggregate [15], that the tensile strength of lightweight aggregate concrete depends on two factors.   Firstly, the influence of the aggregate and secondly the influence of the curing method, which in this case can be discounted as the same curing conditions were used throughout the test and these results also form a relative comparison between the samples.

The influence of the aggregates can be explained by examining the fracture of the specimen. The tensile strength depends on the tensile strength of mortar and aggregate and on the interaction of both components.

On examination of the failed specimens, it was noted that the pull out characteristics of the plastic aggregates showed that the plastic particles remained fully in tact and that all the specimens tested failed with the normal mode of failure for the indirect tensile tests, whereby the sample fails in two sections, approximately down the axis of load application.   This is a clear indication of the failure of the plastic aggregates to form an adequate bond with both the mortar and surrounding aggregates.

The samples tested, containing no plastic, demonstrated signs that a much better bond had developed, this is illustrated by the fact that failure of the conventional aggregates in localised patches in the failed specimen occurred.   This point is clearly illustrated by the results gained in this study; as the plastic content increases, the tensile strength is reduced due the weakened bond in the mix matrix.

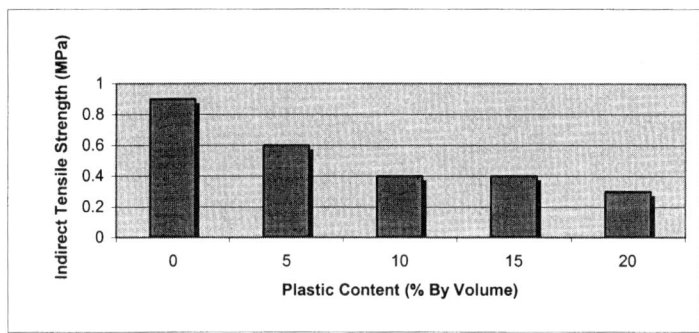

Figure 6 - Bar Chart to Show Indirect Tensile Strength at Varying Plastic Contents

Flexural Strength Results

Again the results (Figure7) show that an increase in the plastic content consequently lowers the flexural strength of the beams tested. It is generally accepted that a direct relationship between flexural strength and compressive strength exists, and is due to the bonding between the particles, due mainly to the surface roughness of the particles.

During testing, all the specimens broke in two with a catastrophic failure. All the specimens tested exhibited the normal failure mode for flexural strength tests – fracture directly in the centre of the beam.

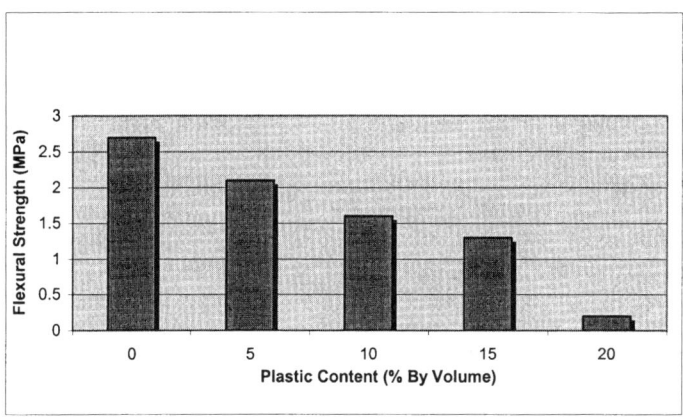

Figure 7 - Bar Chart to Show Flexural Strength with Varying Plastic Contents

Shear Strength Results

As with the results for compressive, tensile and flexural strength, they are all functions of the bond between the plastic and the concrete matrix, which in turn is a function of the aggregate properties.   Again it can be shown that the plastic aggregate is much weaker than the surrounding matrix. Figure 8 indicates that the shear strength for the concrete containing no plastic has mean shear strength of 4.2MPa, whereas the mean strength of the 5% plastic samples was 1.6MPa. Therefore it can be seen that the inclusion of plastic into the mix greatly reduces shear strength. The modes of failure of the samples were as expected for a shear strength failure, i.e. splitting into three sections, along the lines of application of the load.

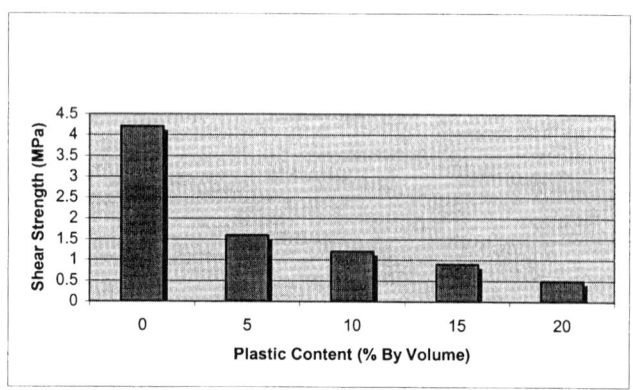

Figure 8 - Bar Chart to Show Shear Strength at Various Plastic Contents Tested at 28 Days

Falling Head Permeability Results

The values of the permeability coefficient, k, were calculated. For each mix design, two samples were tested, with both samples subjected to water pressure heads of 1.0m, 0.8m and 0.5m, and the permeability coefficient calculated for each head using the average time taken for the respective heads of water to permeate each block and the permeability calculated using the average time. The overall mean value of permeability, as found in Figure 9, was then calculated by taking an average of the previously calculated coefficients for each head.

The results in figure 9 clearly show the effect of increasing the plastic content on the blocks' permeability characteristics. It should be noted that the 0% plastic content samples did not allow any water to pass through the block within an hour of testing, at this time the test was concluded, and a 'k' value of 0 assumed. However, it is accepted, that after some time the water would eventually seep through the block, and a 'k' value exhibited. For the purposes of this comparative test this assumption can be justified.

The test results support the theory of the weak interfacial bond between the plastic and cement matrix, in that the inclusion of plastic inhibits the formation of a full cement matrix. The formation of crystalline structures and the bond between them at the microstructure level causes blockages of flow paths through the concrete. Therefore it could be argued that the inclusion of the plastic promotes the formation of micro-cracks and voids within the concrete matrix.

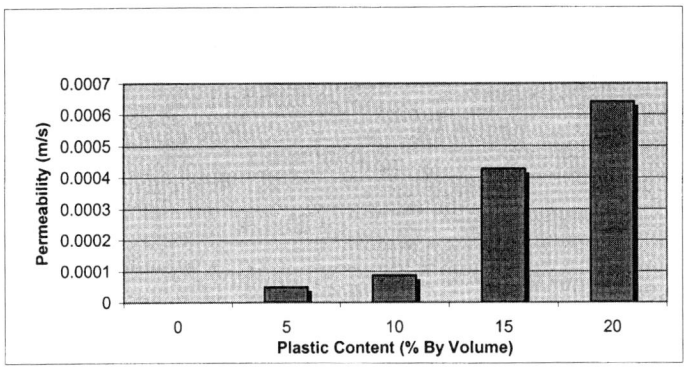

Figure 9- Graph to Show Permeability of Blocks With Varying Plastic Contents

Density of Hardened Concrete

Figure 10, shows the average densities of all the cubes, beams, and cylinders tested for the various mechanical strength tests with an overall average given for each mix design. Figure 10, also shows that the relationship between plastic content and density, as expected.

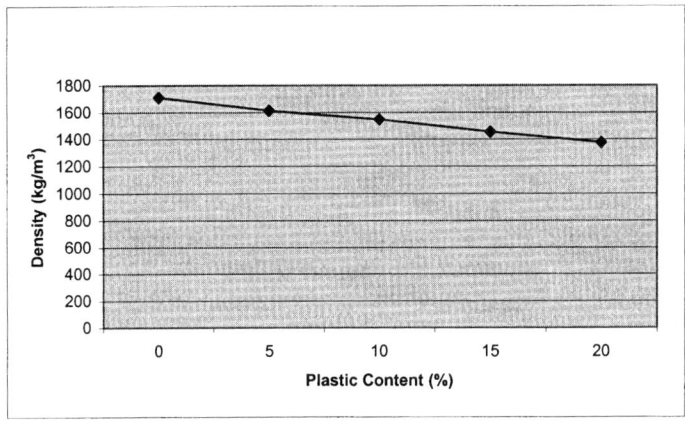

Figure 10. Relationship between Plastic Content and Density

CONCLUSIONS

1) As the level of plastic aggregate replacement increases in the lightweight concrete mixtures, the compressive strength, of the cubes decreased. The results show that a replacement of 5% plastic by volume, significantly reduces the compressive strength properties by approximately half at each age of testing, supporting findings in the literature [3-7].

2) The workability of the concrete mixtures increased fractionally with an increase in plastic content. The impermeable nature of the plastic meant that free water was available to the remaining concrete matrix, therefore increasing the water/cement ratio, which consequently reduces the strength of the matrix.

3) The flexural, shear and indirect tensile strength all decreased with an increase in plastic content. Examination of the failure specimens indicated that the plastic particles had been pulled out from the matrix in all tests analysed. This was taken as evidence of weak interfacial bonding between the plastic and matrix, supporting claims in the literature [3-7]. During mixing it was noticeable that no physical bond formed between the plastic aggregates and concrete mix.

4) The permeability test results provide some useful comparative information; it appears that the inclusion of plastic makes the blocks more permeable. This may be due to the poor bond created between the plastic and the concrete matrix, hence promoting the growth of micro-cracks within the concrete microstructure and hence pathways allowing water permeation.

5) To achieve a commercially viable concrete product containing waste plastic would need a dramatic increase in the strength properties of the concrete blocks. To achieve this would require changing the mix proportions, most notably an increase in cement content and/or pre-treatment of the plastic aggregates. To do so would increase the cost of production, due to the extra materials and resources involved, which would probably outweigh the benefits of utilising a waste product.

REFERENCES

1.   THE WASTE AND RESOURCES ACTION PROGRAMME (WRAP): Stakeholder Update Progress Report: Aggregates, 2002.

2.   ROSSIGNOLO, J.A. AND AGNESINI, M.V.C: Durability of polymer-modified lightweight aggregate concrete, cement and concrete composites. Elsevier Science Ltd., 2003.

3.   NEHDI, M. AND KHAN, A: Cementitious composites containing recycled tyre rubber: An overview of engineering properties and potential applications, cement, concrete, and aggregates, CCAGDP, Vol. 23, No. 1, 2001, pp. 3-10.

4.   FEDROFF, D., AHMAD, S., AND SAVAS, B.Z: Mechanical Properties of Concrete with Ground Waste Tyre Rubber, Transportation Research Record, No. 1532, 1996, pp 66-72.

5.   KHATIB, Z.K. AND BAYOMY, F.M : Rubberised Portland cement concrete, Journal of Materials in Civil Engineering, 1999, pp.206-213.

6.   ELDIN, N.N. AND SENOUCI, A.B: Rubber-tire particles as concrete aggregates," Journal of Materials in Civil Engineering, ASCE, Vol. 5, No. 4, 1993, pp 478-496.

7.   FATUHI, N.I. AND CLARK, N.A: Cement-based materials containing tire rubber, construction building materials, Vol. 10, No. 4, 1996, pp 229-236.

8.   TANTALA, M.W., LEPORE, J.A., AND ZANDI, I: Quasi-elastic behaviour of rubber Included concrete, Proceedings, 12th International Conference on Solid Waste Technology and Management, 1996.

9.   HUYNH, H., RAGHAVAN, D. AND FERRARIS, C.F: Rubber particles from Recycled tires in cementitious composite materials, US National Institute of Standards and Technology, Report 5850,1996.

10.   NAIK, T.R., SINGH, S.S., HUBER, C.O, and BRODERSEN, B.S: Use of Post-Consumer Waste Plastics in Cement-Based Composites, Cement And Concrete Research, Vol. 26, No. 10, 1996, pp. 1489-1492.

11.   BRITISH STANDARDS INSTITUTION: BS 1881-116: 1983, Method for determination of compressive strength of concrete cubes.

12.   BRITISH STANDARDS INSTITUTION: BS 1881-117: 1983, Method for determination of tensile splitting strength.

13.   BRITISH STANDARDS INSTITUTION: BS 1881-104: 1983, Method for determination of Vebe Time.

14.   BRITISH STANDARDS INSTITUTION: BS 1881-118: 1983.: Method for determination of flexural strength.

15.   SURREY UNIVERSITY PRESS, HALSTED PRESS: FIP Manual of lightweight aggregate concrete, 2nd Ed., 1983.

USING WASTE MATERIAL IN MASONRY

M Maghrebi

Sepad Khorsasn Co.

Iran

ABSTRACT. In this appear, the destructive effect of waste material on environment is discussed with an emphasis on the usage of the waste in masonry. Today in Middle East, we consider bricks as an undeniable material for constructions even high buildings .The simple production process and routine use in different situations makes it more popular among engineers. But this heavy material with no constructional roll is not reliable in earthquakes and its weight influence on the analysis of buildings. In this paper we focused on a replacement material for bricks which is much lighter and cheaper. Instead of aggregates in the concrete we used a light, cheap and un-recyclable material, which is waste production of a mold factory.We have tested different types of this new foam concrete with different cement content and compared the weight, density, pressure resistance, abrasion resistance . And at last we compared four cases of loads with this new foam concrete and bricks in the analysis of a virtual six floor building.

Keywords: Lightweight concrete, MLC, Environmental masonry, Waste production, Cellular concrete, Foam concrete.

Mojtaba Maghrebi is Senior student of civil engineer, Faculty of Engineer at Azad University of Mashhad , Iran . The Faculty includes Civil Engineer, Mechanic Engineer, Computer science, Electronic Engineer. He is researcher on R&D Department of Sepad khorasan Co. He is editor in Chief of two scientific student`s magazine, EMARAT magazine specially in civil and architecture (best scientific student`s magazine in Iran), and ALGORITHM magazine especially computer science. His research interests include waste and un-recyclable materials, lightweight concrete technology.

INTRODUCTION

Reducing of density and the weight of concrete is investigated in this paper. it's known that concerts weight is the most important factor in aggregate in this study the aggregates of concrete is replaced with lighter material which is addressed FA in this paper. Different types of materials are studied, and FA of them will be discussed in detail here.

FA

FA [Figure 1] is a waste production of a mould factory. FA is light material with a density of $19.5 \, ^{kg}/_{m}{}^{3}$ which is 50 time lighter than water density with low water absorption. FA is found to be quite appropriate to our aim. FA has the property mentioned in ASTM C33.69 and BS 3797-1995 regarding lightweight aggregates.

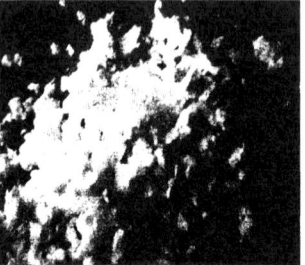

Figure 1.   Picture of FA

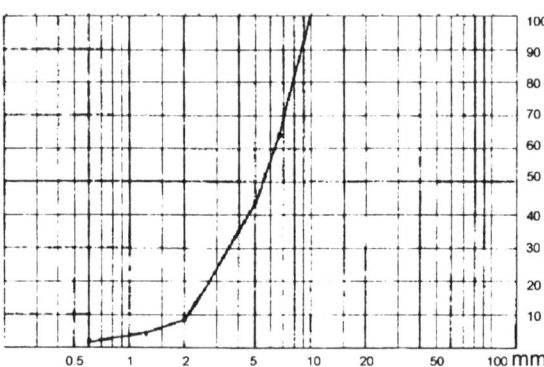

Figure 2.   Griding of FA

The particle size distribution of FA is given in [Figure 2], this figure shows that grading of FA is quite similar to that of gravel. This property helps us to reach a cellular and lightweight concrete .

Materials and detailed techniques used to produce MLC (the registered name of Concrete that made with FA) is as follows.

CEMENT

The lightweight concrete is not so strong enough to resist huge stresses [1] and the strength of MLC is due to percentage of cement used. In this paper the strength of cements ranged between 250-450 kg/m^3 have been tasted. Experimentally to reach an optimum cement percentage.

MICROCYLIS

Microcylis is used in milk to enhance, the strength , durability of concrete, and to weekend the water absorption[2].

Microcylis with the density of 250-300 kg kg/m^3 has be used which has 90% sylisiumoxyde cement ASTM C124 cold recommends different amount of microcylise. Which depends on the aim of experiment. In this investigation cement are mixed with an extra 7% microcylis is used.

WATER

Reology property of lightweight concrete is different to that of other concrete in use [3] this is mostly due to existence of a cellular in lightweight concrete [4].In this study the percentage of water used is between 0.4-0.6 cement weights.

MIXTURE

In most codes the percentage of lightweight concrete mixture is based on volumetric ratio of cement and lightweight aggregate [5]. The volumetric ratio which is used here is different from those recommended by the other codes.
The volumetric mixture formula, equation (1), used in this study is according to the following equation.

$$A+B+C+D = 1\ m^3 \tag{1}$$

Where :

Cement (Kg/m^3 250 - 450) = A

Water (0.4-0.6 Cement weight) = B

Microcylis (7% Cement Weight)C =

FA 1- (A+B+C) D =

COMPUTED RESULT

4 different case are modeled in this study for compare the effects of MLC and brick on the structure. The steel structure is constructed in 6 stories Building each have 288 m^2 [figure 3]
.

: The cases are as follows

1-1:Case 1 – roof with joists & partitions with bricks
:Case 2 – roof with joists & partitions with MLC
2-1 : Case 3 – Composite roof & Partitions with bricks
2-2 : Case 4 – Composite roof & Partitions with MLC

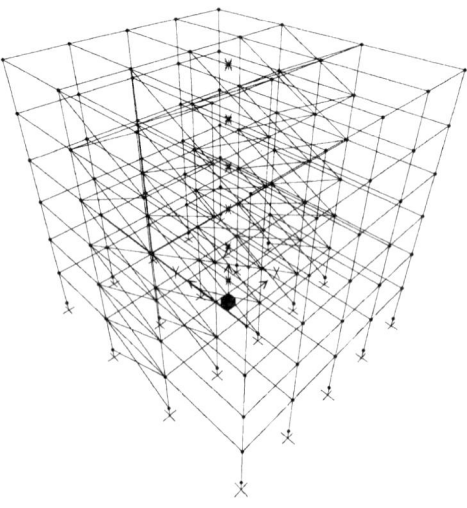

Figure 3.   Simulated steel structure

The following table [Table1] shows the effects which are simulated on the building and some random section [Figure 3] in 4 case .

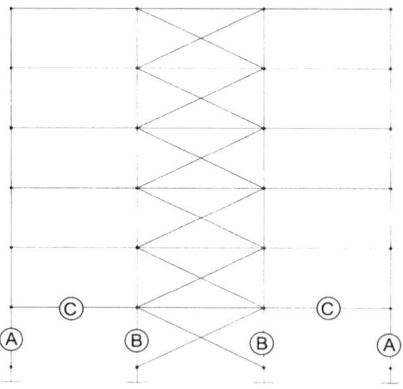

Figure 4.   Random Sections

	A	**B**	**C**
CASE 1	3IPE16	3IPE20	2IPE24
CASE 2	3IPE12 +	3IPE18	2IPE22
CASE 3	3IPE14	3IPE18	2IPE22
CASE 4	3IPE12	3IPE16	2IPE18

Table 1:   Compare of Random Section

DISCUSSION

the extra cement content increase the Strength as predicted [Figure 5] 1

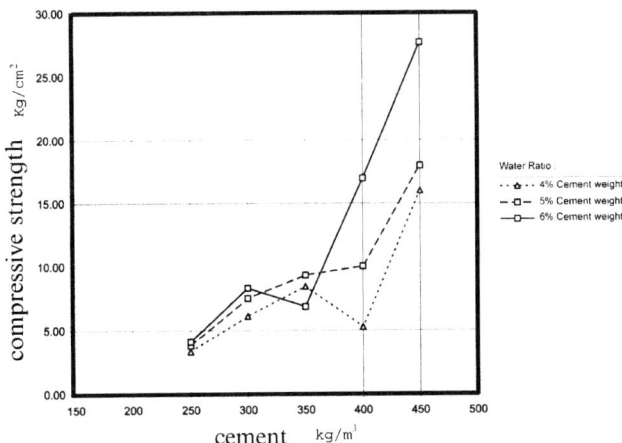

Figure 5. 28 day Compressive Strength

2- the optimum water cement percentage was found to be 0.6 [Figure 5] Since the density FA is 50 time lighter than water. The water mixture percentage most be taken into account accurately.

3- MLC is useful for partitions and it is not appropriate to bear pressure.

4- MLC has not only advantage in technical properties but also is a quite commercial product .

5- Due to the lightweightness of MLC it cause much less Immolates in disaster like earthquakes.

REFERENCE

1   TAM C.T. , Relationship between Strength and volumetric Composition of Moist-cured cellular concrete , Mag Concr Res. 1987

2.   SVENKERUD . P.j, FIDJESTOL .P, and ARTIGUES Texsa . J.C . Micrisilica Based Admixture for Concrete in Admixtures of Concrete : Improvement of Properties , Proc .Int. Symposium , Barcelona , Spain , Ed. E. Vazques , Champan and Hall , London 1990

3.   ACI 213R-87 , Guide for Structure lightweight aggregate concrete , ACI Manual of Concrete Practice , Part 1 : Materials and General Properties of Concrete , 27 pp . (Detroit , 1994)

4.   ACI 304.5R-91 , Batching, mixing , and job control of lightweight concrete , ACI Manual of Concrete Practice , Part 2 : Construction Practices and Inspection Pavements , 9 pp. Detroit , Michigan , 1994

5.   NEVILE ,A.M. , Properties of Concrete , Longman House, Burnt Mill , Harlow, London 1995

SESSION FOUR:

WAY FORWARD AND DEVELOPING SUSTAINABLE MARKET

APPRECIATION OF RISKS INVOLVED IN USE OF INDUSTRIAL BY-PRODUCTS IN BUILDINGS

S B Desai

Butler & Young Limited

United Kingdom

ABSTRACT. Research related to sustainable concrete construction has encouraged development of composite binders, comprising Portland cement and industrial by-products, and c oncrete m ixes w ith p artial r eplacement o f n atural a ggregate w ith r ecycled aggregate. Compared with concrete made with Portland cement and natural aggregate, such concretes would involve an increase in the number of constituents with different sources and with variable control on their production. Certain measures may become necessary, therefore, during construction or post-construction stages, in the interest of ensuring the desired performance of the structure. At the same time, UK industry is resorting to various measures for improving quality of construction, including procedures for compliance with the Construction (Design and Management) Regulations. These procedures include records of risks perceived by the designer and any actions essential for achieving the desired performance of structural concrete and other products incorporated in the building. Such records are expected to lead to reduction or elimination of risks through adequate measures, either during construction or during the post-construction stage. This paper seeks to promote awareness among engineers regarding their role in the processes leading towards achieving safety of construction and expected performance of buildings.

Keywords: Industrial by-products, Recycled aggregate, Risk assessment, I dentification of hazards, Performance of structural concrete.

Dr S B Desai is a Chartered Structural Engineer, Fellow and Member of the Council of the Institution of Structural Engineer and Fellow of the Association of Building Engineers. He is also an Honorary Visiting professor at the University of Surrey. His 44 years of experience include responsible positions with the public sector and with private sector organisations in the construction industry. His fields of experience include civil engineering works (design, construction and project management), UK Building Regulations, British and European Codes of Practice (loadings and concrete), preparation of specialist guidance documents and research in concrete construction. His main research interests include sustainable concrete construction and various aspects of structural reinforced concrete, e.g. design of flat slabs, shear d esign at n ormal and h igh t emperatures, i nfluence o f c onstituents o f c oncrete o n i ts tensile and shear strength, etc.

INTRODUCTION

At the 2002 UN World Summit on Sustainable Development, the participating governments recognised the importance of industry's contribution to sustainable development and pledged to enhance corporate environmental and social responsibility and accountability. Many industries in the UK are participating in the Global Reporting Initiative, a worldwide non-profit organisation. The UK Government has demonstrated its support to sustainable construction and sponsored many research projects concerning environmental issues, reduction in waste, use of industrial by-products and recycling of products of demolition.

Sustainable concrete construction implies prudent use of natural ingredients as constituents of concrete, so that they remain available for future generations to meet their needs. Increasing awareness of this principle has led to research aimed at optimum and prudent use of cement and its replacement with industrial by-products, e.g. pulverised fuel ash (PFA) and ground granulated blast-furnace slag (GGBS). Another development concerns recycling products of demolition, which is being considered seriously by the industry. Any reduction in use of cement in modern concrete mixes should result in saving in energy and reduction in demand on sources of natural materials required for manufacturing cement. Increased use of industrial by-products and recycled materials should assist in reducing the demand for waste disposal sites. Furthermore, modern mix design techniques could result in improved durability of concrete that is attributable to improvement in its microstructure, reduction in voids and increase in resistance to ingress of moisture, gases, etc. Such attributes of different concretes could also improve their engineering properties, e.g. increase in the ratio of tensile strength to compressive strength, better bond between steel and concrete and, in the end, enhanced shear strength[1].

These developments should be beneficial to the concrete industry in the long term and they should contribute towards improving image of concrete as high-tech construction material. However, the new mix designs would essentially increase the number of constituents in the process of making concrete, which are required to be mutually compatible for production of a compound product, i.e. structural concrete. Concrete mixes made with these constituents should, therefore, require more careful analyses of materials compared with those according with the previous practice. The earlier concrete mixes used constituents with strict control during their manufacture and the constituents would be derived from primary products with well-established sources. The same cannot be said about all industrial by-products used as binder components and recycled aggregate used to partially replace natural aggregate. It is probable, therefore, that the resulting concrete could perform unsatisfactorily, if it contains constituents that are not mutually compatible or if they are produced with unsatisfactory quality control and testing regime. Furthermore, increasing pressure on availability of land for construction has led to building on reclaimed ground, grounds with soft soils and brownfield sites containing imported wastes or wastes remaining from their previous industrial use. In the absence of any thorough geotechnical investigation, foundation concrete may be exposed to undesirable environments and the situation may be worsened due to any problems with the concrete constituents. BCA publication on "Concrete resistant to chemical attack" gives very useful guidance on the subject [2].

While the principle of sustainable concrete construction remains indisputably important for the future, its practical application must be accomplished with due caution and higher level of technological skills. Professor Bishop has observed in his paper (on the subject of professional liability) that construction has become more technically problematic than it used to be [3]. Performance requirements are more demanding and there is much "involuntary innovation" as new materials and products are pressed into use. He has added that some construction techniques are much more demanding on the control of accuracy. Professor Bishop's comments in his paper are to be taken in the context of its subject but they represent real life concerns directly relevant to the role of an engineer. It is often experienced that the financial constraints (fees and project costs) are such that there is no room to tolerate even a minor error and perfection is expected irrespective of the level of skills that is available and affordable. Moreover, developers and clients have the cost as the priority consideration and no real enthusiasm for technical innovations or complexities.

These circumstances should not rule out any room for modern concrete technology solutions. However, some choices made by engineers, as competent practitioners, may adversely influence the end results in spite of their exercising of reasonable skills, care and diligence. Such choices should be discussed with the clients in advance and agreements should be reached with full knowledge of the implications and recorded. Further along the line, the design teams should adopt measures that would reduce or eliminate any risks and any potentially residual hazards should be recorded according to the procedures forming parts of the Construction (Design and Management) Regulations or CDM Regulations.

BACKGROUND FOR STRUCTURAL CONCRETE DESIGN IN THE UK

In the UK, general rules for design of reinforced concrete elements were developed during the first decade of 20th century, which did away with various empirical methods existing in the latter part of 19th century. In general, the basis was elastic theory and permissible stresses under w orking l oads. T hese m ethods w ere r eviewed a nd refined o ver t he n ext f orty years resulting in the metric units version of CP114 in 1969 [4].

In its earlier versions, CP114 gave mix proportions by volume and prescribed compressive strength of concrete made with cement, fine aggregate (sand) and coarse aggregate (crushed stone or gravel), mainly in proportions of 1:2:4. The associated strengths were expressed in terms of "preliminary" or "works" cube strengths. The "works" cube strength was meant to serve the purpose of basic control mechanism and it was presumed to be approximately 75% of the "preliminary" cube strength. The Code prescribed some 350 kg/m^3 of cement for structural concrete, which would give an expected 28 N/mm^2 preliminary cube strength and 21 N/mm^2 works cube strength respectively. Such cement content may seem high compared with the present practice.

CP114 introduced further grades of concrete in its later versions, along with "design mixes" giving freedom of choice to make concrete with different strengths using graded aggregate and cement content to suit the required strength. However, caution was exercised through a recommendation "to allow for unavoidable variations", which meant that the mean strength of a designed mix had to exceed the specified works cube strength by "twice the expected standard deviation". Standard deviation had to be based on 40 cube test results subject to a minimum of 3.5 N/mm^2 and, in the absence of previous information, 7 N/mm^2. This meant that a margin of 14 N/mm^2 was expected initially for mix design purposes. Such recommendations show that, in the earlier days, authors of codes of practice relied heavily on such large margins to compensate for any uncertainties in the process of making concrete and in the structural design models.

CP114 went only as far as recommending that the quantity of water should be sufficient, but not more than sufficient, to produce dense concrete of adequate workability for its purpose. Concrete was required to be just sufficiently wet to be placed and compacted without difficulty. Durability was governed by the provision of cover and the works cube strength was selected to suit the conditions of exposure, for example, stronger concrete for more severe exposure conditions. Most engineers of the earlier times believed, perhaps with some justification, that "good concrete structures" depended on cement, cover, compaction and curing, and mix design techniques and other sophistication were not so important.

During the 1950s, some commercial developments began to influence the construction practice. First, cements became more reactive and it became possible to develop higher compressive strengths at earlier ages [5]. At the same time, for certain strength of concrete, cement content could be lower and water-cement ratio higher than the earlier practice. Table 1 shows representative values of 100 mm cube strengths (not air-entrained concrete), which illustrate the influence of changes in cement manufacturing with time.

Table 1. Influence of changes to manufacturing of cement on strength of concrete

W/C Ratio	100 mm CUBE STRENGTH (N/mm^2)			
	1950	1960	1970	1980
0.4	46	52	65	72
0.5	36	45	51	56
0.6	28	35	40	44
0.7	22	27	31	35
0.8	17	21	24	27

This trend raised a question among the experts about the durability of structural concrete, if it continued to be linked solely to the compressive strength. It was realised that the water-cement ratio affected the pore structure, the main factor governing permeability of concrete and its durability. It became apparent, therefore, that a maximum limit on water cement ratio had to be a part of the durability specifications.

After its reign for some 22 years, CP114 was replaced by CP110, followed by the current code of practice BS 8110 in 1985 [6]. This code retained the previous practice of using compressive strength as the input parameter in structural design of concrete and as a measure of its quality, without any regard to the constituents of the concrete mix.

Further developments include special concrete mixes using combination of Portland cement and binder components, for example, PFA and GGBS. These developments contributed to economy and sustainability of concrete construction but they also introduced some considerations requiring careful attention, e.g. rate of strength development, influence of curing, compatibility between binders and aggregate, chemical properties of constituents of concrete (especially with recycled aggregate), etc. Another notable advance was emergence of high strength concretes, with additives and incorporating various fibres. However, questions have been raised concerning their performance under exposure to fire, e.g.

explosive spalling caused by vapours trapped behind the impermeable cover region concrete. Designers have to ensure, therefore, that the special mix concrete elements would perform satisfactorily under various design situations arising during the service life of the building.

ENGINEER'S DUTIES ACCORDING TO CDM REGULATIONS

It is fair to say that an engineer has always been expected to issue a warning against any potential hazard, as a part of his or her ethical duty as a professional. In a different context, Neville has described the lessons from history regarding use of high-alumina cement outside the limitations for its use and without the relevant precautions [7]. It is not intended to draw any parallel between the case of high-alumina cement concrete and that of the modern mix design concrete. However, a designer often may not have direct control on construction (e.g. curing of concrete) or the post-construction usage or maintenance of a building. It makes it essential, therefore, that the designer should be aware of any potential "risks" that may arise during the life of the building structure.

CDM Regulations have a broad definition of a "designer", which includes architects, engineers and contractors carrying out design work. CDM requires that designers should satisfy themselves that there is at least one way to construct the building safely and in a reasonably economical manner. This leads to the expectation that the designers should eliminate or reduce potential hazards and record the information about residual hazards that may have implications for the construction, maintenance and dismantling of the project.

It must be clarified here that CDM does not require engineers to stifle their creativity or to choose "the safest form of construction", for example, conventional concrete mix. At the same time, CDM takes account of the fact that all engineers may not have detailed knowledge of the construction process or control on maintenance and that they would not be able to take into account any unforeseeable hazards. An engineer is not even obliged to produce a "Design Risk Assessment" for every medium or large project. Risk assessment requires an evaluation of probability of occurrence of hazards, which may be impracticable within the scope of an engineer's responsibility for a specific project. All that is expected of an engineer is that he or she should prepare a list including any potential hazards and measures to be taken to eliminate or reduce the hazards. Such records may primarily focus on justifiable design assumptions and attributes of various construction products, and they may also involve declaration of limitations on performance of some structural elements during the working life of the building. In effect, CDM Regulations have a system whereby a record of the topics illustrated above is kept in the "Health and Safety File". The contractor would then take account of any points and actions relevant to the construction stage. The file would finally be handed over to the clients at the time of completion of the works, along with other documents, for post-construction actions as appropriate.

EXAMPLES OF POTENTIAL HAZARDS IN CONCRETE CONSTRUCTION

General
It is a common practice to list hazards and identify the order of risk they may represent - low, medium and high. This assessment may be based only on judgement and views taken by the engineer in the context of the project and not necessarily on any statistical data on probability of occurrence of the hazards. An engineer is also expected to record his or her view on any measures and actions that may be taken during the construction and post-construction stages. An example of the engineer's identification of hazards is shown in Table 2.

Table 2. Identification of potential hazards from an engineer's point of view

DESCRIPTION/LOCATION	RISK	AVOIDANCE/ CONTROL MEASURE	ACTION/COMMENTS
Chemical attack on pile cap concrete	Low	Control on calcium carbonate in the mix and analysis of underground water	Tests on limestone aggregate and limestone filler.
Deterioration of blockwork in walls below ground level	High	Monitoring and analysis of underground water	Do not use open-textured blockwork.

Choice of Aggregate

Selection of aggregates usually depends on commercial considerations. Choice of local aggregates may afford economy of construction, saving of energy in transportation and good control on procurement. The designer should be aware of any implications for performance of concrete made with the chosen aggregate. For example, concrete made with limestone aggregate may perform well at normal temperatures but not under exposure to fire or high temperatures. Similar situation may arise with the use of fibres or other additives in concrete, which may improve performance of concrete at normal temperatures but undergo permanent and irreparable damage with exposure to high temperatures.

Choice of aggregate in foundation concrete is closely related with soil conditions. In some instances, soil investigation is not done very thoroughly and aggressiveness of the soil is not clearly established. For instance, the soil may contain some chemicals at the time of investigation and they may not require any specific provisions. However, as a result of the construction, water may flow from elsewhere and raise the water level. Foundation concrete with certain levels of "calcium carbonate equivalent" (e.g. limestone and excessive gypsum in the cement) has been known to have suffered damage by chemicals that were not present in the ground at the time of construction but were produced by reaction of existing chemicals with the migrated water. Apparently, such reaction, which would produce "Thaumasite", requires combination of many circumstances, e.g. cold temperatures (lower than $15°$ C), presence of very wet conditions (mobile water), sources of carbonates, sulfates and magnesium contents in the ground, etc. The resulting Thaumasite excessively weakens the concrete, which may be below ground and not noticed, and lead to failure of the foundations.

In another instance, walls below ground level were built with open texture concrete blocks, which appeared quite in order for the site with some underground water. The ground was slightly sloping and contained gravel that allowed some storage of water. The situation was changed by the construction, resulting in a flow of the underground water that eroded the reactive limestone. If the water had remained stationary, it would probably have been saturated with the alkaline content with time and the damage might not have occurred.

GGBS Concrete

It is a common experience that the hessian-cured cubes give strength that is about 90% of the water-cured cubes. However, for the low water-binder ratio mixes, particularly mixes containing GGBS, this difference can be very significant. While curing is beneficial to all concrete construction, it is most important for development of strength in GGBS concrete.

The writer has reported tests on beams where air-cured cubes (placed by the side of test beams) w ere u sed t o d etermine t he c ompressive s trength o f b eam c oncrete [1]. T his w as crosschecked with results of tests on water-cured cubes, which were noted to be about 10% higher for PC concrete but about 30% higher for GGBS concrete.

Tests carried out by Nakamura et al on concrete specimens with GGBS have shown a similar trend and revealed some characteristics of such concrete associated with different curing methods [8]. They concluded that air-curing without any water-curing could be a substantial impediment in development of strength. Irrespective of slag-fineness and water-binder ratio, 91 days strength of air-cured specimens was found to be only about 60 to 70% of 28 days control strength of water-cured specimens. They recommended that a minimum of six days water-curing would be required for slag concretes to reach strengths obtained by continuous water-curing. If such curing regimes were to be necessary for using slag concrete in buildings, its use may not be very practicable.

Concrete Society Technical Report 40 comments on the effect of temperature during the early life on strength development of GGBS concrete [9]. Exposure to modest thermal cycling is believed to have a beneficial effect but, for concrete with 70% GGBS, standard cured strength at 28 days is observed to be 32.5 N/mm^2 compared with 43.5 N/mm^2 for PC concrete with cement content of 300 kg/m^3. Further research is required to clarify such dependence on curing conditions, if GGBS concrete is to be used in common structures.

If GGBS concrete is used in any part of the building or a bridge connecting the car park and the building, a designer should record the explicit curing regime essential for achieving the expected performance of the structure, both from the point of view of strength development as well as that of the requisite durability.

Recycled Concrete Aggregate (RCA)

Recycling concrete aggregate is important for reducing demand on primary sources of natural aggregate and for easing the pressure on sites for disposal of products of demolition. At present, RCA use is limited mostly to road construction and plain concrete. Rapid disposal of waste from demolished concrete structures has tended to be the preferred practice of the UK construction industry. Production of RCA for use in structural work would require careful and systematic processing, grading and testing of aggregate, essential for ensuring its acceptability in structural concrete. However, costs of such operations are considered to be too high at present.

There is enough evidence to show that RCA can be used to replace some 30% of the coarse aggregate in some grades of concrete meant for less demanding performance, e.g. concrete suitable for drives and parking for dwellings, external paving, foundations in non-aggressive soils, blinding, kerb bedding, etc. RCA can also be used in some concrete grades meant for internal reinforced concrete members. Further work may suggest that use of RCA concrete can be extended to higher grades of concrete. However, in such instances, a designer has to appreciate deficiencies of RCA concrete regarding workability and the problems presented in the form of harshness of the mix, together with porosity and inadequate durability of the hardened concrete. It would be most inappropriate to increase water content and the cement content (to maintain the water/cement ratio) for the purpose of overcoming these difficulties. Such measures w ould jeopardize the basic principles of sustainability. Addition of PFA is understood to be a successful solution for overcoming some of the problems, provided that its compatibility with other concrete constituents is established. Another important issue, related to identifying any potential hazard, would be the specification for testing the RCA. RCA should not be used in any significantly important structural concrete elements, unless it is chemically inert and without any inherited impurities and problems such as alkali-aggregate reaction or chloride attack. Furthermore, special design considerations may be required before indulging in any proposal to use RCA concrete in members sensitive to creep and shrinkage.

CONCLUSIONS

Engineers are expected to collaborate with other disciplines, so that a building is constructed safely, e conomically a nd w ith d ue r egard t o i ts p erformance d uring i ts e xpected l ife. It i s essential, therefore, that the design assumptions are justifiable for the assembly of all products and for achieving safety and serviceability of the building. Simultaneously, it is essential to eliminate or reduce any potential hazards and provide information on issues that are likely to present residual hazards, including measures essential for long-term performance of the assembly of products. This information is used by the contractor for the construction stage and included in the Health and Safety File, a part of the hand-over documentation, for the benefit of those responsible for post-construction maintenance. CDM Regulations have a structured procedure to achieve these objectives, which are not different in principle to those expected from professional engineers as a part of their ethical duties.

REFERENCES

1. DESAI, S. B.: Influence of constituents of concrete on its tensile strength and shear carrying capacity, Magazine of Concrete Research, London, February 2003, pp 77-84.

2. BRITISH CEMENT ASSOCIATION: Concrete resistant to chemical attack, one of the series of publications "Specifying concrete to BS EN 206-1/BS 8500, Crowthorne, Berkshire, October 2001.

3. BISHOP, D.: Professional liability, insurance, and their impact on practice – an overview, The Structural Engineer, London, November 1990, pp 425-427.

4. CP114: 1969: T he s tructural u se o f reinforced c oncrete i n b uildings, British S tandards Institution, 1969.

5. POMEROY, C. D.: Concrete durability: From basic research to practical reality. International Conference on Concrete Durability. April 1987.

6. BS 8110: PART 1: Structural Use of Concrete: Code of Practice for design and construction. British Standards Institution, 1997.

7. NEVILLE, A.: Should high-alumina cement be re-introduced into design codes?, The Structural Engineer, 2 December 2003, pp 35-40.

8. NAKAMURA, N., SAKAI, M. AND SWAMY, R. N.: Effect of slag fineness on the engineering properties of high strength concrete. Proceedings of a conference on 'Blended cements in construction'. University of Sheffield 9 - 12 September 1991. Elsevier Applied Science, London.

9. CONCRETE SOCIETY TECHNICAL REPORT No 40: The use of GGBS and PFA in concrete. Concrete Society, Crowthorne, Berkshire.

AN INVESTIGATION OF THE BEHAVIOUR OF DIFFERENT FIRED-CLAY PRODUCTS IN AUTOCLAVED CEMENT: FIRED- CLAY PRODUCT BLENDS

D S Klimesch

M Gutovic

S Ray

University of Technology, Sydney

Australia

ABSTRACT. High pressure steam curing or autoclaving has proven extremely versatile for the manufacture of cement-based building products incorporating waste materials such as fly-ash, blast furnace slag, and more recently fired-clay products. With regard to the latter, it was found that the mineral composition, which differs from that of sands that are normally used, affects both microstructure and mechanical properties of autoclaved cement: fired-clay product mixtures. In this article, experimental data for autoclaved cement: fired-clay product blends, made with different amounts and types of fired-clay products, are presented. A combination of techniques including X-ray diffraction, differential thermal analysis, and electron microscopy were used to ascertain differences in mineralogy and chemical reactivity amongst the fired-clay product types and their effects on subsequent hydration product formations.

Keywords: Construction demolition, waste, Fired-Clay Product, Autoclaving, Tobermorite, Recycling.

Dr D S Klimesch is a Research Fellow in the Faculty of Science at the University of Technology, Sydney, Australia. Her research interests include recycling and utilisation of both waste materials and various mineral sources for the development of sustainable, advanced cement-based building products cured under hydrothermal conditions.

Mr M Gutovic is a doctoral candidate in the Faculty of Engineering at the University of Technology, Sydney, Australia. His research topic is concerned with the fundamental understanding of the behaviour of clay-brick waste in autoclaved calcium silicate based building products.

Associate Professor A Ray is an internationally recognised expert in inorganic materials analysis. He has established an active multi-disciplinary research team in cement-based materials at the University of Technology, where he is an executive member of the Centre for Built infrastructure Research, a Key University Research Strength.

INTRODUCTION

There is a great impetus worldwide towards recycling construction and demolition (C&D) waste and other industrial by-products and their utilization as renewable construction materials. Clay-brick waste (CB) rich in Al_2O_3 and SiO_2, is generated in abundance worldwide from demolition of buildings and during the manufacturing process and is a significant component of C&D waste. In recent years, the construction industry has shown considerable interest in the utilization of CB as a sub-base material for road construction, car parks, backfill for landscaping, drainage material and as aggregate for low strength concrete [1,2].

High pressure steam curing or autoclaving is an established manufacturing process for cement-based building products affording rapid strength development and producing a binder with superior chemical stability and physical properties. It is a proven and versatile process for utilizing waste materials such as fly ash and blast furnace slag for the production of dense concretes, lightweight or aerated concretes (AAC), thermal insulation boards, and fiber-reinforced cement boards [3]. Previous work by the present authors has demonstrated that the formation of 1.1 nm tobermorite, the principal binder of most autoclaved calcium silicate based building materials, is enhanced by the addition of finely ground CB and C&D waste containing CB to mixtures of ordinary Portland cement (OPC) and quartz sand [4,5]. This article examines the effects of different types of CB in OPC-CB blends on strength development and their effects on hydration product formations using a combination of analytical techniques. This study is part of a more comprehensive research project on OPC-quartz-CB and OPC-CB blends cured under hydrothermal conditions.

EXPERIMENTAL

The following raw materials were used:
  (a) Goliath cement (OPC) containing: SiO_2 20.0%, CaO 64.2%, Al_2O_3 4.5%, Fe_2O_3 3.7%, SO_3 3.5%, and having a fineness index (Blaine) of 350 m^2/kg, was produced by Australian Cement, Sydney, NSW, Australia.
  (b) Different types of crushed CB and passing 2.38 mm, were obtained from major clay-brick manufacturers in Sydney, Australia. Three types of CB denoted CB1, CB2, and CB3 were used in this study. The dry ball milling method of pre-screened crushed material was employed to produce ground CB having a mean particle diameter of 22 micron and a particle size distribution comparable to that of quartz sand used in industry. Table 1 depicts the major oxides of the different CB types as determined by X-ray fluorescence (XRF). The mineralogical compositions were determined qualitatively using X-ray diffraction (XRD). Crystalline phases were identified using JCPDS powder patterns and are summarized in Table 2, in decreasing order of abundance.

OPC-CB blends were prepared using a water-to-total solids ratio (w/s) of 0.35 as this yielded comparable workability for all the mixtures. CB was added at 0, 10, 20, 30, 40, 50, 60, 70 and 80 mass % as cement replacement. The bulk OPC:CB and Ca/(Al+Si) atom ratios for the different blends are given in Table 3. Mechanical mixing was conducted in accordance with ASTM C 305-99 [6]. Pastes were cast into stainless steel moulds and consolidated on a vibratory table, followed by a 24 hour curing period in a moist cabinet. Demoulded specimens were autoclaved for 7.5 hours, 6 hours of which were at 180°C under saturated steam. After autoclaving specimens were allowed to air dry for 2 days, followed by compressive strength testing according to ASTM C109-02 [7]. Three specimens per mix were loaded under compression until failure using a Tinius Olsen testing machine. Compressive strength results are reported to the nearest 0.1 MPa.

Table 1. Major oxides of different types of CB used in this study

OXIDE, MASS %	CB1	CB2	CB3
SiO_2	53.36	71.82	65.30
TiO_2	1.25	0.82	0.81
Al_2O_3	41.44	18.68	18.79
Fe_2O_3	1.40	4.05	7.10
MgO	0.23	0.95	1.34
CaO	1.04	0.45	1.56
Na_2O	0.27	0.11	0.62
K_2O	0.56	2.36	2.11

Table 2. Major crystalline phases in different CB in decreasing order of abundance

CB1	CB2	CB3
mullite	quartz	quartz
cristobalite	mullite	mullite
quartz	hematite	hematite
andalusite	cristobalite	Rutile

Table 3. Bulk OPC:CB and Ca/(Si+Al) atom ratios of different OPC:CB blends

CB, MASS %	OPC:CB RATIO	Bulk Ca/(Si+Al) Atom Ratio		
		CB1	CB2	CB3
0	-	2.72	2.72	2.72
10	9.00	1.88	1.94	1.97
20	4.00	1.36	1.43	1.47
30	2.33	1.00	1.07	1.11
40	1.50	0.74	0.80	0.84
50	1.00	0.55	0.59	0.62
60	0.67	0.39	0.42	0.46
70	0.43	0.27	0.29	0.32
80	0.25	0.17	0.18	0.20

Fragments of the broken paste specimens were retained for further evaluations by XRD, scanning electron microscopy (SEM), and thermal analysis. SEM was conducted on carbon coated specimens at an accelerating voltage of 8kV using a Jeol 6300F, fitted with energy dispersive X-ray microanalysis system (EDS). XRD analyses on powdered specimens were carried out using a Siemens D5000 diffractometer and copper $K_{\alpha 1}$ radiation from 3 to 60° 2θ at 0.02° 2θ per second. Thermal analyses on powdered samples were preformed using a TA-instruments SDT 2960 simultaneous differential thermal and thermogravimetric analyzer (DTA-TGA) at a heating rate of 10°C/min under flowing air (20mL/min) from 20° to 1000°C. The amount of reacted CB was determined from the mass remaining after dissolution in 2N hydrochloric acid and in 10% NaOH [8].

RESULTS AND DISCUSSIONS

Strength Development
For all CB types, compressive strength increased with increasing amounts of CB up to 40 to 50 mass % additions, and then smoothly decreased with any further additions (Figure 1). Specifically, the maximum strength observed was in the following order:

$$CB1 \approx CB2 > CB3$$

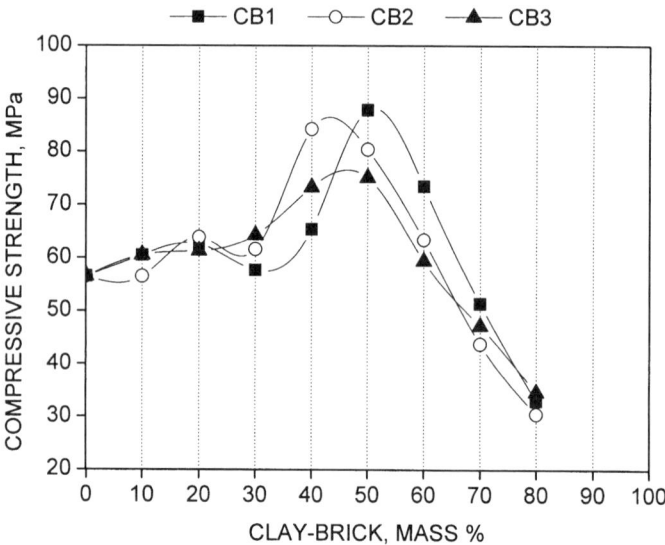

Figure 1. Average compressive strength for autoclaved OPC:CB blends.

In the case of CB1 and CB3, maximum strengths of 87.8 and 75.1 MPa, respectively, were reached at 50 mass % addition. In contrast, for CB2 maximum strength of 84.2 MPa was reached at 40 mass % addition. For comparison, companion control specimens made with OPC and ground quartz sand had compressive strengths of 86.1 and 75.3 MPa for 40 and 50 mass % additions, respectively. The variations in strength development among the different CB types as well as their amounts, may be explained, at least in part, by the nature of the phases formed, their morphology, and the nature of the different CB types.

Phases Formed and Microstructure

From a combination of XRD, DTA-TGA, and SEM analyses, the low strengths for 0 to 30 mass % additions for all CB types could be associated with Portlandite (CH) and alpha-dicalcium silicate hydrate (α-C_2SH) presence, the latter is known to be a low-strength giving phase [3]. With larger amounts of CB (greater than 30 mass %), both CH and α-C_2SH were fully consumed, and large amounts of calcium silicate hydrate (C-S-H) and crystalline 1.1nm tobermorite (tobermorite) dominated the matrix with traces of hydrogarnet also being observed. For additions greater than 50 mass % of CB2, and greater than 60 mass % for CB1 and CB3, tobermorite amount and its crystallinity decreased significantly, and this was reflected in a sharp decrease in the strength (Figure 1). In other words, additions of CB beyond the strength maxima acted only as a diluent, and the strength decreased. Alternatively, a decrease in the bulk OPC:CB ratio (Table 3), results in reduced availability of C aO f rom c ement a nd f ewer h ydration p roducts a re f ormed. It i s a lso n oteworthy, t hat hydrogarnet presence decreased significantly beyond 40 mass % CB additions and then finally disappeared. This reinforces previous observations, that hydrogarnet presence is favoured by lime rich environments. A decrease in OPC reduces the formation of lime-rich phases, resulting in a decrease in hydrogarnet amount [9].

Differences in tobermorite morphology for the different CB types at additions of 50 mass % were also observed. For example, for the CB1 specimen tobermorite morphology was comprised of dense crystalline plates (Figure 2). In contrast, for the CB2 specimen, a lath to split-type morphology was prevalent (Figure 3). The observed splitting of the tobermorite laths may signify a change in crystal form or structure of the binder. This observation and the fact that XRD revealed a lower crystallinity for this morphology when compared to the plate type morphology for the CB1 specimen, supports previous findings that a change in crystal form can been associated with strength loss [10]. It follows that a dense crystalline plate-type morphology is associated with higher strength. A more detailed account of microstructural attributes and mechanical properties of different OPC-CB blends is currently underway, and we will report our findings in future publications.

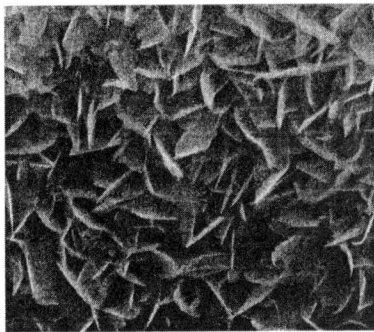

Figure 2. Tobermorite having a plate-type morphology for 50 mass % CB1 as OPC replacement.
Field of view = 4.3μm x 4.3μm

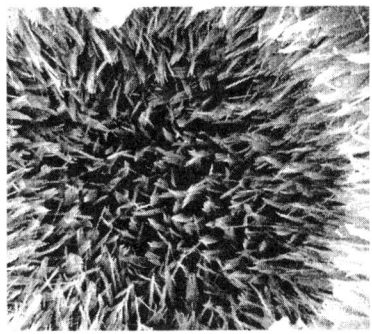

Figure 3. Tobermorite having a lath to split-type morphology for 50 mass % CB2 as OPC replacement.
Field of view = 32μm x 32 μm

Chemical Composition and Mineralogy of CB types

The chemical and mineralogical make-up of the three different CB types varied considerably and this was reflected in the compressive strength development of autoclaved blends (Tables 1 and 2). Specifically, the highest compressive strength was obtained for CB1, rich in Al_2O_3 and SiO_2, and having mullite and cristobalite as dominant crystalline phases. In contrast, CB3 had the highest amount of quartz and this was reflected in lower maximum strength when compared to CB1 and CB2. To recapitulate, the abundance of mullite, cristobalite and quartz for the different CB types, as well as maximum compressive strength decreased in the following order:

Mullite:	CB1 → CB2 → CB3
Cristobalite:	CB1 → CB2
Quartz:	CB3 → CB2 → CB1
Strength:	CB1 → CB2 → CB3

From the viewpoint of strength development, the inverse relationship between mineralogical make-up and the compressive strength is, therefore, apparent. In other words, there exists a good correlation between the compressive strength and the amount of mullite, cristobalite, and quartz, with higher amounts of both mullite and cristobalite in combination with lower amounts of quartz favouring higher strength development. Overall, the present data reinforce earlier findings that optimum strength development is a function of the nature of the siliceous source [3]. Future aims of this research will be to establish the effect of mineralogy on mechanical properties by conducting multivariant analyses.

CONCLUSIONS

From the findings presented in this study the following conclusions are made:

1. The use of clay-brick fines in combination with OPC for the production of hydrothermally cured calcium silicate based materials is a viable option for the future.

2. Differences in chemical and mineralogical composition of clay-brick fines are important factors in relation to strength development, with a combination of high amounts of both mullite and cristobalite and lower amounts of quartz favouring higher strength.

2. High compressive strengths corresponded to the maximum amount of tobermorite formed. There exists a good correlation between the total sum of tobermorite and C-S-H present and the compressive strength.

4. There exists a good correlation between tobermorite morphology and the compressive strength, with a dense crystalline plate-type morphology favouring high compressive strength.

ACKNOWLEDGMENTS

The authors acknowledge the Australian Research Council for supporting this research via grant DP 0211950. In addition we would like to thank stuff of Microstructural Analysis Unit in the Faculty of Science, UTS.

REFERENCES

1. BAKOSS, S.L., SRI RAVINDRARAJAH, R., Recycled C&D Materials for Use in Roadworks and other Local Government Activities: Scoping Report, Centre for Build Infrastructure Research, University of Technology, Sydney, 1999, pp 1-68.

2. KLIMESCH, D.S., SRI RAVINDRARAJAH, R., BAKOSS, S.L., New Technologies for use of Recycled Brick as Construction Material: Final Report to NSW Waste Boards, Centre for Build Infrastructure Research, University of Technology, Sydney, 2001, pp 1-66.

3. TAYLOR H.F.W., Cement Chemistry, 2nd ed., Thomas Telford, London, 1997, p 459.

4. KLIMESCH, D.S., RAY, A., Incorporating brick fines in cement based building materials: Proceedings of the International Conference on Composites in Construction, Figueiras et al. (EDS), Porto, 2001, pp 47-49.

5. KLIMESCH, D., GUTOVIC, M., RAY, A.: Brick waste a supplementary cementing material in autoclaved building products. In: Dhir RK, Newlands MD, Csetenyi LJ eds. Role of Cement Science in Sustainable Development, Thomas Telford, London, 2003 pp. 283-290.

6. AMERICAN SOCIETY FOR TESTING AND MATERIALS: ASTM C 305-99, Standard Practice for mechanical mixing of hydraulic cement pastes and mortars of plastic consistency, Philadelphia. 1999.

7. AMERICAN SOCIETY FOR TESTING AND MATERIALS: ASTM C 109/C 109M-02, Standard Test Method for Compressive Strength of Hydraulic Cement Mortars (Using 2-in. or [50-mm] Cube Specimens), Philadelphia. 2002.

8. KLIMESCH, D.S., RAY, A., Metakaolin additions to autoclaved cement-quartz pastes: evaluation of the acid-insoluble residue, Advances in Cement Research, 9, No. 36, October 1997, pp 157-165.

9. KALOUSEK, G.L., Tobermorite and Related Phases in the System $CaO-SiO_2-H_2O$, Journal of the American Concrete Institute, Vol. 51, June 1955, pp 989-1011.

10. AL-WAKEEL, E.I., EL-KORASHY, S.A., EL-HEMALY, S.A. AND UOSSEF N.: Promotion effect of C-S-H-phase nuclei on building calcium silicate hydrate phases, Cement and Concrete Composites, 21, 1999, pp173-180.

DETERMINING ECONOMIC VIABILITY OF PRECAST CONCRETE PRODUCTS MADE WITH RECYCLED CONSTRUCTION AND DEMOLITION WASTE

J Gradwell

Enviros Limited

R G Tickell, S G Millard, M N Soutsos, J H Bungey, N Jones

University of Liverpool

United Kingdom

ABSTRACT. This paper outlines the findings from an on-going project to investigate the case for using recycled construction and demolition waste (C&DW) as aggregates in pre-cast concrete building blocks. Currently, C&DW is typically processed for low value end uses. The project aims to investigate both the economic case and the technical capability of using C&DW in higher specification applications. The research to establish the economic case has involved surveying the producers, current end users and potential future users of C&DW. This has included an assessment of the practicalities of using C&DW and the current climate for the construction sector in the North West.

The data gathered will be used to inform a cost model that will compare various scenarios for manufacturing pre-cast blocks using C&DW.

Keywords: Construction and demolition waste, Recycled aggregates, Precast concrete blocks, Sustainable construction, Economic viability.

Ms. J. Gradwell is an Environmental Consultant with Enviros Ltd.

Mr. R. G. Tickell is a Senior Lecturer in the Department of Civil Engineering, University of Liverpool.

Dr. S. G. Millard is a Senior Lecturer in Department of Civil Engineering at the University of Liverpool.

Dr. M. N. Soutsos is a Lecturer in the Department of Civil Engineering, University of Liverpool.

Prof. J. H. Bungey is a Professor in the Civil Engineering department of the University of Liverpool and Head of the Structures and Materials Group.

Dr. N. Jones is a Research Associate at the Department of Civil Engineering, University of Liverpool.

INTRODUCTION

Construction and demolition waste (C&DW) accounts for approximately 17% of the annual UK waste stream, amounting to a total of 70 million tonnes of waste material [1, 2].

Around 30% of the 70 million tonnes arising is reused in low-grade applications such road sub-base construction, engineering fill, or landfill engineering where some crushing and separation of materials such as metal and wood is required [2]. Only a small percentage is recycled for high specification applications and this tends to come from easily identifiable source, such as railway ballast. Higher value uses for the majority of C&DW have not in the past been thought possible because of the heterogeneous nature of the material compared with primary aggregates.

Around 220 million tonnes of natural aggregates are quarried each year in the UK [4]. Because of the environmental impact of quarrying, the Government has used introduced legislation - such as the aggregates levy, introduced in April 2002 - in the UK to minimise C&DW production and maximise the use of alternatives to primary aggregates [5]. As C&DW is largely inert, consisting of soil, brick, glass, concrete and plaster etc., landfill charges for its disposal have traditionally been low.

This study provides an assessment of the economic and technical aspects of using C&DW. The research to establish the economic case has involved surveying the producers, current end users and potential future end users of C&DW. The work includes an assessment of the practicalities of using C&DW and the current climate for the construction sector in the North West by establishing likely future demand and the impact of planning controls and other contributing factors, such as Government funding, legislative and financial pressures.

CURRENT QUANTITIES AND USES FOR C&DW

C&DW Arisings

The most comprehensive research to date on C&DW arisings has been undertaken by Symonds for the Office of the Deputy Prime Minister (ODPM) in 2002 [4]. This research surveyed the industry and users of C&DW and extrapolated the results to estimate the total arisings. This showed that there are more than ten million tonnes of C&DW arisings in the North West each year. Almost half is hard C&D waste that is currently used as an aggregate in road building, engineering or land restoration.

The precise requirements for C&DW that could be used in precast concrete blocks will be determined by the laboratory research, but of the estimated 4.5 million tonnes of C&DW that could be available in the North West the actual amount that could be used in block production will be affected by:

- The location of the demolition site and the processing site;

- The level of contamination arising from it use, materials of construction or caused by poor segregation during demolition; and

- Local demand for the material that varies depending on current development and infrastructure projects.

An Example of the Current Market for C&DW

The landfill sector in one of the markets currently taking C&DW in the UK. It has traditionally taken large quantities of C&DW in various forms and to serve various purposes, such as;

- 'Void fill' – materials that are not economically recyclable, i.e. 'soft strip' including carpets, plasterboard, insulation materials, wood, glass etc. Landfill companies charge a gate fee and landfill tax per tonne to tip this material.
- Roadway construction – by their nature, landfill sites are constantly changing. C&DW has been a preferred material to make temporary access roads.
- Cover – demolition waste containing a high percentage of fine organic material (e.g. soil) is often screened by demolition companies and used by the landfill companies as cover.

POTENTIAL FUTURE MARKET - CONCRETE BLOCK MANUFACTURING

The concrete block is a commodity product that is manufactured to British Standard (BS) 6073. According to the Concrete Block Association (CBA) the market is supplied by over 50 manufacturers a cross t he U K, w ho b etween t hem p roduces i n e xcess o f 6 0 m illion m2 o f aggregated concrete building blocks per year.

In summary, the market for precast concrete blocks is very competitive. The basic block is a commodity product and therefore the profit margin is low. The sector is dominated by large multi-nationals that generally own quarrying operations; as the raw materials used to manufacture these blocks, be it virgin aggregate, lightweight or manmade materials are costly to transport. Most manufacturers are located close to or at the quarry site. In the UK there are approximately 100 plants producing a wide variety of blocks for the construction industry. Despite the wide variety of designs of block manufactured the Standard less than 20 k g b lock c ontinues t o d ominate s ales o f b uilding b locks, a ccounting for 7 o f e very 1 0 blocks sold.

The market has developed to expect good quality and consistency beyond the British Standard specifications. The Standard block that is disguised by render or paint is the cheapest precast block product on the market. The Standard block is produced in large volumes, with a clearly defined and growing position in the market place.

The total square metres of precast concrete blocks produced have increased by almost 30% over the last ten years. Using DTI data on the construction output it has been estimated that the North West is 10% of the UK precast concrete block output. Using this assumption it can be estimated that the North West produces 10% of the dense aggregate blocks in the UK. This suggests that the current market for aggregate for dense block production in the North West is approximately 5-600,000 tonnes annually and growing.

A conservative estimate of the market for aggregate used in precast concrete blocks in the North West suggests that at least 500,000 tonnes of dense aggregate is consumed each year. This compares with the 4.5 million tonnes of hard C&DW arisings identified by the Symonds report. This suggests that there is the opportunity to replace some of the virgin aggregate consumed in the North West with C&DW without unbalancing the existing markets for C&DW that exist currently.

Manufacturing a Standard block using C&DW is the focus of the laboratory work. One problem with producing a product to compete in this marketplace is the level of competition for sales and the minimal profit margin per block – thus requiring them to be sold by the thousand to cover operating costs and to see a viable profit.

Transportation rather than the processing costs appear to be a key consideration when producing blocks as there need to be sufficient customers within a viable haulage distance.

CURRENT CONSTRUCTION CLIMATE

Research has been undertaken on the construction activity and effects of regional planning guidance. The findings indicate that the market for the construction industry has been healthy and is forecast to maintain output and in some areas continue to grow. All sectors of construction, except public housing and industrial, have seen significant increases in activity levels and continued growth is expected, particularly in the areas of private commercial, public non-residential and public housing repair, maintenance and improvement.

In addition significant amounts of public monies are being spent and are expected to be spent on construction related activities over the coming decade.

PRACTICALITES/ TECHNICAL FEASIBLITY OF USING C&D WASTE IN CONCRETE BLOCKS

Some of the practicalities that need to be addressed in order to use C&DW in the manufacturing of precast concrete blocks are outlined here. Further issues such as flow smoothing, supplying to meet demand and storage also need to be addressed.

Demolition & Processing Equipment

The equipment used by demolition contractors to demolish and process C&DW has remained largely unchanged for the past decade. Demolition contractors typically use jaw crushers to produce a crushed concrete and masonry of a particle size of 75mm.

To produce C&DW of the particle size needed to manufacture precast concrete blocks will require the use of a cone or impact crusher which will process material more slowly or it will be necessary to pass C&DW through a jaw crusher several times to achieve the required particle size.

Screening

Screening after the material has been crushed is necessary if any control is to be exerted over the particle size of the finished product. Material is typically moved by conveyor from the crushing equipment. As all this equipment tends to be mobile rather than static plant it is possible to introduce variations in the plant layout depending on the requirements of the end product.

The particle size of material required to manufacture precast concrete blocks will require demolition contractors to invest in new screens in order to produce the correctly graded material.

British Standards
The new European Standards for aggregates (BS ENs) that replaced existing British Standards (BS) in January 2004. The mandate behind the new BS EN standards is to change the approach to using aggregates, not to define the use of primary aggregates for particular end uses, but rather to expect them to have certain performance characteristics for particular end uses. This change in approach means that recycled and manufactured aggregates are as valid for any end use provided they meet the performance requirements.

Contamination
The threat of contamination in the C&D waste is one of the biggest barriers to its use in construction products. The perception is that C&D waste will be difficult to process, contain some contamination that will affect the handling and properties of the end product and will always be consider inferior to virgin materials.

The presence of other construction waste materials that may contaminate the hard C&D waste fraction is common. These contaminants include organic materials, such as wood, bitumen, glass and metal. This contamination in C&D waste needs to be removed during processing or excluded that is to be used for manufacturing precast concrete blocks will affect the strength, bonding and lifetime of the product.

Consistency, Colour and Surface
Consistency of the raw material is a key issue for the research elements of the project to address. The aspects of consistency that are important from the manufacturer view point are:

- Handling properties – cement and water requirements to achieve the properties required by relevant standards;

- Colour – effects of variations in mix of masonry and concrete in the input C&D waste;

- Surface finish – difference between the finish achieve by different variations in mix of masonry and concrete in the input C&D waste.

Manufacturers have stressed the fact that customers are used to purchasing a product manufactured from a very consistent raw materials. This has implications for even the cheapest precast blocks, as they tend to be consistent in terms of colour and surface finish.

Even the addition of cement may not completely mask the colour variations in the input raw material and the use of pigments to hide to variations may add excessive cost burdens.

Processing Site Issues
There are many factors that will affect a manufacturing operation using C&DW to produce precast concrete blocks. The issues include – environmental legislation covering handling waste, hazardous materials, air pollution and statutory nuisance and other factors such as health and safety.

Suitable methodologies for dealing with the practical issues identified will continue to be determined and tested during this research phase.

ECONOMIC CASE FOR USING C&D WASTE IN CONCRETE BLOCKS

The economic case for using C&DW to replace virgin aggregates will be affected by the findings of the on-going laboratory research.

The factors that will affect the economic viability of using C&DW in the manufacturing process include:

- The price of the finished block – the market suggests it need to be cheaper than a conventional block manufactured with virgin aggregates. This is necessary to surmount "conservatism" by the marketplace.

- Positive financial factors such as Landfill tax and Aggregates Levy that will be avoided by using C&DW to manufacture a product.

- Negative financial factors, such as transportation costs, the requirements of legislation, the need for demolition contractors to invest in new equipment and the additional processing costs which may arise from variations in the raw material that increase maintenance costs, downtime, processing cost due to more product line change, etc;

Other important factors that may be positive or negative include:

- Maintaining market confidence;
- Maintaining market share; and
- Shareholder pressure.

COST MODEL

A production cost model has been developed to compare the identified costs of five production scenarios for manufacturing precast concrete blocks using C&DW. The five scenarios are:

1. Demolition followed by on-site sort and crushing, transportation to a remote block manufacturer, manufacturer and sell.

2. Demolition followed transport to a regional site for sorting and crushing waste with a) manufacture at regional site then sell or b) transport to remote block producer, manufacture and sell.

3. Demolition followed by on-site sort and crushing, then manufacturer on site and sell.

4. Demolition followed by alternative use.

5. Model based on current use of natural aggregates (control).

The laboratory findings will impact directly on the various manufacturing arrangements for each scenario, affecting aspects such as the recipe required to manufacture a block that will meet the BS 6073 block specification, for example the quantity of virgin to C&DW aggregate used.

CONCLUSIONS

Research has taken place to establish the amount of C&DW available in the waste stream, the make-up of the precise concrete block marketplace, the current level of construction activity. Work has been undertaken and is also ongoing to establish all the practical considerations when u sing C &DW t o manufacture p recast c oncrete b locks, t o i dentify t he a ll t he f actors affecting the economic viability of using C&DW and in the manufacture of precast concrete blocks.

In the laboratory the factory procedure for fabricating concrete blocks using natural aggregate has been successfully replicated. The laboratory research is ongoing to determine the viable recipes for manufacturing a block that will meet the BS 6073 specification. The basis for the cost model has been determined and further work is on-going to collect all the necessary data and to model the various scenarios within the range of recipes that are achievable.

ACKNOWLEDGEMENTS

The a uthors w ish t o t hank t he O NYX E nvironmental T rust a nd t he F lintshire C ommunity Trust Ltd. (AD Waste Ltd.) for funding this project. Thanks are also extended to the following industrial collaborators for their assistance: Clean Merseyside Centre, Marshalls Ltd., Forticrete Ltd., Liverpool City Council, Liverpool Housing Action Trust, RMC Readymix Ltd., W. F. Doyle & Co. Ltd., DSM Demolition Ltd., and Environmental Advisory Service – Merseyside.

REFERENCES

1.   Estimated Annual Waste in the UK by Sector (Million Tonnes). (2000), Environment Agency.

2.   Williams, P. T., *Waste Treatment and Disposal*. (1998): John Wiley & Sons Limited.

3.   Symonds Group Ltd., A., COWI & PRC Bouwcentrum. *Construction and Demolition Waste Management Practices, and Their Economic Impacts*. in *Final Report to DGXI, European Commission, Feb. 1999*. (1999).

4.   Symonds Group Ltd., *Survey of Arisings and Use of Construction and Demolition Waste in England and Wales 2001,for the Office of the Deputy Prime Minister*.

5.   Department of Environment Transport and the Regions, Planning for the Supply of Aggregates in England - a Draft Consultation Paper. (2000). p. 70.

6.   HM Treasury, *Consultation on the Objectives of Sustainability Fund under the Aggregates Levy Package*.

STRATEGIC C&D WASTE MANAGEMENT TOOL FOR SUSTAINABLE URBAN RENEWAL

M Petersen

Golder Associates

United Kingdom

ABSTRACT. Urban renewal and regeneration works typically generate large quantities of construction and demolition wastes. Considering it is possible to recycle more than 80% of demolition wastes from typical buildings, strategic waste management offers a significant opportunity to improve the sustainable development performance of the project.

Golder a ssociates h ave developed a s trategic w aste m anagement t ool t hat a ssists p lanners, developers and consultants in identifying mechanisms whereby the expected wastes generated can be handled and processed, either on site or in the local urban area, for use in the construction works.

Starting with a prognosis of the demolition wastes expected, the strategic demolition waste management tool helps to identify demolition and recycling methods which will increase the recyclability of the wastes, taking *financial, environmental, technical and social* parameters into account.

Keywords: Demolition wastes, Recycling, Urban renewal, Sustainability, Demolition, Land development.

Martin Petersen is a Project Manager at Golder Associates UK and has worked within the field of decommissioning as both a contractor and consultant for over 15 years. A significant portion of this work has been on large scale, complex decommissioning works, for example airports, hospitals, industrial plants refineries and nuclear facilities, both in the UK as well as in Europe and Asia. Martin is a member of the Institute of Demolition Engineers, the Chartered Institution of Wastes Management and the Institute of Explosives Engineering.

INTRODUCTION

Sustainable Development within the construction industry has seen considerable progress in the past years, with a growing emphasis on the need to mainstream sustainability into practice. There are several initiatives underway, for example the London Sustainable Construction Project and research projects with best practice guidelines by such organisations as CIRIA and BRE. These are being supplemented by such guidance notes as "Six Steps to Sustainable Development for Housing Associations" as prepared by the Housing Corporation[i].

However, there is need for a more systematic approach to the integration of sustainability into land development as a whole, whereby a holistic overview can be taken in order to make informed decisions on which environmental benefits to seek, with which financial costs and how to engage the local community in the works.

The aim of the tool presented in this paper is to provide such a holistic and systematic process for sustainable demolition waste management, which is a key component of sustainable land development projects. The tool focuses primarily on the initial works comprising demolition and site clearance for ensuing construction works.

From the results of the tool, the users are able to assess, along sustainability parameters, various options for waste management. This enables decision makers to design and implement the works in line with the sustainable development targets for that specific project.

OPPORTUNITIES FOR SUSTAINABLE WASTE MANAGEMENT

A significant challenge within sustainable land development is to reach an *optimal balance* between the three key sustainability parameters; economical, environmental and social.

The environmental and economic factors are normally quantifiable and have been assessed in various past and ongoing development works, with case studies available for projects using increased recycled materials whilst achieving cost savings[ii].

The environmental and economic parameters must now be integrated with improved social impacts on the community (both local and national), in order to attain true sustainability in the works. There are opportunities for this as detailed in Table 1.

APPLICATION OF STRATEGIC DEMOLITION WASTE MANAGEMENT TOOL

Types of Development Works

The tool has been designed to deal with large-scale land development works within urban renewal, regeneration and redevelopment sites. The land development works comprise demolition and site clearance of the present built environment, with recycling and handling of the demolition wastes to feed back into the subsequent construction works planned for the site.

Table 1. Opportunities for sustainability in land development works

INDICATOR		OPPORTUNITY FOR SUSTAINABILITY
Social	✓	increased employment generation through utilising local skills
	✓	skills training and accreditation
	✓	preservation of historic buildings leading to increased community value
	✓	active participation by local stakeholders
	✓	business development opportunities for small scale recycling operations
	✓	potential for re-use of demolition materials as artistic value in new development
Economic	✓	use of recycled and secondary materials can be cheaper than primary materials (including Aggregate Levy Tax)
	✓	reduced waste disposal costs due to on-site recycling (including non payment of landfill tax)
	✓	reduced haulage costs for export of wastes and import of aggregates from on-site recycling
	✓	value of social and community gains
Environmental	✓	increased recycling and re-use of wastes from the demolition and construction works
	✓	reduced energy consumption from reduced haulage and raw materials extraction processes
	✓	conservation of finite resources
	✓	reduced traffic to/from site leading to reduced risks (traffic accidents, nuisance to neighbours, schools etc.)
	✓	reduced emissions from reduced haulage requirements

Target Audience / Users

The tool is aimed at the strategic level of planned development works, in order to provide the Local Authority, Planners and Developers with an overview of the expected costs from various site clearance, demolition and recycling options.

Through running various scenarios, from 100% disposal of wastes to maximum recycling of wastes and full local community engagement, the Developers can use the tool to find the optimal balance between the three main sustainability indicators.

The tool is thus aimed at the following users, each with their own aims:

✓ *Local Authorities;* to gain better understanding of opportunities to ensure sustainable development is integrated into the land development works and thus set appropriate sustainability targets.

✓ **Planners;** to gain insight into the potential environmental, economic and social impacts from the planned development and set realistic benchmarks and targets for the development, in line with the expectations of the Local Authority.

✓ **Developers;** to understand the costs and waste handling issues arising from the planned development, as well as actions required in order to meet those sustainability targets set by the Local Authorities.

✓ **Contractors;** to receive practical guidance on how to meet sustainability targets set, as well as a tool for achieving cost benefits through increased recycling, local community engagement and reduced negative environmental impacts.

HOW THE TOOL WORKS

Introduction

This section presents the use of the tool with an illustration of the outputs.

The program is predominantly a *mass-flow analysis program* and has been developed in Microsoft Access.

In the program two types of accounting are considered; recycling economy, selected environmental indicators as well as social impact opportunities. The following quantitative indicators are included:

- Energy consumption (in MJ, divided into fossil fuels and electricity)
- Kilometres covered by truck (divided into city traffic and highway traffic)
- Substituted materials (aggregates and steel)
- Waste (deposit on landfill)
- Cost

The following material types can currently be evaluated in the program:

- High quality concrete
- Low quality concrete
- Bricks
- Mixed concrete and bricks

The different handling options included in the program are, for:

- Concrete: Aggregates for new concrete, road bases, sub bases, fill and landfill
- Brick: Aggregates for road bases, sub bases, fill and landfill
- Steel: Recycling, fill (as part of concrete), landfill (as part of the concrete)

Inputs

The tool is based on a model with three main parameters, which when combined will provide the possibility of building scenarios:

- *Addresses:* The site where the land development works are planned, including data on waste quantities, expected materials consumption for the new construction and local community skills audit.

- *Processes:* The purpose of treating a material (e.g. selective demolition)

- *Equipment:* The machinery used to treat the material in a process (e.g. demolition equipment and crushing and sorting plant).

During the input of data, both the equipment and the processes are assembled into processes and the processes are allocated to an address.

Outputs

After all material flows have been defined the results for each parameter are displayed with a summary page illustrating the net result. The net result is an addition of all costs and savings for the different material flows in a scenario (including subtracted effects from substituted materials), as well as the environmental impacts and social opportunities for that specific scenario.

The key benefit from the outputs from the tool, is the opportunity to run several scenarios in order to find the optimal balance between economic, environmental and social parameters for the waste management aspects of a specific land development project, in accordance with the aims of the project as a whole.

Implementation

Once the optimal and most cost-effective balance between the three key parameters has been agreed upon within the Project Design Team and/or Client, the concept is to be implemented, as described below.

Sustainable Waste Management Strategy

The outputs from the tool form the basis for an informed decision on the waste management issues surrounding the implementation of the land development works and thus contribute to a Sustainable Waste Management Strategy for the specific project.

Such a strategy should ideally be prepared in partnership with the key stakeholders (including the planners, local authorities, contractors and local community), where the implementation of the strategy is monitored by an authorised person.

For the implementation of the strategy, specific clauses can be included within the Tender Documentation, which ensure that the sustainability of the works are sought. These can, for example, include recycling targets for the project, requirements for local skills employment and access to structures before demolition to allow local community removal of re-usable / recyclable for subsequent sale. A further component of the strategy may relate to the training of site engineers and personnel in the required methodologies and principles, as well as identified potential labour in the local community.

In addition, the tender evaluation criteria can be formulated to consider the key sustainability indicators for the specific works.

CASE STUDY

The following case study presents the use of the strategic demolition waste tool on an actual project in Norway during 1999 – 2001. Please note that the social dimension as included in the revised tool were not available at the time, thus only the environmental and economic parameters were used for this case study.[iii]

Redevelopment of Fornebu International Airport, Oslo

Fornebu airport was the main airport of Oslo Norway since 1939. The new airport "Oslo Airport Gardermoen" opened on 8th October 1998, where after the Fornebu airport was planned for re-development as housing, office and recreation facilities. The amount of asphalt left at Fornebu was approximated to the total amount of 425,000 tonnes.

In order to create a base for strategies for contracts and co-ordination with the proceeding building projects, an analysis was performed in 1998 for The Norwegian Directorate of Public Construction and Property (Statsbygg), concerning how asphalt waste at Fornebu should be used, taking into consideration both economic, environmental and social aspects.

Options Analysed

Based on the needs in forthcoming projects at Fornebu, focus was set on the following recycling options (scenarios):

I	Aggregates for the production of warm asphalt
II	Aggregates for the production of cold asphalt
III	Road base (without adding new bitumen)
IV	Sub-base (without adding new bitumen)
V	Filling material
VI	Use as a road surface in its existing form (using the former landing area as roads)

A *zero-alternative* for calculating/comparing effects from different options comprised leaving the asphalt in-situ without use. The option to take the material to a landfill was never analysed in detail, since it was concluded at an early stage that this would be too expensive and would have a negative environmental impact.

In addition to these alternatives on site the effect of transportation to other sites was investigated. The sites and forms of transport chosen were:

- Construction site in a nearby project by truck
- Regional site by barge
- North Germany by barge.

Selection of Parameters

The environmental parameters were chosen with background in the Norwegian political agenda, policies in Statsbygg, and by examination of Life Cycle Analyses performed for other building projects with the same kinds of material.

The different recycling options were analysed according to the following parameters:

Headlines for qualitative discussion;
- Minimising production of residual products (load on landfill, pollution)
- Minimising the risks in working environments (calculating cost to keep a certain level)
- Minimising impact from dust, vibrations and noise emissions
- Opportunities for local community employment

Quantitative calculations for;
- Use of energy (MJ)
- Use of materials (oil and gravel/crushed rock)
- CO_2 emissions
- SO_2 emissions
- NO_X emissions
- Transport km by truck
- Costs

RESULTS

The results for the different parameters were presented separately for the project management to use as a background to discussions. An attempt was also made to use LCA weighting-models to match the selected investigated emissions.

The analysis indicated that all options for local recycling were feasible with respect to costs compared with the "zero-alternative" of leaving the asphalt without use. The best options when considering economic and environmental aspects jointly were presumed to be: use as existing road system, recycling as cold asphalt, or road bases and sub bases locally at Fornebu.

Most energy could be saved using the material as warm asphalt (if the inherent energy were counted), and most oil resources could also be saved. Use as warm asphalt in top layers would, moreover, mean that the quality of the aggregates probably would be best utilised (closing the cycle), and the future utilisation of the asphalt products simplified. However, other parameters such as costs, the selected emissions and transportation did not point out warm recycling as feasible.

The following graph (Figure 1) illustrates a typical result from the strategic demolition waste management tool where other results concern energy consumption, transportation requirements etc.

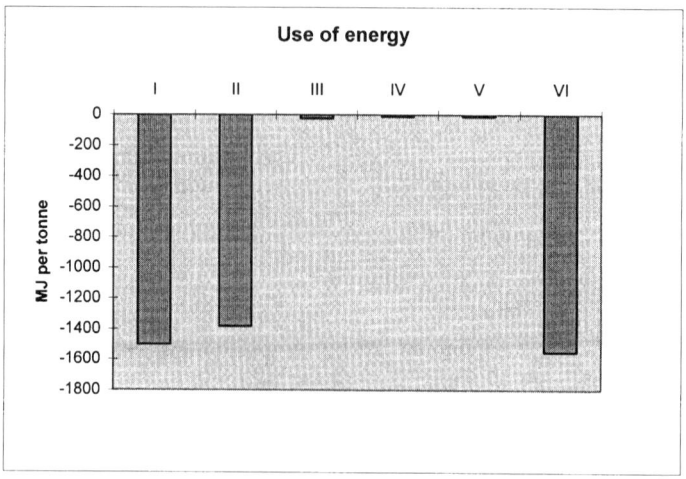

Figure 1. Use of energy for different applications on site at Fornebu

CONCLUSIONS

This paper has presented a strategic demolition waste management tool which assists Local Authorities, Planners, Developers and Contractors in achieving cost benefits through increased engagement of local communities and reduced negative environmental impacts in land development works.

The tool allows the user to run various waste management scenarios for the planned land development works, from off-site disposal of all wastes to maximum recycling of demolition wastes and engagement of the local community. The optimal balance can thus be found, which meets the sustainability targets set by the Local Authorities, as well as provides best value for the works.

REFERENCES

[i] HOUSING CORPORATION. *Six Steps to Sustainable Development for Housing Associations* Report on mainstreaming sustainable development by Beyond Green (14/01/2004) at www.housingcorplibrary.org.uk.

[ii] For case studies and guidance on use of recycled materials in construction, refer to WRAP's web site at www.wrap.org.uk and www.aggregain.co.uk as well as CIRIA's work at www.ciria.org.uk.

[iii] TORRING M. *Management of Concrete Demolition Waste*. PhD Thesis from NTNU Trondheim, Norway. 2001

ASK THE FELLOWS WHO CUT THE HAY

A Pears

Kotuku

United Kingdom

ABSTRACT. The construction industry consumes and wastes extensive resources. Numerous campaigns have been initiated to reduce this waste, but their impact on small and medium sized business – the heart of the industry – has been minimal. This project collected data on current building activity and waste management practices by cycling over 450 miles in four London Boroughs. From this information key stakeholders were identified – clients, architects, builders, merchants, building control officers, and waste handlers – and interviewed either in focus groups or individually. What emerges is a clear picture of why current policy is so ineffective and what could be done to improve it. . The title is from a book on agricultural practices in the 19c East Anglia; if you want to know what happens in the fields, you "Ask the fellows that cut the hay".

Keywords: Resource Efficiency, Sustainable Construction, Construction and Demolition Waste, Recycling, Reuse, Waste management.

Andrew Pears is a graduate of York University and has run his own Construction Company for 15 years. He is director of Kotuku, an environmental body he formed in 2001. It carries out research and develops initiatives to improve resource efficiency in the construction industry particularly for small and medium sized companies.

ASK THE FELLOWS WHO CUT THE HAY . . . .

A Study to Determine Attitudes to Waste, Resource Management, and the Use of Recycled Materials in Construction in four London Boroughs of Lambeth, Wandsworth, Hammersmith and Fulham and Kensington and Chelsea

INTRODUCTION

Construction is not a resource efficient activity. To add 275 million tonnes of stock to the built environment requires over 420 million tonnes of material resources, 8 million tonnes of oil, and generates nearly 3 tonnes of waste for each man woman and child in the UK. It is estimated that 13 million tonnes of building materials, delivered to site are sent away unused. In London construction sends to landfill twice the entire municipal waste stream. This position is not sustainable. Yet despite initiatives, research, and fiscal incentives sustainable resource management has had little impact on the industry, and particularly on the 90% of building firms that employ less than eight people.

This project is a first step in developing specific local initiatives to improve the sustainable nature of construction in London. The project recognises the complex nature of construction procurement and the large number of stakeholders responsible for opinion making and decision implementation. To find answers to questions about resource management it targets the heart of the industry by talking to the people who have the day to day responsibility for making construction work. It will "Ask the fellows who cut they hay".

PROJECT AIMS

The project will provide answers to four questions. These are focused at a local level, with a particular interest in small and medium sized companies.
1. Why are practitioners in the construction industry so resource inefficient?
2. What would need to happen to improve current practice?
3. How do people in the industry view the environmental agenda and recycled and recovered materials?
4. What factors would increase the specification and use of such materials?

PROJECT OUTPUTS

1. An analysis of the types and levels of building work currently taking place in four London boroughs.
2. Identification of levels of waste production, recycling and reuse.
3. An assessment of current levels of knowledge on resource efficiency and sustainable construction.
4. An assessment of the effectiveness of current campaigns to reduce waste in construction.
5. An understanding of the value currently put on environmental issues in construction.

6. Identification of effective and productive initiatives to improve resource efficiency particularly in small and medium sized companies.

7. The setting of a datum from which to measure the effectiveness of future campaigns.

PROGRAMME

The project comprises the following phases:

a. An analysis of types and volumes of building work carried out in the Boroughs. Principal categories will be Public and Private Sector, new build, repair and refurbishment, and infrastructure. (Completed).

b. Identification by visual survey of all current building sites in the boroughs. (Completed).

c. Selection of industry stakeholders. (Completed).

d. Research through focus groups and individual interviews to determine the attitudes of these stakeholders. (Currently running) Principal members will be;

    a) Clients – Public sector / Private Sector
    b) Architects
    c) Construction Companies
    d) Builders Merchants
    e) Waste Handlers
    f) Building Control Officers.

e. Compilation of a report analysing the research and setting a practical and action orientated agenda for targeting waste minimisation and green procurement in the boroughs. (Awaiting research results).

f. Distribution, dissemination and promotion of results to industry stakeholders and the wider community. (The Kingston Conference will be the official launch of the research findings).

FUTURE STRATEGY

Kotuku sees the project as the starting point for a realistic and relevant campaign to influence waste minimisation and green procurement by the construction industry in London. The target boroughs are also participants in the Reth!nk Rubbish Western Riverside campaign, and we see it as relevant to explore and understand the links between domestic and commercial waste initiatives. The report will identify local issues and help determine practical initiatives.

These could include:

1. An industry and locality specific website
2. A directory of green products and suppliers
3. Site based waste management initiatives
4. Education campaigns
5. A dedicated information and advice centre
6. A conference targeted at opinion formers and decision makers in the Borough
7. The development of communication networks in the Boroughs
8. Sector specific information sheets
9. Develop a Green target code
10. Free training modules

Key to a successful campaign is the setting of a datum from which to measure change and the effect of new initiatives.

CONCLUSIONS

Waste should be an important issue for everyone, and particularly those whose work produces it. This focus has identified why an agenda that seems so relevant, and is at the forefront of so much policy, has had so little impact. But it has gone further than that. From this understanding it becomes clear what type of campaign is needed, and what initiatives will meet with the greatest approval by a sceptical industry. If our proposals are adopted they will make a direct contribution towards improving resource efficiency and reducing waste in the London construction industry. This will be to the benefit of the community, business and the environment.

ACKNOWLEDGMENTS

Kotuku is grateful for the support of the London Development Agency and London Remade in funding this project.

REFERENCES

1. SMITH, R. A, KERSEY, J.R., and GRIFFITHS, P.J. The construction Industry Mass Balance: resource use, wastes and emissions. Viridis Report VR4 2002.
2. COVENTRY, S., WOOLVERIDGE, C. Environmental good practice on site. CIRIA 1999.
3. DETR: Building a Better Quality of Life: a strategy for a more Sustainable Construction: April 2000.
4. LISNEY, B., RILEY, K., BANKS, C. From waste to resource management: Hampshire Natural Resources Initiative, Sep 2003.

5. Biffa: Future Perfect: An analysis of Britain's Waste Production and Disposal Account: Biffa winter 2002/3.

6. Strategy U nit C abinet O ffice: W aste n ot, W ant not: A strategy f or t ackling t he w aste problem in England: November 2002.

7. W S Atkins on behalf of NetRegs: SME-nvironment 2003.

8. The Sustainable Construction Task Group: The UK Construction Industry: progress towards more sustainable construction 2000 – 2003: October 2003.

9. THOMAS, C., YOXON, M., SLATER, S., : Public attitudes towards waste and recycling: Rethink Rubbish Western Riverside Campaign Evaluation Research: Open University: July 2003.

10. www.bre.co.uk

11. www.londonbuildingcontrol.org.uk

12. www.constructingexcellence.org.uk

13. www.capitalwastefacts.co.uk

14. www.envirowise.gov.uk

15. www.ciria.org/recycling

16. www.defra.gov.uk

17. www.environment-agency.gov.uk

18. www.environment-agency.gov.uk/netregs/

19. www.lbhf.gov.uk

20. www.lambeth.gov.uk/

21. www.wandsworth.gov.uk

22. www.rbkc.gov.uk

23. www.lda.gov.uk

24. www.aggregain.org.uk/

25. www.wrap.org.uk/

26. www.lsx.org.uk/

27. www.londonremade.com/

28. www.ice.org.uk

GREEN CONCRETE OR THE ALTERNATIVE USE OF WASTES

S Weber

S Pommerenke

C Schwambera

University of Applied Science, Stuttgart

Germany

ABSTRACT. Waste management and recycling play an increasing political, social and economical role in Germany. Increasing costs for transport, incineration and storage have to be considered. As the waste sector will face enormous challenges new recycling strategies from industry are required. On the other hand at the Kyoto Conference in 1997 Germany agreed to reduce the CO_2 emission by 21% to the level of 1998 by 2005 [1,2]. Cement and concrete have an important role to play in enabling Germany to fulfil its obligation to reduce the CO_2 emissions. Nevertheless the potential of traditional reduction methods for concrete, such as very expensive new technologies for cement and concrete production, the substitution of cement by fly ash and the minimization of the cement content is limited.

Because concrete is used in the majority of buildings in the structural elements, the use of waste is imminent and can lead to a substantial reduction of CO_2 emissions of concrete while reducing the mentioned costs of waste management. The paper will show the optimisation procedure for concrete mixtures using crushed concrete as aggregates and substituting cement with ash from the combustion of sewage sludge or by waste incinerator slag and low quality fly ash at the same time. The main mechanical properties of such "green" concretes are presented. Chemical analysis will give further information about the effects of such "green concrete" on the environment.

Keywords: CO_2 emissions- Kyoto, Green concrete, Recycled aggregates, Residual products, Concrete mix design, Ecological effects, Sewage sludge incineration ash, Compressive strength, Durability, Life cycle assessment

Professor Dr S Weber is teaching Building Materials, Concrete technology and Maintenance and Repairing of Concrete at the University of Applied Sciences, Stuttgart, Germany. Her interest is focused on the mechanical and physical properties of cement-based materials.

Dipl.-Ing. (FH) S Pommerenke has made his diploma theses on these topics at the University of Applied Science, Stuttgart, Germany.

Dipl.-Ing. (FH) C Schwambera has made his diploma theses on these topics at the University of Applied Science, Stuttgart, Germany.

INTRODUCTION

To reach the goals of the environmental protection agency as agreed at the Kyoto conference, it is vital to reduce emissions worldwide. The Kyoto protocol allows emission trading, which will start in 2005. Both methods of emissions trading (Joint Implementation and Clear Development Mechanism) will be helpful to establish new technologies or products to save energy and natural resources. Companies and industries will get fiscal advantages if they produce green types of concrete.

The German government decided on its own initiative to reduce emissions by up to 25% by 2005. Especially the cement industry with the high volumes produced can contribute to reducing the emissions. For that matter it is indispensable to develop new cement- based products, thus new types of concrete.

It is known that the type and the amount of cement have a major influence on the environmental properties of concrete. So it should be possible to produce concrete with less cement. One known way of minimising the cement content in concrete is to partly replace the cement by pozzolanic materials. Commonly fly ash was used as a substitution either for the cement directly or for the Portland cement clinker, when producing blended cements like CEM II A-S or CEM II B-S. However a limitation of the utilisation of fly ash is given by DIN EN 450 [3]. Additionally a higher content of ordinary fly ash has no corrective influence on the mechanical properties of concrete.

On the other hand in Germany there is a real problem regarding the residual products from sewage treatment plants. The annual yield of sewage sludge in Germany of approximately 3.75 tons was usually disposed of in landfills. This is nowadays and the residual products from sewage treatment plants have to be incinerated. Normally this happens in a mono-combustion unit and the incineration ash is disposed of in special deposits leading to increased costs. A part of the increasing costs of the water supply and purification of sewage, of the incineration of sewage as well as for the transport and for the storage of the ash could eventually be reduced by using this sewage sludge incineration ash (SSIA) in concrete.

In accordance with the way of producing types of concrete with lower CO_2 emissions it will also be necessary to consider concretes produced with recycled aggregates. According to the German Standard DIN 4226 Part 100 [4] a certain amount only of the aggregates can be replaced by recycled aggregates. To produce "green concretes" with recycled aggregates in a simple and economic way, the recycled aggregates should totally replace the natural aggregates. Basically, green types of concrete will only be produced in large quantities if they comply with the following aspects of performance:

- Method of production (cheap and similar to conventional concrete)
- Workmanship (workability, strength development, curing)
- Mechanical properties, such as compressive strength, bending strength and splitting tensile strength
- Durability (corrosion protection, frost resistance)
- Environmental aspects, i.e. CO_2 emissions, energy, recycling, etc.

However, the challenge is to develop a green concrete that fulfils all of those properties.

MATERIALS

For all concrete mixtures a Portland Cement CEM I 32.5 R according to EN 197-1 [5] was used. This cement is widely used in Germany. Sand and gravel from the Upper Rhine Valley selected in four size fractions 16/32, 8/16, 2/8 and sand 0/2 mm were used as natural aggregates (NA). The aggregates have a grain-size distribution acc. DIN EN 12620 [6] representing an A/B 32 curve. The used recycled aggregate is a recycled concrete aggregate (RCA). According to the German standard DIN 4226 Part 100 [4] it is not allowed to use grain sizes smaller than 2 mm and to use recycled aggregates exclusively. It was selected in the fractions 16/32, 8/16, 2/8 and has a similar grain size distribution to the natural aggregates. Fly ash is a standardised by-product from a coal combustion unit acc. EN 206-1 [7]. Sewage sludge incineration ash (SSIA) is a residual product from sewage treatment plants and was added as a suitable substitute for ordinary fly ash. SSIA is extracted by a mono-combustion unit. SSIA differs from normal fly ash in having a higher CaO content and lower SiO_2 and Al_2O_3. For some mixtures a superplasticizer type FM 40 was used.

CONCRETE MIX DESIGNS AND OPTIMISATION PROCEDURE

In Table 1 the proportions of the concrete mixtures are given.

Table 1. Concrete mix proportions [kg/m³]

COMPONENT	DESCRIPTION	A1	B1	B2	B3	B4	B5	B6
Cement	CEM I 32.5 R Schwenk	288	272	272	288	288	240	217
Fly ash	SFA	95	136	136				
Water		163	163	212	163	212	196	163
Equiv. w/c ratio		0.5	0.6	0.6	0.5	0.6	0.5	0.5
Superplasticizer	FM 40			6.8		8.3	5.0	
Residual Product	SSIA				95	95	172	217
	0/2 mm			374.8		380.1		
RCA	2/8 mm			322.1		326.7		
	8/16 mm			299.5		303.8		
	16/32 mm			437.8		444.1		
	0/2 mm	419.0	415.5		419.0		391.8	405.3
NA	2/8 mm	435.8	430.0		435.8		406.5	420.5
	8/16 mm	377.8	331.0		377.4		353.2	365.3
	16/32 mm	555.4	548.8		555.4		518.8	536.7

Concrete Type A1 with normal aggregates was designed as a reference concrete representing a typical German concrete containing the maximum amount of fly ash.

When replacing cement by fly ash (f), the DIN 1045 -2 [8] admits a maximum replacement of cement according to:

$$f < 0.33 * cement \quad (1)$$

Thus the maximum amount of fly ash is 95 kg/m³.

The required compressive strength of 35 N/mm² was tested at the age of 28 days on 150 mm cubes, demoulded at the age of 24 hours and stored in water at 20°C.

In a first optimisation procedure the amount of fly ash was increased, and then the normal aggregates were completely replaced by recycled aggregates. The concrete type B1 shows an increased amount (136 kg) of fly ash and slightly reduced cement content. Concrete type B2 has the same amount of cement and fly ash as concrete type B1 but uses recycled aggregates instead of natural aggregates.

In a next optimisation procedure sewage sludge incineration ash was added, replacing cement. Concrete Type B3 and B4 have the same cement content as the reference concrete, but contain sewage sludge incineration ash to a level of 95 kg/ m³. Considering that for the SSIA there is no limitation, a maximum replacement as for fly ash was performed acc. (1).

For concrete type B5 and type B6 the optimisation procedure lead to a further reduction of the cement content while increasing the amounts of SSIA. All concretes using RCA needed a higher amount of water, thus showing an increased w/c ratio. When calculating the equivalent w/c ratio an activity factor of 0.4 for fly ash (2) and 0.5 for SSIA (3) was added [9].

$$\frac{w}{c}eq = \frac{w}{c + 0,4f} \quad (2) \qquad\qquad \frac{w}{c}eq = \frac{w}{c + 0,5s} \quad (3)$$

To improve the workability of the fresh concrete mix superplasticizer was added.

TESTING PROGRAMME

Compressive strength tests were performed on cubes 150 x 150 x 150 mm according to European standard code DIN EN 12390-3 [10] were performed. The cubes where tested at concrete ages of 7 and 28 days. The results are given in Table 2.

In order to test the durability the ph-level of the fresh concrete was measured.

At the age of 84 days the carbonation rate was determined on standard cubes of 150 mm length. The cubes were cut in half and sprayed with phenolphthalein. The depth of carbonation was measured.

Frost tests were performed on cubes of 100 x 100 x 100 mm using demineralised water according to German Standard E DIN EN 12390-9 [11]. The cubes were demoulded at the age of 24 hours and stored in water. The test program covers periods of 7, 14, 28, 42 and 56 days. The authoritative result is the scaling after 56 days. In Germany there are no requirements for that scaling. The results of the frost tests are shown in Table 2.

Water permeability was measured for the concretes type A1, B1, B2 and B3 on cubes of 150 x 150 x 150 mm and for concretes type B4, B5, B6 on prisms of 200 x 200 x120 mm at the age of 28 days. It was performed acc. to EN 12390-8 [12]. The results are given in Table 2.

Table 2. Test results

TEST	Unit	A1	B1	B2	B3	B4	B5	B6
STRENGTHS								
Compressive 7 days	N/mm²	34.4	28.3	23.2	35.9	27.5	22.2	15.3
Compressive 28 days	N/mm²	45.6	41.2	30.8	50.7	38.3	34.6	23.1
DURABILITY								
pH-level	-	11.5	11.0	11.0	12.0	12.0	13.5	13.0
Frost resistance								
Scaling 7 days	M.-%	0.2	0.1	0.0	0.0	0.0	0.0	0.0
Scaling 14 days	M.-%	0.3	0.1	0.1	0.0	0.0	0.1	0.0
Scaling 28 days	M.-%	0.4	0.1	0.1	0.0	0.1	0.1	0.1
Scaling 42 days	M.-%	0.4	0.1	0.2	0.1	0.1	0.1	0.1
Scaling 56 days	M.-%	0.4	0.1	0.2	0.1	0.1	0.6	0.2
Depth of Carbonation	mm	3.0	2.0	2.0	0.5	0.5	1.0	1.5
Water permeability	mm	26	43	44	83	28	99	28

TESTING RESULTS

Due to a different water requirement of the very different aggregates leading to different w/c ratios, a direct comparison of the compressive strength is not possible. Based on the variety of the water contents, the different evolution in compressive strengths is actually inevitably. Nevertheless from Table 2 it can be seen that the compressive strength at the age of 7 days for concretes with NA is higher than that of the concretes using recycled aggregates. The increase in compressive strength at the age of 28 days is relatively the same for all concretes.

To guarantee the passivity of reinforcement in concrete a pH-level above the critical level of 9 is mandatory. The pH-levels of all green concretes tested are between 11.5 and 13.5. The highest level was observed on type B5 produced with the higher amount of SSIA.

The scaling of the "green concretes" after 28 days is constantly lower than that of the reference concrete A1. The results of the 56 day scaling differ. The concrete type B5 containing 172 kg SSIA showed the highest scaling. Best frost resistance was achieved by the concrete type B3 followed closely by concrete type B4. When adding SSIA to the mixture no influence of the type of aggregates (natural or recycled) on frost resistance can be observed.

The carbonation depth at an age of 84 days performed spraying with phenolphthalein shows a carbonation depth of a maximum of 2 mm for the "green concretes". Regarding the carbonation rate there is no significant difference between the concretes using natural aggregates and those using recycled aggregates.

The permeability of concretes with recycled aggregates indicates well-performed results. Concretes with normal aggregates showed a higher water penetration and thus an increased permeability.

DISCUSSION

The performed tests of compressive strength showed lower values for the concretes produced with RCA. There are two primary reasons for this behaviour. On the one hand there is the failing of the crushed aggregate while the cement based matrix remains more or less intact. On the other hand due to the capillary suction on the surface of the recycled aggregates, supplementary water has to be added to the mixture. This water might increase the porosity of the concrete and result in lower compressive strength. In spite of reducing the cement content it is possible to start by simultaneously tuning the compressive strengths and the accompanying content of SSIA. The optimum for the SSIA/cement- ratio was between 0.4 for recycled aggregates and 0.45 for natural aggregates. These ratios result in maximum compressive strengths, being higher than those achieved by the reference concrete. By a further increase in SSIA the strength development decreases but up to an approximate ratio of 0.6 the compressive strength is still higher than for a concrete produced with 100 % cement. Referring to this ratio it is possible to reach a reduction of cement by almost 30 % while no decrease of the compressive strength will occur. Using these ratios it is also possible to improve the workability of concretes produced with RCA and achieve the required compressive strengths.

The pH-levels of all green concretes tested were between 11.5 and 13.5. The highest levels could be determined on concretes produced with major contents of SSIA. In fact the decreasing of the cement content had no influence on the pH-level. On the contrary, depending on the amount of SSIA, it enables us to keep up or even to increase the pH-level.

The frost resistance tests showed that when using SSIA the type of used kinds of aggregates have no impact on frost resistance performance.

There are no significant differences between concretes with NA and RCA, but the content of residual products, especially SSIA, has a positive influence on the rate of carbonation. Therefore the depth of carbonation will not reach the steel during the average useful life of construction units. This has to be verified by the $\sqrt{t}$-formula as well.

The water permeability of concretes produced with RCA showed the lower value. From the point of view of concrete technology there is no suitable explanation for this behaviour, or for the increase in permeability of the concrete with fly ash and the decrease of permeability of concretes with an increasing SSIA content. Further examinations are required.

EFFECTS ON THE ENVIRONMENT AND CO_2 EMISSIONS, LIFE CYCLE ASSESSMENT

In particular environmentally friendly building materials have to be tested for their eco friendliness themselves. Especially harmful substances, e.g. chloride, sulphur and organic compounds or even heavy metals should not be elutriated by rain or ground water. The content of

- Harmful substances in conventional concrete is very low. In general concrete produced with fly ash or SSIA show an increasing content of different heavy metals [13].

The realistic process of exhausting depends first on the dilution capacity in water and then on the diffusion of such materials through the concrete. These processes are depending on the porosity of the concrete.

As the contents of heavy metals in SSIA are slightly higher than in FA, additional examinations have to be performed to verify if there is any endangering of the environment when using them in concrete. The chemical analysis performed be the producers of SSIA showed a lower content of chrome, nickel and quicksilver (mercury) then in fly ash, instead the content of cadmium and lead is twice and zinc six-time higher than those in FA. During dilution tests with water increased traces of inorganic materials were detected only in the first two days [12].

In conclusion the level of harmful substances in green concrete were initially general too low to permit higher amounts of zinc or other heavy metals to be washed out of the cement matrix. Finally the measured exhausting of zinc for example will have no negative influence on the environment [13].

In order to analyse the total effects on the environment it is necessary to examine a life cycle assessment for that product. That assessment looks at the whole life cycle that means the extraction and preparation of raw materials, general production, distribution (including transport), usage and consumption as well as the waste disposal.

Although some CO_2-emissions are initiated during the combustion of sewage sludge; these emissions will not be considered because it is an external process to produce SSIA. By reducing the amount of cement it is possible to reduce the CO_2 emissions of concrete by up to approximately 26%. This reduction can be reached by using a high amount of SSIA (172 kg) and simultaneously decreasing the cement content to 240 kg/m^3.

POSSIBLE FIELDS OF APPLICATION

The compressive strengths and durability performances will permit the use of "green" types of concrete in various areas of application. Even some special applications, such as concrete with low water penetration depth can be produced using SSIA.

Because of the fast hardening, green concretes with SSIA are perfectly suitable for prefabricated elements. For using these types of green concrete on building sites it is necessary to slow down the hardening process by using chemical additives.

CONCLUSIONS

The first results of this research program show that it was possible to produce concretes with a high environmental benefit.

The main aim in producing such new types of concrete, e.g. "green concrete" was to minimize the CO_2 emissions. The most effective way is to reduce the cement content replacing it partially with SSIA. In Germany approximately 1.13 Tons CO_2 (in CO_2-equivalents) were generated by producing 1 ton of cement. The reduction of the CO_2 emissions in concrete can be as high as 30 %. It is realistic to assume that the goals of the Kyoto conference can be achieved.

SSIA as a residual product is a very good substitute for parts of the cement and for other pozzolanic materials such as ordinary fly ash especially in large quantities. Using suitable combinations of recycled aggregates and SSIA it is possible to produce high-quality concretes, eliminating as well the normally negative effects on compressive strength when using RCA instead of NA. The produced concretes have at least the compressive strength of conventional concretes. With some mixtures higher values were obtained as well. The compressive strengths can be verified and calculated in advance by different SSIA/cement - ratios, similarly to water/cement-ratios used in concrete technology.

The durability performances such as frost resistance, permeability and carbonation rate are similar to standard reference concretes. This concretes show better eco friendliness and can be produced at reasonable costs. Due to suitable combinations of different cement contents, varying amounts of SSIA, the use of RCA or NA it is possible to produce "green concretes" for different areas of application.

REFERENCES

1. PROTOKOLL von Kyoto zum Übereinkommen der Vertragsnationen über Klimaänderungen, Artikel 3 Abschnitt1 und Anlage B: Quantifizierte Emissionsbegrenzungs- oder reduktionsverpflichtung

2. EU Institutions press releases; Memo/02/46 (04/03/2002): Das Kyotoprotokoll undder Klimawandel

3. DIN EN 450: Verwendung von Flugasche (Tabelle S2)

4. DIN 4226: Gesteinskörnung für Beton und Mörtel;
    Teil 100: Rezyklierte Gesteinskörnung

5. DIN EN 197-1: Anforderungen an mech. und physik. Eigenschaften von Normzement

6. DIN EN 12620: Gesteinskörnungen für Beton und Mörtel

7. DIN EN 206-1: Beton: Festlegung, Eigenschaften und Konformität

8. DIN 1045-2: Anwendungsregeln zu DIN EN 206-1

9. Damtoft S. J; Glavind, M; Munch-Petersen, C.: Danish Centre for Green Concrete. In: Features in Proceedings of CANMET/ACI International Conference, San Fransisco, September 2001

10. DIN EN 12390-3: Prüfung von Festbeton; Druckfestigkeit von Probekörpern

11. E DIN EN 12390-9: Prüfung von Festbeton; Frost- und Frost-Tausalzwiderstand, Abwitterung

12. DIN EN 12390-8: Prüfung von Festbeton; Wassereindringtiefe unter Druck

13. BRAMESHUBER, W., HOHBERG, I., UEBACHS, S.: Auslaugung von Frischbeton; Schriftenreihe des BTB Heft 11, Jahrgang 2002, p. 611-615

14. BRAMESHUBER, W., BROCKMANN, J., RANKERS, R.: Emissionen von umweltrelevanten organischen Bestandteilen aus Betonen mit organischen Betonzusatzstoffen. Forschungsbericht Nr. F 626, Institut für Bauforschung, Aachen, 1999.

HI-VALUE PROJECT: PRODUCTION OF HIGH ADDED VALUE MATERIALS FROM CLEAN COAL GASIFICATION BY-PRODUCTS

F Cioffi

R Ippoliti

C Maria-Pia

Contento Trade srl.

Italy

ABSTRACT. In some c ountries, the increase of natural g as prices is a sign of a possible comeback for coal in the power generation industry. Shunned for years, due to its poor environmental performance relative to natural gas, coal fired power plants are beginning to look like a feasible prospect for new capacity build, and as a result, coal could make a strong comeback within the next decade. If natural gas prices continue to rise and coal prices remain flat, advanced coal technologies become competitive. IGCC (Integrated Gasification in Combined Cycle) technology could be an economic option for new power-generation schemes and if emission standards become more stringent, IGCC could be more important relative to other coal technologies. However economic incentives for companies to move toward IGCC plants are scarce. Obtaining high additional benefits from Gasification By-Products (GBPs) may enhance these incentives. Two main activities are presented in this project to reach this aim:

    i)   Extraction of high value metal (Ni, V, Ge and Ga) from fly ash.

    ii)   Production of glass fibres, lightweight aggregates, hydraulically bound mixtures and aggregates, from slag, and Contento Trade is working on this

Keywords: Slag, Coal gasification, Binder, Aggregate

Dr Flavio Cioffi is the director of the company and is responsible of this project for Contento Trade srl. Italy.

Dr Rebecca Ippoliti is a researcher working for Contento Trade company in the environmental field.

Dr Contento Maria Pia is the Scientific Responsible of Contento Trade company. Italy.

INTRODUCTION

This work was developed within an European project funded by the EC (Coal Research Agreement ECSC n. 7220-PR145) "Production of high added value materials from clean coal gasification by-products (HIVALUE). The coordinator is Elcogas from Puertollano, Ciudad Real, Spain and Contento Trade srl is one of the project partners involved in slag valorization. Slag and fly ashes are the main by products of IGCC treatment for coal gasification and the project aims at their valorization.

The project is still running and at the moment only the slag application on ceramic materials and hydraiulic binders has been developed , while the application related to lightweight aggregates and vitreous fibers will be studied afterwords.

HYDRAULIC BINDER PRODUCTION AND TESTING

The material that will be developed in this research is a slag binder for hydraulically bound mixtures according to prEN 14227-2. These mixtures are largely used in Europe for road foundation layers and embankments and have been included in the CEN mandate for construction materials.

The mixtures are normally made of foundry slag and their grain size distribution is chosen and regulated on the basis of the slag reactivity and fragmentation resistance. As reported in the following table, Puertollano slag has a chemical composition very similar to one of the most used foundry slag, that is water quenched and thus amorphous and highly reactive:

Table 1. Comparison of slag composition

COMPOSITION	BLAST FURNACE SLAG	IGCC SLAG
SiO_2	27-41%	55%
Al_2O_3	7-20%	28.39%
CaO	30-50%	6.08%
MgO	<20%	0.94%

The lack of calcium oxide can cause a reduction in the hydraulic reactivity of the slag but the higher Si content can increase the pozzolanic reactivity of the slag. A perfect slag is highly reactive and easily crushable: the results obtained with Puertollano slag show that the reactivity is lower than the standard reactivity of water quenched foundry slag while the fragmentation resistance is higher.

Several tests have been done in order to try to increase the reactivity of the slag, using chemical activators and thermal activators. Best results have been obtained by mixing Puertollano slag with a standard water quenched foundry slag and optimized activators: in some cases the mechanical properties of these mixtures were higher than the mechanical properties of the pure foundry slag, as showed in the following Figures 1 to 3.

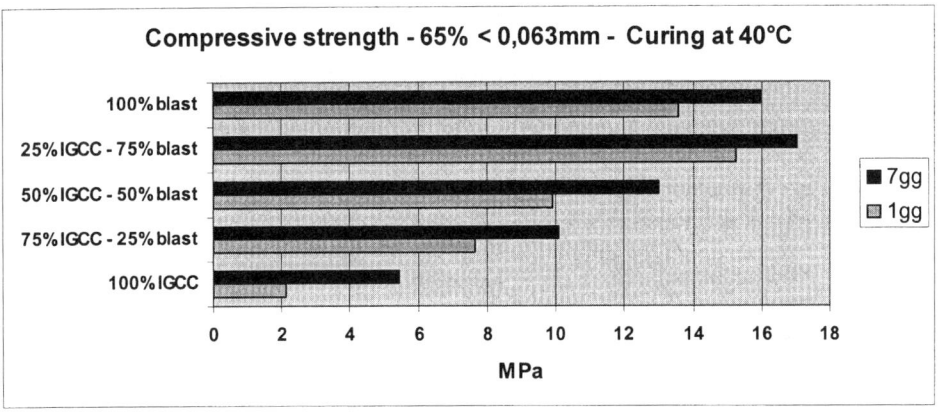

Figure 1. Compressive strength behavior of mixtures of crushed slag (65% passing the sieve of 0,063mm) cured at 40°C, 100% R.U.

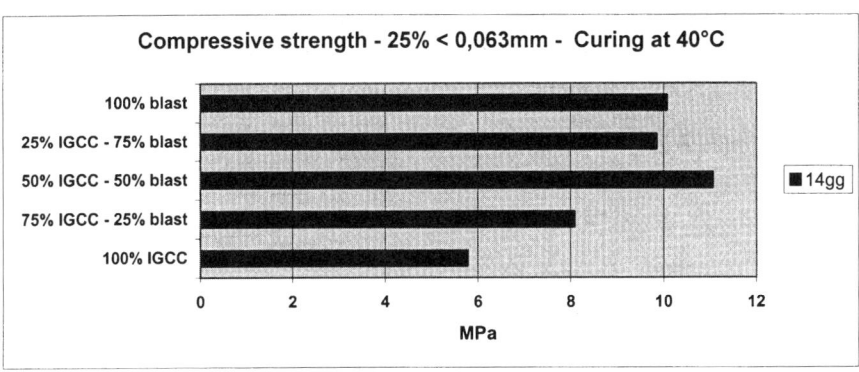

Figure 2. Compressive strength of mixtures of crushed slag (25% passing the sieve of 0,063mm) after 14 days of curing at 40°C, 100% R.U.

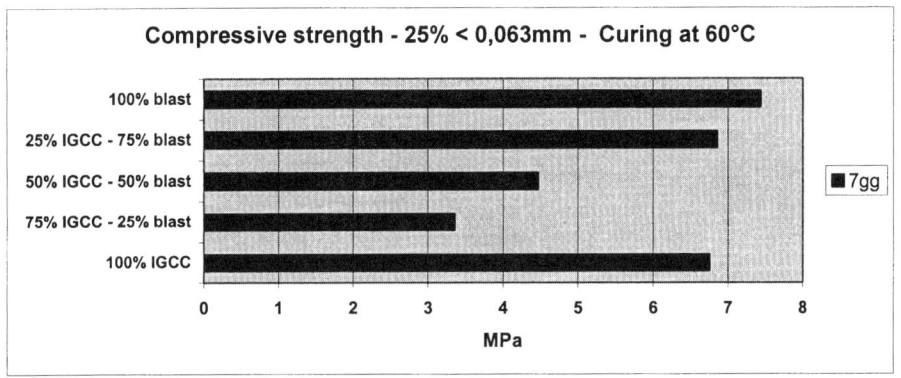

Figure 3. Compressive strength of mixtures of crushed slag (25% passing the sieve of 0,063mm) after 7 days of curing at 60°C, 100% R.U..

The obtained results show that it is possible to substitute about 50% of a good quality foundry slag with the same amount of Puertollano slag. The implications of this substitution towards the physical, chemical and mechanical behavior of the mixture will be analyzed in real scale specimens in next period of the research.

Another material that has been also developed in this research is a slag binder for concrete mixtures according to EN 206. The binder itself is not comparable with those listed in EN 196 and EN 197 and the hardening process is really different but the physical, chemical and mechanical properties of the obtained conglomerates are expected to be similar to the best available Portland cements.

The main components of the new hydraulic binder: Puertollano slag, alkaline activator, silica binder (optional). At present the new binder has a setting temperature of about 80°C and a hardening time of about 24 hours. Setting temperature and hardening time can be reduced using some silica based admixtures. The addition of silica binder to the mixtures improves mortar workability and reduces water absorption of the hardened mortar but seems not to modify its mechanical properties.

It is also possible to produce mortars avoiding completely the use of additional silica binders. In order to minimize the cost of the binder, some test have been performed on Puertollano slag simply grinded with a hammer mill equipped with a 0,500mm grid passing the sieve of 0,125mm. The obtained results are presented in the Tables 2 to 4:

Table 2 Samples preparation

SAMPLE PREPARATION	CONDITIONS
Binder/aggregate ratio	1 : 3
Water/binder ratio	1 : 2
Curing temperature	60°C
Curing time	72h
Consistency	Fluid

Table 3.Target physical properties

PHYSICAL PROPERTY (hardened mortar)	TARGET	PRESENT RESULTS
Density	>2000 kg/m3	2100 kg/m3
Water absorption	< 5%	8 %

Table 4. Target mechanical properties

MECHANICAL PROPERTY (hardened mortar)	TARGET	PRESENT RESULTS
Bending strength (MPa)	> 8,0	8,4
Compressive strength (MPa)	> 60	50

The obtained results demonstrate that it is possible to obtain a conglomerate with physical and mechanical properties very close to those of conventional concrete, using a binder made for 85% with Puertollano slag slightly grinded. The rapidity of the hardening process can be very interesting for the building industry but the durability of the obtained conglomerate must be verified.

CERAMIC MATERIALS PRODUCTION AND TESTING

The main target of this working phase was to test and optimize the chemical stability of the obtainable ceramic materials. In order to do that we developed a rapid test for chemical stability, based on the EN99 standard for boiling water absorption.

In the EN 99 a sample of the material to be tested should be immersed for two hours in boiling water in order to determine the weight increase and thus the water absorption. A vapour cooler was used in order to recovery and recycle all the water evaporated during boiling and we analyzed the chemical composition of this water, looking for the Si and Na leachate. A maximum Si leaching of 100 mg/kg and a maximum Na leaching of 300 mg/kg, referred to the dry mass of the tested sample before boiling, were fixed as optimal targets.

The in deep study carried out raised to the classification of several parameters affecting chemical stability of the obtainable ceramic: the most important are given in Table 5.

Table 5. Process parameters influencing leaching

PARAMETER	Influence on Si Leaching	Influence on Na Leaching
Ratio slag/silica binder	***	***
Liquid/solid ratio of the mixture	***	***
Catalyst dosage	**	****
Curing temperature	*****	*****
Specific surface of the specimen	***	**
Water absorption of the specimen	***	**
Density of the specimen	***	**

Therefore, it is possible to produce both pressed and extruded samples of ceramic materials characterized by a Si leaching below 80mg/kg and Na leaching below 250 mg/kg with a curing temperature below 500°C.

In this period, the possibilities to improve the physical and mechanical properties of the new ceramic material were also investigated. The study is still running but the shaping process and the curing process have been identified as key points to optimize the properties of the final product. The present situation of the study is summarized in Table 6.

Table 6. Present best results for extrudable ceramic material

PROPERTY	TARGET	PRESENT RESULTS
Binder %	< 25	24
Water absorption acc. EN 99 (%)	< 3	7
Indirect Tensile strength acc. EN 13286/42 (MPa)	> 10	6

Table 7. Present best results for pressable ceramic material

PROPERTY	TARGET	PRESENT RESULTS
Binder %	< 17	16.7
Water absorption acc. EN 99 (%)	< 0.5	8
Indirect Tensile strength acc. EN 13286/42 (MPa)	> 5	3

The chemical stability of the ceramic materials obtained at about 500°C has been verified. For this purpose some samples of ceramic materials and also some samples of untreated Puertollano slag have been tested in a sequence of leaching test in boiling demineralised water in a boiler equipped with a system for the recovery of evaporated water. After 2 hours, 48 hours and 72 leaching water was separated from the solids and analyzed while new fresh water was added. After 144 hours the test was stopped. Figures 4 and 5 show results obtained.

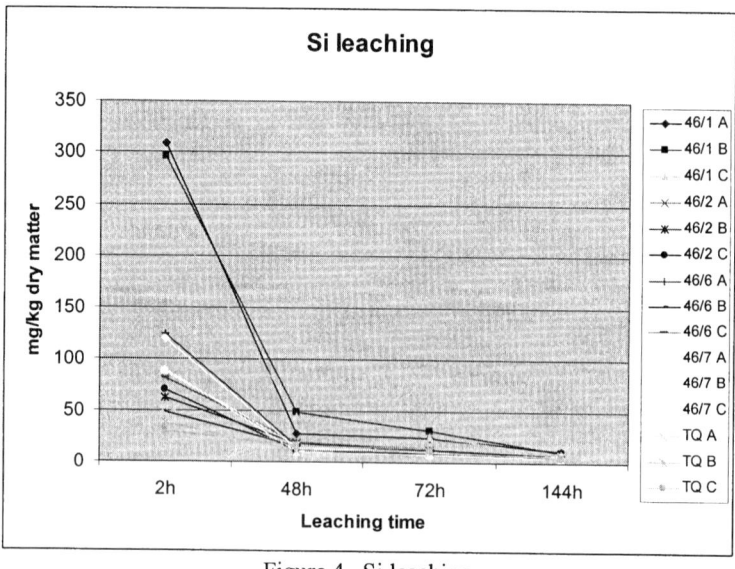

Figure 4. Si leaching

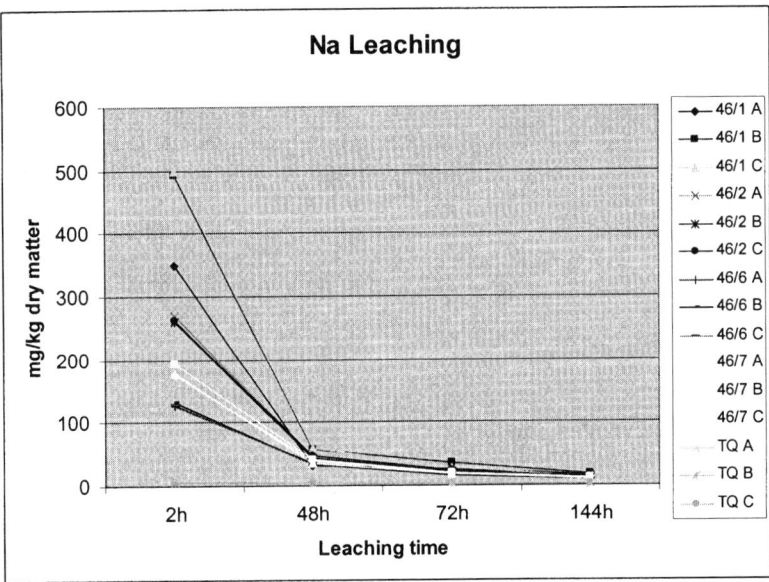

Figure 5 Na leaching

These data confirm that the light release of silica and soda from the produced ceramic material is not due to the decomposition of the structure, but to the presence of traces of components of the raw materials that did not complete the reaction and so are still soluble. Already after 72 hours of boiling the release is negligible. Moreover long term leaching tests show that the release of Si and Na from the untreated slag (samples coded wit TQ...) is very similar to the release from the obtained ceramics.

Even in the ceramic materials the introduction of a new alkaline activating agent in the mixture has been studied, and even for these materials the results were very interesting. The baking temperature of the samples has been reduced to 90°C and to reduce the costs has been used the Puertollano slag simply grinded with a hammer mill equipped with a 0,500mm grid and passing the sieve of 0,125mm. The characteristics of this new ceramic material has been adapted to the products that are objectives of this activities, that are:

- Low cost extrudable ceramic material, to be used for pipes and profiles production
- Low cost pressable ceramic material, to be used for tiles production

Even in this case some preliminary samples have been produced, using different curing conditions, summarized in Table 8. On new preliminary samples, some tests similar to those already performed have been done, to verify the water absorption, the mechanical properties and the chemical stability. For the physical properties and the mechanical ones the objectives previously decided (for the produced ceramics at 500°C) have been maintained and have been subdivided by finished products, while regarding the chemical properties the unified objective has been risen on the basis of the results of the long lasting leaching tests. Table 9 gives results obtained.

Table 8. Condition for low cost samples preparation

SAMPLES PREPARATION	CONDITIONS
Silica binder/binder ratio	< 25%
Curing temperature	< 90°C
Curing time	< 7 days

Table 9. Leaching from low cost ceramic materials

PROPERTY	TARGET	PRESENT RESULTS
Si leaching during EN 99 (mg/kg s.s.)	300	240
Na Leaching during EN 99 (mg/kg s.s.)	900	1700

Table 10: Properties of low cost extrudable ceramic material

PROPERTY	TARGET	PRESENT RESULTS
Activated binder %	< 25	24
Water absorption EN 99 (%)	< 3	12
Compressive strength (MPa)	> 80	40

Table 11 Properties of low cost of pressable ceramic material

PROPERTY	TARGET	PRESENT RESULTS
Activated binder %	< 17	15
Water absorption EN 99 (%)	< 0,5	5
Compressive strength (MPa)	> 60	35

CONCLUSIONS

The obtained results demonstrate that Puertollano slag is useful for the production of good quality ceramic materials, both using moderate heating treatments (500°C) and using steam conditioning at 90°C.

On the basis of the laboratory tests it is possible to foresee that the production costs will surely be competitive compared to the costs of traditional ceramic products. These aspects, the market preliminary evaluations and the verification of other important parameters as the durability of the new products will be studied in the future activities of the research.

NEW COMPLEX ECOTECHNOLOGY FOR OIL-DEMOLISHED WASTE

L B Svatovskaya

O Y Trunskaya

D V Gerchin

E V Rusanova

N I Yakimova

E A Dziraeva

E I Makarova

N B Krylova

Railway University- St Petersburg

Russia

ABSTRACT. One base is suggested to be for improving technology and utilization of the different nature waste, and such kind of base can be energy-transform. In the paper is being given an example of complex technology for improving purification metal details from oil (aim №1) and using oil pollutions solution for improving properties construction material (aim №2). The waste for the first aim includes Na_2SiF_6 and both technology of the phosphate materials and bricks have been used for utilization oil-content solution. In the end the good purification and the good properties of construction material have been received.

Keywords: Oil pollution, Metal details, Energy, Thermodynamic parameters, Phosphate materials, Bricks, Properties.

Professor, Dr L B Svatovskaya is the Head of Engineering Chemistry Department, Railway University, St Petersburg, Russia. She specializes in the chemistry of binders and ecology.

Dr N I Yakimova is Lecturer in Engineering Ecology Department, Railway University, St Petersburg, Russia. Area of scientific interests is economics.

Dr O Y Trunskaya is Lecturer in Engineering Ecology Department, Railway University, St Petersburg, Russia. Area of scientific interests is ecology.

Mrs E A Dziraeva is Post-graduate in Engineering Chemistry Department, Railway University, St Petersburg, Russia. Area of scientific interests is ecology.

Dr D V Gerchin is Lecturer in Engineering Chemistry Department, Railway University, St Petersburg, Russia. Area of scientific interests – dry mixes and concretes.

Mrs E I Makarova is Post-graduate in Engineering Chemistry Department, Railway University, St Petersburg, Russia. Area of scientific interests is ecology.

Mrs E V Rusanova is Post-graduate in Engineering Chemistry Department, Railway University, St Petersburg, Russia. Area of scientific interests is ecology.

Mrs N B Krylova is Post-graduate in Engineering Chemistry Department, Railway University, St Petersburg, Russia. Area of scientific interests is ecology.

INTRODUCTION

There is a great problem of the transport to use oil-demolished waste, for example after metal details washing and a lot of oil-polluted water challenge for new using. New directions are suggested in this paper [1-7] – in the technology of the phosphates material obtaining and in the raw mix for forming building ceramics. Both the first and the second objections are based on the natural and technical resource and the common idea is the next. Any substance holds energy (enthalpy, $\Delta H^{\cdot}_{298}$) and it is possible to take that energy by means of the reaction thermodynamic allowed ($-\Delta G^{\cdot}_{298}$). It is possible to lock oil-pollution in the artificial stone together with heavy metal ions reactions (numbers 3-6, table 2) or to use inside energy instead of outside energy in the build brick according reactions number 7-8. Properties of the obtained phosphates materials are not only good enough but are improving. In this paper the example of the utilization oil pollution solution is being shown.

MATERIALS

The taken consideration materials are being shown in the Table 1. The spontaneous reactions are known to go under condition of the negative parameters $\Delta G^{\cdot}_{298}$. In the Table 2 reactions 1 and 2 have been the base of the new washing mail for oil-polluted metal details; reactions 3-6 have been the base of the new technology of the utilization oil polluted solutions and reactions 7-8 have been the base of the ceramics technology of the utilization oil polluted solutions.

Table 1. Energy – nature-and-artificial substances

№	RESOURCE	ENTHALPY, $\Delta H^{\cdot}_{298}$
1.	Clay contained waste ($Al_2O_3 \cdot 4SiO_2 \cdot 2H_2O$)	- 5764,68
2.	Fe (II) – oxide	-272,2
	Fe (III) – oxide	-821,9
3.	Fe, Zn-contained dusty waste (FeO, Fe_2O_3, ZnO)	-821,9
4.	Phosphorous acid (H_3PO_4)	-1289,26
5.	Sand (SiO_2)	-901,9
6.	Oil-production waste ($C_{10}H_{40-42}$)	-456
7.	Na_2SiF_6 (waste)	-2849,72

Table 2. Changing of the thermodynamic parameters of the reactions

№	REACTIONS	$\Delta H^{\cdot}_{298}$	$\Delta G^{\cdot}_{298}$
	Washing mail for oil-polluted materials		
1.	$2Na_2CO_3 + Na_2SiF_6 + 2H_2O = 6NaF + SiO_2 \cdot 2H_2O + 2CO_2$	-21	-107,8
2.	$2H_2O_2 = O_2 + 2H_2O$	-196,2	-233,4
	Reactions for the phosphoric artificial stone obtaining to conserve heavy metal ions		
3.	$FeO + 2/3H_3PO_4 + 1/3H_2O = 1/3Fe_3(PO_4)_2 \cdot 4H_2O$	-394,1	-236,9
4.	$ZnO + 2/3H_3PO_4 + 1/3H_2O = 1/3Zn_3(PO_4)_2 \cdot 4H_2O$	-332,2	-184,4
5.	$CuO + 2/3H_3PO_4 = 1/3Cu_3(PO_4)_2 \cdot 3H_2O$	-442,4	-302,0
6.	$Al_2O_3 \cdot 2SiO_2 \cdot 2H_2O + 6H_3PO_4 = 2(SiO_2 \cdot 2H_2O) + 2Al(H_2PO_4)_3 + H_2O$	-417	-87
	Oxidation reactions oil demolished production during bricks obtaining		
7.	$C_{20}H_{42} + 30{,}5O_2 = 20CO_2 + 21H_2O$	-12441,4	$\Delta H^{\cdot}_{298}$ per
8.	$C_{21}H_{44} + 32O_2 = 21CO_2 + 22H_2O$	-13108,5	tone -44,3·10^6

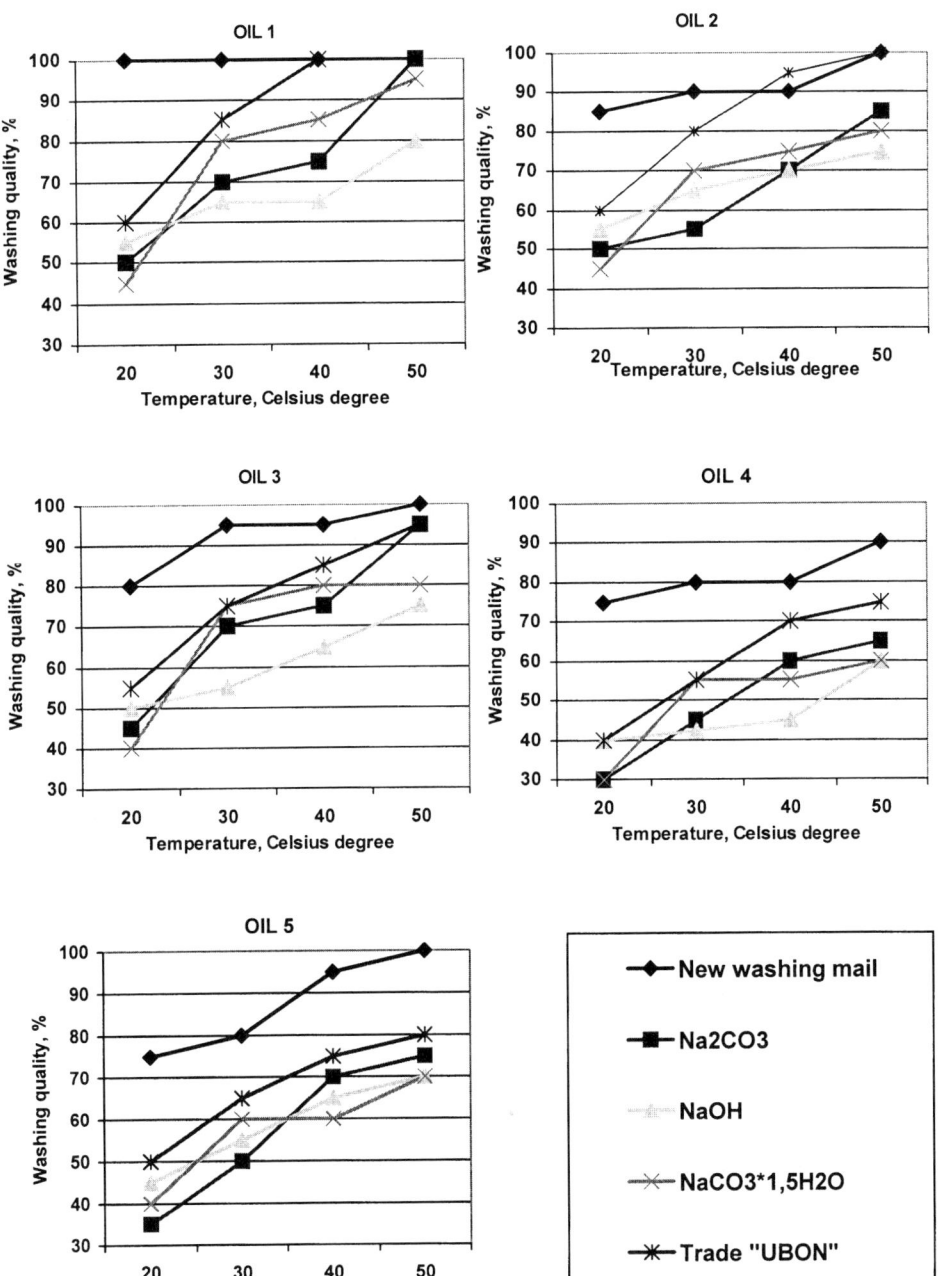

Figure 1. Relationship between washing quality and the temperature, °C.

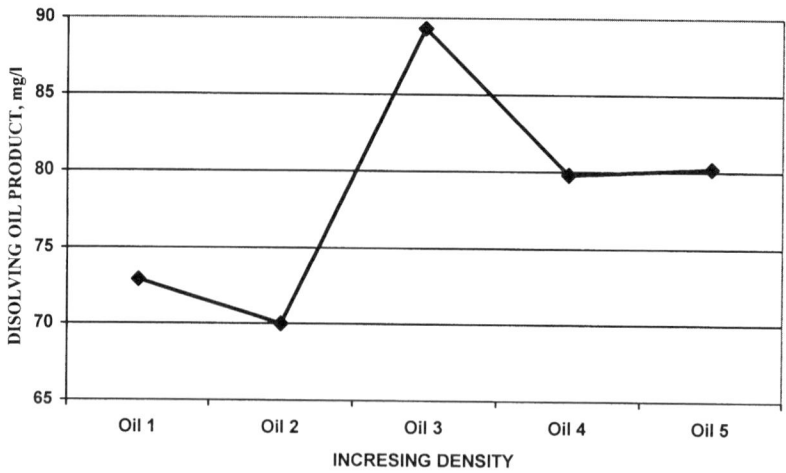

Figure 2. Dissolved oil-productions in the polluted solution

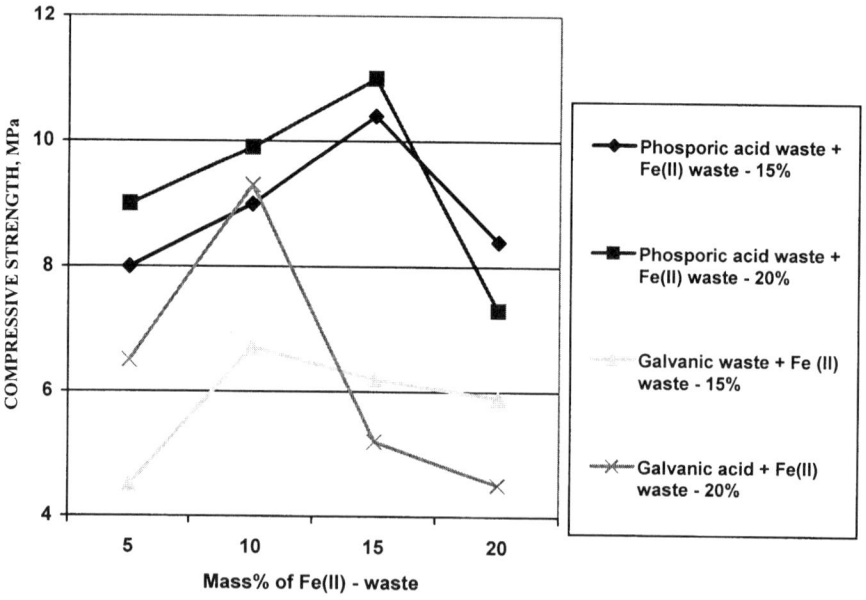

Figure 3. Relationship between compressive strength
of phosphates materials and mass % of Fe-waste

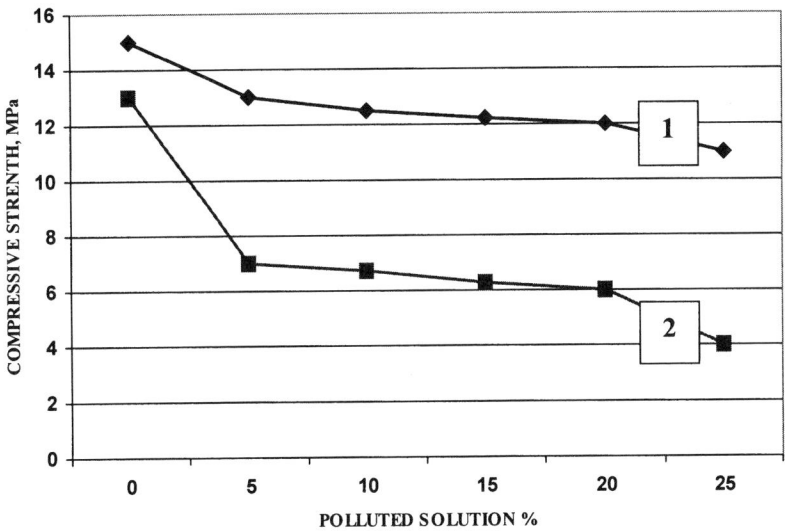

1 – Only polluted solution;   2 – The same and Fe (III)-waste;

Figure 4. Relationship between compressive strength
and amount of the polluted solution %

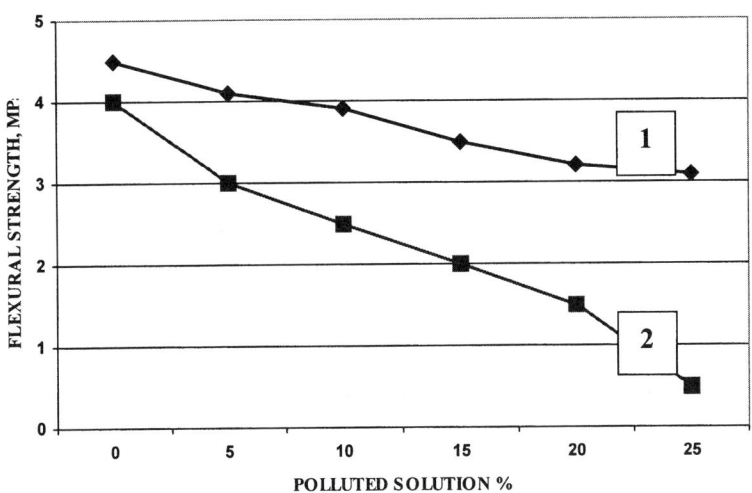

1 – Only polluted solution;   2 – The same and Fe (III)-waste;

Figure 5. Relationship between flexural strength
and amount of the polluted solution %

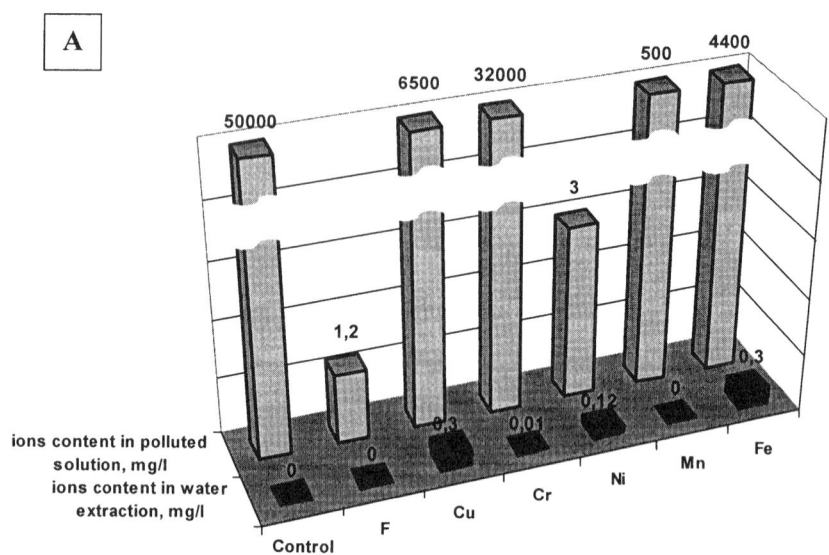

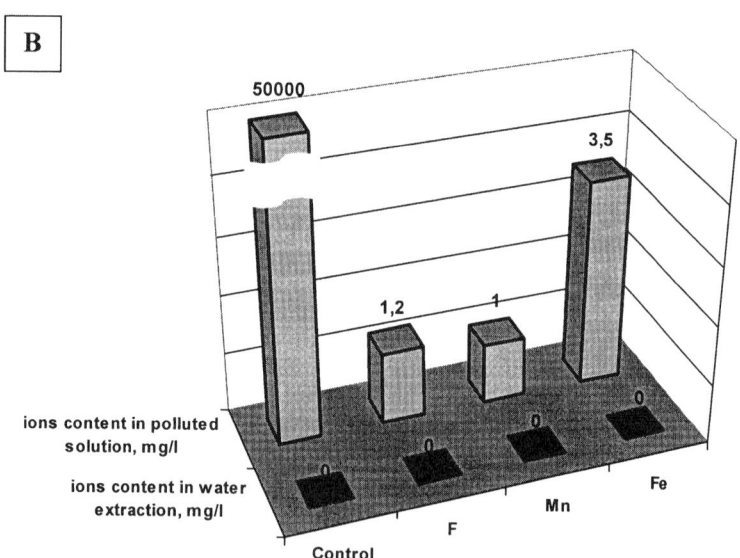

Figure 6. Water extraction from materials: A – phosphate; B - ceramic;

RESULTS AND DISCUSSION

On the Figure 1 is being shown the new quality of the metal purification by means of new washing mils based on reactions 1 and 2, table 1, the large is the number the large is density of the oil pollution. The result is being given comparing to the known washing mils. The consist of dissolved oil production after purification of metal details is being shown on the Figure 2. Before our work that solution had been taken for conservation into the proving ground. But according to our research (reactions №2-8 in the Table 2) the others technology of the utilization was suggested can be used. The main ecology idea had been that it has been possible to use a great number waste together (table №1). The first technology has been phosphate substances from Table 1 and oil polluted solution. On the Figures 3 and 4 is being shown the results of the adding oil – and other waste during phosphates material obtaining, using substances from Table 1. On Figure 5 "A" you can see, that all of using waste are locked and the base of the oil product localization is ability of the new products in phosphate system to oil waste ad- and absorbing. The second technology has been ceramics one, and it has been curved out that it is possible to add dusted of pouter oil-polluted solution in raw mix for brick obtaining together with Fe (III) waste. The material has all properties according Russian standard. On Figure 5 "B" you can see, that all dangerous substances is locked in the material.

PRACTICAL APPLICATION

On the fig 6 you can see the pure special metal details for transports and the examples of pure phosphate material for deco ring is being shown on the fig 7.

CONCLUSIONS

1. Using one waste group (for example Na_2SiF_6) the new washing solution has been obtained and the high quality of washing has been proofed.

2. Using the oil polluted solution after metal details washing and the great number the other waste the pure phosphate material has been obtained and ceramics material as well.

ACKNOWLEDGMENTS

The authors would like to acknowledge the scientists of the department for discussing problems.

REFERENCES

1. SVATOVSKAJA, L.B. and others: Thermodynamic and electron aspect properties composite materials for building and ecology, Building Published, St. Petersburg, 2003.

2. SVATOVSKAJA, L.B.: Modern natural scientific bases in the science of the constructive materials and ecology, St. Petersburg, 2000.

3. SVATOVSKAJA, L.B., MASLENNIKOVA, L.L., SOLOVYEVA V.Y.: Basic approaches for creation of the new complex technologies for biosphere purification, St. Petersburg state university of Transport, St. Petersburg, 2003.

RECYCLING OF COMPOSITE PLASTICS IN CONSTRUCTION AND DEMOLITION WASTE BY PYROLYSIS

A M Cunliffe

P T Williams

University of Leeds

United Kingdom

ABSTRACT. Composite plastics, comprising a polymer matrix moulded around fibre reinforcement, are present in construction and demolition waste in the form of shower trays, baths, utility meter boxes and interior panels, amongst others. However, composites present major problems to existing recycling technologies. This study aimed to investigate the viability of pyrolysis, or thermal degradation, of the plastic component of composites as a means of recycling this waste. Three thermoset composite wastes were pyrolysed at final temperatures between 350 and 800°C. The oils and waxes derived were then analysed by gas chromatography/mass spectrometry. Industrially useful chemicals such as styrene and phthalic anhydride could be recovered from a polyester/styrene co-polymer based composite. The styrene concentration increased with final pyrolysis temperature between 400 to 600°C, whilst the concentration of diphenyl compounds decreased. The oils derived from both a phenolic resin composite and an epoxy resin-based waste contained significant concentrations of useful phenolic compounds. However, in the case of the phenolic resin, pyrolysis also resulted in the production of a significant amount of water. The pyrolysis temperature had little observable effect on the composition of the epoxy-derived oil.

Keywords: Pyrolysis, Composite, Plastic, Oil, Recycling.

Dr. A.M. Cunliffe is a Research Fellow in the Energy and Resources Research Institute, University of Leeds, with research interests in the thermal processing of waste materials by pyrolysis and incineration. After completing his doctorate at Leeds on the recycling of scrap tyres, he worked as a Research Associate at Cardiff University. He is currently an Incorporated Engineer and an Associate Member of the Energy Institute.

Prof. P.T. Williams is Professor of Environmental Engineering in the Energy and Resources Research Institute at The University of Leeds and has a research background in both applied chemistry and process engineering. He has published more than 200 academic papers in the area of environmental engineering, including waste incineration and waste and biomass pyrolysis. He is a member of the Editorial Boards of the journals, Environmental Technology and of Fuel.

INTRODUCTION

Composite plastics comprise a resin matrix moulded around a reinforcing fibre mat or roving, sometimes with inorganic resin fillers and with other additives. In construction and demolition waste, composites are present in the form of shower trays, baths, water tanks, utility meter boxes and interior panels, amongst others. Some of the most widely used matrices are thermoset polyester/styrene copolymers, phenolic resins and epoxy resins. The reinforcement may be glass, carbon and/or polymeric fibre, while fillers include calcium carbonate, aluminium trihydrate and antimony oxide.

The complex nature of composite materials, coupled with the thermoset resins often used, has presented major problems to existing recycling technologies. There has been interest in using the emerging technology of pyrolysis to recycle composite wastes [1-3]. Pyrolysis thermally degrades the plastic component to produce an oil and gas product. The oil may be used as a liquid fuel or returned for processing at the refinery [2], whilst the gas could be burnt to provide the energy requirements for the pyrolysis system. The fillers and reinforcing fibre can also potentially be reclaimed or reused [1, 4-6]. In this work, oils derived from the pyrolysis of three common composites are characterised using gas chromatography/mass spectrometry (GC/ms).

MATERIALS AND METHODS

Pyrolysis

Commercially available composites were selected to represent unsaturated thermoset polyester/styrene copolymer, phenolic resin and epoxy resin based materials. Details of the samples are given in Table 1. The composite samples were cut into approximately square pieces of sides 10-15mm and pyrolysed in a bench-scale reactor that has been described previously [3]. Samples were heated at $10°C$ min^{-1} to final pyrolysis temperatures between 350 and 800°C, with a holding time at the final temperature of 60min. The reactor was continuously purged with nitrogen (0.2L min^{-1}). The condensable products recovered were stored in a laboratory refrigerator to prevent degradation or sample loss.

Table 1. Composite samples

SAMPLE CODE	SAMPLE DESCRIPTION
PE	Thermoset polyester/styrene resin with calcium carbonate and aluminium trihydrate fillers (36wt %), glass fibre reinforcement (12wt %)
PH	Phenolic resin with magnesium oxide and calcium carbonate fillers (41wt %), glass fibre reinforcement (31wt %)
EP	Epoxy resin, inorganic fillers (30wt %) glass and carbon fibre reinforcement (45wt %)

Condensable Product Analysis

Samples (0.02g) were drawn from the non-aqueous (oil) fraction of the condensable products and made up in 10cm^3 dichloromethane. The sample was briefly sonicated to assist dissolution into the solvent. In the case of PE, the products were also sonicated prior to sampling to put the wax observed in this sample into emulsion. A Hewlett Packard 5280 gas chromatograph was used for analysis, coupled to a HP 5271 ion trap detector. During the course of analysis, two columns were employed i) Restek RTX-5ms, 30m x 0.25mm i.d. ii) Zebron ZB-Wax mixed polarity, 30m x 0.25mm i.d. The latter column was found to give good separation of polar species, such as phenol derivatives. The column was held at an initial temperature of 40°C for 5min, then heated at 5°C min^{-1} to 245°C, with 15min final hold time. External standards were employed to verify the identity of the major peaks by retention time and to provide a semi-quantitative analysis.

The PAC group contains the largest number of known or suspected carcinogens of all chemical groups [7]. In addition, the presence of PAC in fuels can favour the generation of particulates upon combustion [8]. For these reasons, it is important to determine the PAC content of the pyrolysis oils. In order to produce a cleaner sample to analyse for the presence of PAC, the condensable products were separated by liquid chromatography into chemical class fractions. Full details of the separation technique employed have been published elsewhere [3]. The benzene fractionation residue was made up to 10cm^3 in DCM and analysed by GC/ms using the Restek RTX-5 column. The column oven was held at an initial temperature of 100°C for 2min, then heated at 4°C min^{-1} to 320°C, with 3min final hold time.

RESULTS AND DISCUSSION

Polyester Composite PE

The condensable products derived from PE comprised of a yellow/brown, low viscosity oil, which accounted for approximately 80vol % of the total condensable yield, a pale yellow, waxy solid (15vol %) and water. The moisture was derived from the decomposition of aluminium trihydrate filler and the oxygen present in the ester bonds. Table 1 gives the concentrations (as wt % of the non-aqueous fraction only) of the main species positively identified in these products by GC/ms analysis, in relation to the final pyrolysis temperature. It may be seen that the two most important products were styrene and phthalic anhydride. Styrene is directly derived from decomposition of the polystyrene element of the resin, whilst phthalic anhydride is the acid anhydride derivative of ortho-phthalic acid, which was used in the manufacture of the polyester element of the co-polymer resin under study. Previous studies have similarly found that phthalic anhydride is a characteristic decomposition product of ortho-phthalic polyester resins [9-11]. It has also been reported that phthalic anhydride is amongst the first decomposition products evolved from such resins. In this study, phthalic anhydride was the only non-aqueous component identified in the condensable products derived from an experimental run at 350°C (results not shown). The other major hydrocarbon species identified, including α-methylstyrene and 1,3-diphenylpropane, would also be expected from the thermal degradation of the polystyrene element of the resin. With regard to the influence of final pyrolysis temperature, increases were seen in the styrene and 4,7-methanoindene concentrations between 400 and 600°C. However, the concentration of 1,3-diphenylpropane in the oil showed an opposite trend with increasing temperature.

Several products comprising two aromatic rings linked by an aliphatic group were found in addition to those outlined previously, including bibenzyl, stilbene, 1,2-diphenylcyclopropane and three diphenylbutene isomers. These diphenyl products are styrene dimer-variants directly related to the decomposition of the polystyrene element of the co-polymer [9]. The concentration of the diphenyls showed a general decrease as the final pyrolysis temperature increased from 400 to 600°C, then stabilised as the temperature increased further. This trend, mirroring the trend seen earlier for styrene, indicated that the dimer-variant species were increasingly being cracked to form monomer-variant single ring aromatic species as the temperature increased. The only PAC species identified in the benzene fraction were 2-phenyl naphthalene ($660-1190\mu g$ g^{-1} of the pre-fractionation oil) and a methylated derivative ($920-1390\mu g$ g^{-1}). The presence of dibutyl phthalate in the benzene fraction ($0-520\mu g$ g^{-1}) indicated that the glycol used in the polyester resin formulation under study was butanediol [9]. Table 2 shows that butyl ester of benzoic acid was also found during the whole oil analysis, providing further evidence for this conclusion.

Table 2. Main species identified in PE-derived oils

IDENTIFIED SPECIES	FINAL PYROLYSIS TEMPERATURE (°C)			
(wt %)	400	500	600	800
1,2-Diphenylcyclopropane	0.22	0.22	0.18	0.17
1,3-Diphenylpropane	2.91	2.29	1.98	2.02
4,7-Methanoindene	0.00	0.91	2.94	2.31
alpha-Methylstyrene	3.17	2.21	3.00	1.85
Benzoic acid, butyl ester	0.75	0.59	0.65	0.37
Phthalic anhydride	7.86	12.84	4.61	5.66
Styrene	6.81	10.87	15.02	14.68

Styrene, α-methylstyrene and phthalic anhydride are useful industrial chemicals. Styrene is employed in the manufacture of a range of polymers including polystyrene, styrene-butadiene rubber, ABS and thermoset polyesters. Modified polymers and alkyd resins are manufactured using α-methylstyrene [12]. In addition to use as a modifying agent in polyester resin manufacture, phthalic anhydride can also be used as a cross-linking agent for epoxy resins, as a catalyst for the polymerisation of olefins and as a dehydrating agent for alcohols [12]. It may be seen that pyrolysis has the potential to enable a 'closed-loop' recycling scheme for styrene and phthalic anhydride back into polyester/styrene thermoset resins.

Phenolic Resin Composite PH

With the exception of the 400°C sample, which appeared to consist almost exclusively of water, thermal decomposition of PH generated yellow/brown oil (50vol %), water (45vol %) and wax. Table 3 gives the concentrations of the main species identified by GC/ms analysis of the oil fraction using the mixed polarity column. The products were almost exclusively phenol and methylated phenol derivatives, most notably 2-methyl phenol and 2,4-dimethyl phenol. Phenol has a variety of industrial uses, including the production of phenol-formaldehyde resins, bisphenol A and nylon intermediates [12]. Methyl phenols are also important chemicals, being employed in the manufacture of phenolic resins, insecticides and flavouring and fragrance compounds.

With regard to the influence of final pyrolysis temperature on the composition of the condensable products, the only organic species identified in the condensable product generated at 400°C was glycidol, with the remainder being water. This indicated that hydroxyl groups were being stripped away from the aromatic structure of the polymer, forming moisture, but that the structure itself resisted decomposition at 400°C. Table 3 shows an increase in the concentration of all the phenolic derivatives between 400 and 600°C, showing that the remaining polymer matrix was being increasingly broken down into monomer-derivatives. The concentrations of phenol, 2-methylphenol and 2,4-dimethyl phenol decreased between 600 and 800°C, whilst the other phenols identified remained stable. It may be that the apparent decrease was due to further cracking of the phenols to form lighter hydrocarbon gases and liquids. However, it is more likely that the decreases seen were due to unrepresentative sampling, given difficulties encountered when trying to draw off the organic components from the highly aqueous.

Table 3. Main species identified in PH-derived oils using the mixed polarity column

IDENTIFIED SPECIES	FINAL PYROLYSIS TEMPERATURE (°C)			
(wt %)	400	500	600	800
2,4,6-Trimethyl phenol	0.00	3.05	6.07	6.32
2,4-Dimethyl phenol	0.00	22.06	34.88	17.69
2,6-, 2,5- or 2,3-Dimethyl phenol	0.00	4.05	6.43	7.44
2-Heptadecanone	0.00	0.78	1.45	0.63
2-Methyl phenol	0.00	12.29	19.69	11.50
Glycidol	0.43	1.27	1.41	1.23
Phenol	0.00	11.98	21.43	11.99

Four PAC were identified in the benzene fraction of the oil generated at 600°C, phenanthrene (790µg g^{-1} whole oil), 2-methylanthracene (640µg g^{-1}), 1-methylphenanthrene (270µg g^{-1}) and 1,4-dimethyl phenanthrene (130µg g^{-1}). With regard to the health and safety implications of these compounds, phenanthrene is classified as having weakly neoplastic effects, whilst 1-methylphenanthrene shows mutagenic properties [12]. However, 2-methylanthracene is classified as non-carcinogenic. Other hydrocarbons identified in this fraction were 5-methyl-1-phenyloctane, decylbenzene and undecylbenzene, at concentrations of 110, 230 and 120µg g^{-1} respectively.

Epoxy Resin Composite EP

The epoxy resin based composite EP produced viscous, dark brown/orange oil with some water (10vol %). The main products identified in the oil fraction by GC/ms analysis using the mixed polarity column are given in Table 4. As was the case with PH, the oils derived from EP contained high levels of industrially-useful phenolic compounds. In addition, benzyl alcohol finds application in the perfumery and food flavourings industries and is also employed as an anti-microbial preservative in cosmetics and pharmaceuticals [12]. Although the concentration of phenol and methylated derivatives were lower in the oil derived from EP than in those from the pyrolysis of PH, it is important to note that the aqueous fraction of EP was considerably smaller. In addition, the proportion of the mass balance represented by the condensable products from EP was around three times greater than was the case for PH [3]. It may be seen from Table 4 that the final pyrolysis temperature had no observable influence on the composition of the oils derived from EP, suggesting pyrolysis was complete at a considerably lower temperature than was the case for PH. Hence the pyrolysis of epoxy-based composites would seem to offer a significantly more attractive route for the recovery of phenolic chemicals than the pyrolysis of phenolic resins.

In contrast to the phenolic oils, GC/ms analysis of the benzene fraction of the EP-derived oils using the RTX-5 column identified no PACs. Indeed, the benzene fraction was exclusively comprised of oxygenated aromatic compounds, including methylene bisphenol isomers (200-2100ppm). Such compounds were directly derived from the bisphenol compounds commonly used in the manufacture of epoxy resins, such as Bisphenol A. It is highly probable that the epoxy-derived oils contained higher quantities of bisphenolic compounds [13, 14], but that these did not elute not under the analytical conditions employed with the mixed polarity column.

Table 4. Species identified in EP-derived oils using the mixed polarity column

IDENTIFIED SPECIES	FINAL PYROLYSIS TEMPERATURE (°C)			
(wt %)	400	500	600	800
2-Ethyl phenol	1.12	0.97	1.09	0.89
2-Ethyl-6-methyl phenol	4.80	4.28	4.61	4.30
2-Methyl phenol	2.14	2.29	2.41	2.53
4-Ethyl phenol	0.74	0.64	0.80	0.77
4-Methyl phenol	1.62	2.05	2.09	2.06
Benzyl alcohol	3.86	3.80	3.47	3.73
Phenol	13.07	11.19	13.10	12.18
Xanthene	0.63	0.26	0.59	0.62

CONCLUSIONS

1. The pyrolysis of composites based on a thermoset polyester/styrene matrix yielded oil containing significant proportions of industrially useful chemicals. Most notably, styrene and phthalic anhydride could be used in the manufacture of fresh polyester/styrene copolymer resins.
2. Epoxy resin based composite yielded significant amounts of phenol and methylated derivatives, which find many industrial applications.
3. Although similarly useful phenolic chemicals could be derived, the phenolic resin based composite required a high pyrolysis temperature and yielded a large amount of

water, making it the least attractive of the three samples for processing by pyrolysis.
4. The PAC content of all the oils was low, or in the case of EP, not detectable. However, PH generated phenanthrene and anthracene derivatives that have been reported to have detrimental health effects.

ACKNOWLEDGEMENTS

The authors wish to acknowledge the contribution of Nicola Jones and Peter Thomson to this study. The w ork w as f unded b y U K E ngineering a nd P hysical S ciences R esearch C ouncil grant GR/NO 8551, as part of Department for Trade and Industry LINK WMR3 Programme WML 538, Development of a Novel Recycling Process for the Recovery of Plastic Composite Materials. The authors would like to thank the following partner organisations for their support and assistance: Pera Technology; British Plastics Federation; Society of Motor Manufacturers and Traders; Cray Valley; Filon Products; Hepworth Composites; Lotus Cars; Kemlite; NEL; Oakdene Hollins; OSS Group; Reichhold; Shanks Waste Solutions; Swift Caravans.

REFERENCES

1. de MARCO, I., LEGRETTA, J.A., LARESGOITI, M.F., TORRES, A., CAMBRA J.F., CHOMON, M.J., CABALLERO, B., and GONDRA, K.: Recycling of the products obtained in the pyrolysis of fibre-glass polyester SMC, Journal of Chemical Technology and Biotechnology, 69, 2, 1997, pp187-192.

2. TORRES, A., de MARCO, I., CABALLERO, B., LARESGOITI, M.F., LEGRETTA, J.A., CABRERO, M.A., GONZALEZ, A., CHOMON, M.J., and GONDRA, K.: Recycling by pyrolysis of thermoset composites: characteristics of the liquid and gaseous fuels obtained, Fuel, 79, 2000, pp897-902.

3. CUNLIFFE, A.M., JONES, N., and WILLIAMS, P.T.: The pyrolysis of plastic composite waste, Environmental Technology, 24, 2003, pp653-663.

4. WATT, D.F. and XU, J.Z. On the use of pyrolysed SMC as a concrete additive, in Advanced Composite Materials, Proceedings of the 7th Annual Advanced Composites Conference, Detroit, Michigan, USA, September 1991.

5. PICKERING, S.J., KELLY, R.M., KENNERLEY, J.R., RUDD, C.D., and FENWICK, N.J.: A fluidised bed process for the recovery of glass fibres from scrap thermoset composites, Composite Science and Technology, 60, 2000, pp509-523.

6. CUNLIFFE, A.M. and WILLIAMS, P.T.: Characterisation of products from the recycling of glass fibre reinforced polyester waste by pyrolysis, Fuel, 82, 2003, pp2223-2230.

7. LEE, M.L., NOVOTNY, M., and BARTLE, K.D.: Analytical chemistry of polycyclic aromatic compounds. Academic Press, New York, 1981.

8.   HENDERSON, T.R., SUN, J. D., LI, A.P., HANSON, R.L., BECHTOLD, W.E., HARVEY, T.M., SHABANOWITZ, J., HUNT, D.F.: Gc Ms and Ms Ms Studies of diesel exhaust mutagenicity and emissions from chemically defined fuels, Environmental Science & Technology, 18, 6, 1984, pp428-434.

9.   HILTZ, J.A.: Pyrolysis-gas chromatography mass-spectrometry identification of styrene cross-linked polyester and vinyl ester resins, Journal of Analytical and Applied Pyrolysis, 22, 1-2, 1991, pp113-128.

10.   RAVEY, M.: Pyrolysis of unsaturated polyester resin: quantitative aspects, Journal of Polymer Science: Polymer Chemistry Edition, 21, 1, 1983, pp1-15.

11.   EVANS, S.J., HAINES, P.J., and SKINNER, G.A.: The effects of structure on the thermal degradation of polyester resins, Thermochimica Acta, 278, 1996, pp77-89.

12.   BUCKINGHAM, J.: The Combined Chemical Dictionary on CD-ROM. Chapman and Hall/CRC Press, Boca Raton, 2000, pp35000.

13.   BLAZSO, M., CZEGENY, Z., and CSOMA, C.: Pyrolysis and debromination of flame retarded polymers of electronic scrap studied by analytical pyrolysis, Journal of Analytical and Applied Pyrolysis, 64, 2, 2002, pp249-261.

14.   DENQ, B.L., CHIU, W.Y., LIN, K.F., and FU, M.R.S.: Thermal degradation behaviour of epoxy resin blended with propyl ester phosphazene, Journal of Applied Polymer Science, 81, 5, 2001, pp1161-1174.

MUNICIPAL SOLID WASTE MANAGEMENT: AN OPPORTUNITY ANALYSIS AND COMPREHENSIVE BENCHMARKING OF SOLID WASTE MANAGEMENT PRACTICES

J Antony Pulikkottil

R Somasundaram

National Institute of Industrial Engineering- Mumbai

India

ABSTRACT. Management of Solid Waste resulting out of urbanization has become a serious concern for government, pollution control bodies and also public in most of the developing countries including India. The alarming amount of solid wastes produced in Bombay (one of the largest and densely populated metropolitans in the world), coupled with the insufficient and inefficient solid waste management currently implemented here, and highlights the seriousness of the problem. For conducting a Comparative study and benchmarking activity on waste management system, the solid waste and waste sources were grouped into: The Municipal sector, The Commercial and Industrial (C&I) sector and The Construction and Demolition (C&D). For the benchmarking project performance measures are calculated and compared for primary activities like Kerbside collection, Recycling centers, Transfer stations and Landfill sites. The parameters identified to measure the performance of the primary activities are Weights handled by source, by waste stream and by disposal method and operating costs in total and by cost type. The action areas and methods to implement waste reduction identified include Waste and toxicity reduction, Reuse, Landfill & Recycling Model, Composting of yard waste and food waste, Resource recovery through mixed municipal solid waste composting or incineration and Sanitary landfills.

Keywords: Solid waste management, Benchmarking, Critical analysis, Waste management facilities, Waste management network

Joseph Antony Pulikkottil is doing his post graduation in Industrial Management at National Institute of Industrial Engineering, Mumbai. He holds the Bachelor of Technology in Mechanical Engineering discipline from National Institute of Technology, Calicut.

Rohith Somasundaram is doing his post graduation in Industrial Management at National Institute of Industrial Engineering, Mumbai. He holds the Bachelor of Technology in Mechanical Engineering discipline from National Institute of Technology, Calicut

INTRODUCTION

Waste is an unavoidable byproduct of human activities. Economic development, urbanization and improving living standards in cities, have led to increase in the quantity and complexity of generated waste. Inefficient waste management has aggravated this problem which has resulted in the threats on natural resources. This clearly emphasizes the importance of effective and efficient waste management techniques. Considering the rapidly urbanizing world, most of the cities across the globe are suffering from the problem of managing the environmental outcomes of human activities. As we describe the economy of the world through the development of big cities, the solid waste management in those areas has become crucial for human existence in such big cities. Most of the developing cities in Asia particularly face the dual challenge of exploding populations and scarce resources. They will absorb about 2 billion new residents by the year 2030. Given rapid population growth and a concurrent decrease in aid funding from central governments and international donors, city managers have the greatest challenge of finding new and creative ways to provide safe, healthy cities, in which their populations will live and prosper. By 2025, urban waste will more than quadruple. Organic matter forms the bulk of the municipal waste; 36 percent of the waste flow in Organization for Economic Co-operation and Development (OECD) member states is food or garden waste. Organic matter in developing countries accounts for a staggering 50 to 75 percent of the total waste stream. Lack of proper treatment for these waste streams is one of the most serious health issues confronting the world today.

Management of Municipal Solid Waste (MSW) resulting out of rapid urbanization has become a serious concern for these government departments, pollution control agencies, regulatory bodies and also public in most of the developing countries of Asia including India. Mega-cities are perhaps significant point in Asia. There are in the world 22 cities, which have more than five million people. Fifteen of them are in Asia, and eleven of them are in eastern Asia. In a sense we can consider the identity of Asia itself as "very-large cities". Most of these have Dual economies which compose the urban and the land use system. In this paper we have dealt primarily with Mumbai, the most commercially advanced cities in India.

Solid Waste Management in the urban environment

Proper management of solid waste is critical to the health and well-being of urban residents. In most developing cities, several tons of garbage is left uncollected on the streets each day, acting as a feeding ground for pests that spread disease, clogging drains and creating a myriad of related health and infrastructural problems. The urban poor - often residing in informal settlements with little or no access to solid waste collection and often in areas that are contiguous with open dumps - are particularly vulnerable. While urban residents in developing countries produce less solid waste per-capita than in high-income countries, the capacity of their cities to collect, process, dispose of, or re-use solid waste in a cost-efficient, safe manner is far more limited. Municipal SWM efforts often focus on expensive 'end-of-pipe' measures, those involving the collection and disposal of solid waste, yet many of the 'best practices' for SWM improvement are far more accessible and cost-effective opportunities involving waste reduction programs and recycling strategies.

Mumbai, the commercial capital of India is one of the largest and densely populated metropolitan cities in the world. It spreads over an area of 437.71 sq km. with population of about 18 millions. Solid waste generation of the city is the highest by any Indian city with more than 7500 tones a day. Per capita waste generation of Mumbai is among the highest in

Indian cities with about 450 gm per person per day. Though it is not comparable with developed countries whose per capita waste generation goes over 2.5 kg, it is considerably high when compared with many cities in developing countries. The low value of the percapita waste generation is just because of high population in the city. Even though the percapita generation is less, the ill-effects of the high pollution, is felt much more than the developed nations as the population density is very high .The waste generated are mainly in the form of garbage, debris, silt removed from drains , cow dung and also waste matter removed from common house gullies or inaccessible narrow lanes between old buildings. The present study is an attempt to take into account the entire solid waste management in the selected cities and also tries to benchmark the solid waste management system for the city of Mumbai.

CRITICAL ANALYSIS OF THE EXISTING SOLID WASTE MANAGEMENT SYSTEM IN MUMBAI

The municipal authority of Mumbai through its conservancy and transportation functions collects transports and disposes the solid waste generated within the city limits. Conventional MSW management fails to deliver quality services for various reasons, like lack of co-ordination between the concerned parties and the poor efficiency. From the personal interviews conducted with the municipal authorities, it was found that lack of coordination between various functions of the authorities is the major cause for the inefficient waste management system blockage and thereby resulting in the failure. Inappropriate allocation of funds and scarcity of manpower and other resources have resulted in the failure of the system.

From the data given by the Municipal Corporation of Mumbai the waste generated in the city can be broadly classified into three segments. The city produces around 1070 tones of construction waste and silt, around 1700 tones of mix waste (Biodegradable and Recyclable) and around 4730 tones of Biodegradable waste. Further segregation of the waste gave an approximate composition of the waste as:

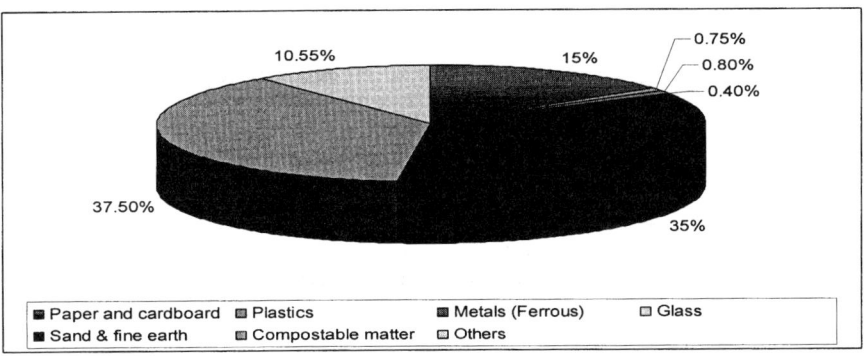

Figure 1. Categorical distribution of waste generated

Looking at the alarming amount of solid wastes produced coupled with the insufficient and inefficient solid waste management currently implemented in the city, there exists an immense potential in this sector. Manual sweeping and collection and inefficient workforce results in low productivity.

The whole study was done in 4 phases.

Phase 1: Identification and classification of waste and waste sources

Solid waste generated in Mumbai can be categorized by its source into three broad sectors, each with distinctive waste types, waste collection systems and issues:

1. The Municipal sector, comprising regular household curbside collections and other wastes collected by local government services such as street sweeping and public place bins. It also captures waste deposited at transfer stations, municipal hard rubbish collections and direct deposits to landfill from home renovations and other activities which accounted for around 35%.
2. The Commercial and Industrial (C&I) sector, comprising business sector waste and also includes wastes arising from state and federal government entities, schools and tertiary institutions which accounted for around 25%.
3. The Construction and Demolition (C&D) sector, comprising waste from residential, civil and commercial construction and demolition activities. (Prime importance) which accounted for about 40%.

Phase 2: Identification of action areas and methods to implement them.

The economic approach has enabled the authorities to think in a new line and strengthen the waste management systems. Major action areas in this phase are

1. Reduction in the amount of solid wastes. Increase awareness among the people by conducting small workshops on the reduction of waste produced and recycling of the waste in the households itself. Reuse and recycle of solid wastes from businesses, homeowners and multi-family buildings could find applications in the energy generation for those groups and thus making them self reliable.
2. Detoxification of the waste stream and initiation of programs that properly manage problem materials and household hazardous wastes.
3. Effective utilization of resource recovery facilities for materials not recycled or reused.

The design, size and the spacing of community bins are inadequate and fail to cater to the associated needs. In the present condition with the available statistics it was found that only one bin is provided for 2500 people on average, and people are left with no option but to dumb their waste in their back yard or in drainage. Unlike the developed cities the waste transportation in Mumbai is based on randomly allocated routes. There is no optimally allocated route for waste collection and transportation and hence results in the inefficient and large amount of waste left in the streets itself. To achieve improved output, a continuous dialogue needs to be maintained between the conservancy and the transport functions, especially at the midlevel operators.

Methods of waste reduction for the above action areas include:

1. Waste and toxicity reduction in the house holds: Households can reduce toxicity by reducing the use of hazardous products. Suggestions for reducing household hazardous materials and selecting less toxic products include (1) instead of pesticides, use beneficial insects, companion planting in the garden, and hand-picking to ward off certain pests, (2) use water-based paint instead of oil-based paint to eliminate the need for paint cleaners; (3) assure proper dilution of condensed cleaners, (4) substitute baking soda and vinegar for many household cleaners or use multi-purpose cleaners rather than a specific cleaner for a specific job; and (5) use the products bought before it expires.

2.   Landfill & Recycling Model: The practice of land filling is becoming less favorable due to the generation of foul odors as communities expand and reside in close proximity to food processing plants. Recycling is currently done in the informal sector by ragpickers who scavenge through neighborhood waste bins, accessible transfer stations, and landfills. Source segregation is not a new concept in India, where households in previous generations separated paper, rags, and metals and sold them to kabadiwalas, men who would buy these items from households. Households also frequently composted vegetable matter in their backyards. Increased urbanization, lack of space, and a cultural shift toward disposable plastics have all decreased household waste segregation. From experimental studies it has been found that collection of the solid waste at the source reduced the effort and hence the manpower needed for the processing of the solid waste by around 70%. This can be achieved by making the people in urban areas aware through educational campaigns and hiring workers to conduct door-to-door collection of segregated recyclables.

3.   Composting is an effective recycling strategy for turning otherwise unrecyclable organic wastes into a valuable resource. It is estimated that approximately 40 to 70 percent of the total municipal solid waste stream could be composted. Some examples of organic material that can be composted include fruit and vegetable scraps, tea bags and coffee grounds, leaves and yard wastes, agricultural crop residues, paper products, sewage sludge, and wood. Composting of yard waste and food waste like Vermi composting and Aerobic composting: Aerobic composting, where the organic waste is decomposed by microbes in the waste, and vermiculture, where worms are used to digest the organic matter in the waste, result in humus, which can be used for soil enrichment. Since vegetative matter constitutes the majority of waste, composting presents dual benefits of resource recovery and reduced need for landfill space. Educational campaigns if conducted would significantly encourage composting at both household and municipal levels. Mumbai currently uses a private sector company to compost 300-350 tonnes of waste per day, and an additional 200 tonnes per day of market waste is composted using vermiculture at different sites across the city. These are really promising as this would greatly reduce the amount of wastes moving to the dump sites. If this quantity could be increased the amount of wastes dumped into the dumping could be reduced significantly.

4.   Resource/Energy recovery through mixed municipal solid waste composting or incineration;

5.   Sanitary landfills: Existing landfills are also required to be modernized if their use was to be continued. The new requirements are similar to those in the developed countries, such as the use of liners, methane recovery systems, daily compaction and cover, fencing to prevent unauthorized access, banning of ragpickers, prevention of uncontrolled leachates, and environmental monitoring.

6.   Construction and Demolition Waste utilization: Primarily this consists of cardboard, clean dimensional wood, wood materials, land clearing debris, concrete, asphalt, metals, concrete masonry units, gypsum wallboard, carpet, insulation, glass, beverage containers.

The benefits identified by the recycling Construction and Demolition Waste included:

1.   Conserves immense space in existing landfills and retard the aggressive need of dump yards for the new generations with in the municipal areas.

2.   Reduces the environmental effects of extraction, transportation, and processing of raw materials. These effects include: greenhouse has emissions, air and water pollution, resource conservation, energy conservation;

3. Reduces project costs through avoided disposal costs and avoided purchases of new materials, This can be a new source of revenue itself;
4. Helps communities, contractors, and building owners comply with state and local policies, such as disposal bans and recycling goals which in turn enhance the public image of organizations that reduce disposal.

Phase 3: Comparative study and benchmarking of the waste management system

This phase tries to study the waste management systems in place in various cities like Hong Kong, Singapore, Melbourne, New York, Minnesota, Victoria, Fitchburg, Utah and various cities of China, Australia and UK. This study is to be followed by a comprehensive benchmarking activity to develop a solid waste management plan for the city of Mumbai.
For this we plan to map the waste management process of various cities across the globe and to adopt the best practices suited for Mumbai. The study will be based both on qualitative and quantitative approach.

BENCHMARKING THE WASTE MANAGEMENT SYSTEM IN THE CITY OF MUMBAI

Benchmarking is the systematic comparison of elements of the performance of an organization against those of other organizations, usually with the aim of mutual improvement. Major points identified from the benchmarking study for Mumbai

Waste management network for Mumbai
The first step in municipal waste management plan is to split the city of Mumbai into 14 waste management sectors .This is essential as the type and composition of the solid waste generated sectors can vary from residential to industrial or commercial wastes. Each of these sectors will operate several small localized Transfer Stations for small concentrated areas and one sector collection station. These individual sector collection stations will coordinate with disposal facility for the entire city of Mumbai- Table 1.

The waste will be collected in two stages

- Primary :from the small localized Transfer Stations to the sector collection center
- Secondary :from the sector collection center to disposal facility

Collection will be handled as

- Primary collection by private haulers
  - For industrial and commercial units collection will be done by professional private haulers
  - For residential areas this will be done by residents committee workers or social unit workers
- Secondary collection by the municipal agency

Collection of special wastes like tires, white goods, and household hazardous material (batteries, fluorescent lamps etc) are done by private parties. This is done once every month using special collection vehicles.

Table 1. Sector wise distribution of waste

SECTORS	TYPE OF WASTE GENERATED	RESIDENTIAL	COMMERCIAL	INDUSTRIAL	TOTAL WASTE
Andheri	Residential/ Commercial/Industrial	346.68	203.32	377.96	927.96
Sakinaka	Commercial/Industrial		203.32	377.96	581.28
Vikhroli	Commercial/Industrial		203.32	377.96	581.28
Airport	Commercial		203.32		203.32
Kurla Ghatkopar Belt	Industrial			377.96	377.96
Bandra	Residential	346.68			346.68
Trombay	Industrial			377.96	377.96
Chembur	Residential	346.68			346.68
Wadala-Mahim belt	Commercial/Residential	346.68	203.32		550
Dadar & Parel	Residential/Commercial	346.68	203.32		550
Mahalakshmi	Commercial		203.32		203.32
Central & Cotton Green	Commercial/Industrial		203.32	377.96	581.28
Marine Drive	Commercial		203.32		203.32
Colaba	Residential	346.68			346.68
					6177.72

Waste Management Facilities Required- Table 2

Table 2. Facilities required in each sector

Sl.No	Sectors	People to be employed in different sectors	Collection centres recommended	Vehicles recommended
1	Andheri	18776	841	361
2	Sakinaka	11762	527	226
3	Vikhroli	11762	527	226
4	Airport	4114	184	79
5	Kurla Ghatkopar Belt	7648	343	147
6	Bandra	7015	314	135
7	Trombay	7648	343	147
8	Chembur	7015	314	135
9	Wadala-Mahim belt	11129	499	214
10	Dadar & Parel	11129	499	214
11	Mahalakshmi	4114	184	79
12	Central & Cotton Green	11762	527	226
13	Marine Drive	4114	184	79
14	Colaba	7015	314	135
		5600		**2400**

Waste Management Group

Working Group on Waste Management Plan under the Waste Reduction Task Force is to be set up. The Working Group, with representatives from the Government and industry, is tasked to develop requirements on waste management plans suitable for use in Mumbai. The following are the basic responsibilities of the group.

- Implementation of waste management programs and effectively controlling them
- Partnership with Community Groups: joining hands with community organizations to organize the Waste Recycling Campaign in Housing Estates. Forming Green groups, youth groups, community centers, property management companies and Environmental Protection Ambassadors.
- Awarding Certificates of Waste wise Logo to industries indicating their environmental support.
- Adopt-A-Street practice– Individuals, community and civic organizations and businesses are required to adopt a section of one or more city streets and required to collect litter at least six times per year. Signs are erected at each location, supplies are provided and trash is collected by the designated organizations.

Recycling

The government has to assure a high level of participation in curbside recycling programs and a quality stream of recyclable materials from these collections. Typically individual householders or waste collectors bring recyclables to a dealer and are paid on the basis of weight and condition of the material received. Materials are bulked up for transportation either directly to recycling factories or to larger dealers.

Residential Recycling:

The City curbside recycling program must provide a five sort system for recyclables using color codes. Newspapers are placed in a green bin. Mixed containers (glass, plastic, aluminum, and tin/steel) are placed in a yellow bin. Mixed paper, paperboard, office paper, and magazines are placed in a paper bag. Polystyrene is placed in a clear plastic bag. Corrugated cardboard is flattened and stacked.

Material Recovery and Reuse

- Market development for processed organics, timber, and construction and demolition rubble and fill material. In non-metropolitan areas, market development initiatives may also be required to develop local markets for lower grades of recyclables such as low-grade cardboard/paper and glass materials.
- Materials specifications for public works projects are to be amended to allow for the use of recycled aggregates in road sub-base and in concrete for minor structures. The Working Group has to monitor and further promote the use of recycled aggregates in public works projects.
- Emphasis on environmental purchasing and use of recycled products by state government departments, agencies and local government.
- Remanufacturing: Involves identifying and supporting local unorganized manufacturing units that can remanufacture goods. Also encouraging the use remanufactured products especially in the government institutions.

- MAX (Metro Area eXchange): MAX tries to match-up businesses that generate unwanted materials with other businesses that can use those materials. The advantages of a materials exchange are that it offers savings:
    1. To the business generating the waste, which avoids paying shipping and disposal costs, and frees up storage space;
    2. To the business receiving the materials, which obtain less expensive feedstock and packaging;
    3. To the community, where processing and landfill capacity is saved; and
    4. To the environment, when fewer raw materials are used.
- Producing energy from waste using technologies such as pyrolysis and gasification from organic waste from residential units and homogenous wastes from industry waste streams (timber, garden and food organics), combined with other sources of prescribed industry wastes (sludge, biosolids).
- One of the primary ways to promote waste reduction, reuse and recycling is through the administration of state assistance programs for waste reduction, recycling and household hazardous waste management. This funding will helped communities to purchase recycling equipment they may not be able to otherwise afford.

Waste Reduction at source

The preferred approach to waste reduction is to employ assertive tools to increase the uptake of cleaner production and product stewardship throughout industry. waste reduction efforts must focus on the following areas:
- Promotion and recognition of voluntary initiatives;
- Conducting an educational outreach program
- Adopt quantity-based user fee disposal programs for commercial and industrial units. This will encourage them to dispose of less waste.

CONCLUSIONS

The proposed system should enable the city to consolidate the existing basic waste infrastructure while anticipating and meeting the challenges of future economic growth.
In the view of the authors one of the greatest benefits this study will provide is the presence of a highly competent group of people who will act as a resource for future solid waste management projects for the city of Mumbai. As we continue to evaluate and modify Mumbai's Solid Waste Management Plan, waste reduction will remain the cornerstone of the solid waste hierarchy. Waste reduction always has and always will make good economic and environmental sense. The purchasing of supplies and equipment by governments, the encouragement of conservation policies and the purchasing of goods by consumers are some of the areas where we need to constantly evaluate the environmental concerns.

REFERENCES

1. Central Pollution Control Board (2002) Management of Municipal solid waste, Ministry of Environment and Forests, Govt. of India.

2. Municipal Corporation of Greater Mumbai (MCGM)(2002), ' Survey of solid waste transported to disposal sited in Mumbai from 24 wards in three shifts' Solid Waste Management Department (Project Division).

3. SUDHAKAR Yedla and SARIKA Kasal (2003) 'Economic insight into municipal solida waste management in Mumbai: a critical analysis', Int. J. Environment and Pollution, Vol. 19, No.5.

4. YELDA, S. and PARIKH, K.S. (1997) 'Economic evaluation of a landfill system with gas recovery for municipal solid waste management: a case study', Int. J. Environment and Pollution, Vol. 15, No.4.

5. New York State Solid Waste Management Plan 2000 – 2002

6. OLLEY Jane and OLBINA Rada (2000), Innovative solid waste management in China. 25th WEDC Conference on Integrated Development for water supply and sanitation, Addis Ababa, Ethiopia 1999.

7. Price Water House Coopers, Water Supply, Wastewater & Solid Waste Performance Benchmarking Studies 2000.

8. Report on sustainability in solid waste management: Singapore's approach ISWA congress Melbourne, Australia 10-14 NOV 03

INDUSTRIAL SYMBIOSIS –
A PRACTICAL INSIGHT INTO BEST PRACTICE

I P Bryan

Waste Per Se

United Kingdom

ABSTRACT. The National Industrial Symbiosis Programme was established in 2003 to promote resource efficiency and waste minimisation and is funded by the Onyx Environmental Trust and a number regional development agency's (RDA's).

The programme works by identifying cross sectoral synergies between non-traditional industrial partners. These potential synergies are identified and evaluated using web based data collection tools.

To date over 500 companies are participating in the programme and over 1 million tonnes of resource have either been diverted from landfill or are in negotiation. Additionally, over 100 jobs have been created or preserved.

This paper describes what Industrial Symbiosis is, how it works and includes case studies relating to the manufacture of building materials including a high recycled content.

Keywords: Industrial symbiosis, Synergy, Construction demolition waste, Recycled aggregates and concrete, Recycled glass, Data collection and processing, Resource efficiency, Waste minimisation.

Ian Bryan is a Chartered Engineer and Programme Director for the National Industrial Symbiosis Programme. His interests include resource efficiency and waste minimisation and he is also a co-opted member of the DTI's Used Tyre Working Group.

INTRODUCTION

Many individual companies or industry associations have made (or perceive they have made) progress on such single issues as energy efficiency, waste minimisation and resource efficiency. Some have experimented with waste exchanges (largely unsuccessfully), energy efficiency clubs and implemented environmental management systems in an attempt to deal with waste and increase profitability.

Industrial Symbiosis (IS) encourages companies to look outside their own physical and sector boundaries, which are inherently limiting, for additional resource efficiencies and market opportunity. By collating and analysing data provided by companies on such parameters as input/output of materials, energy and water and supplementary information on logistics, capacities and human resources, it is possible to undertake a holistic and aggregated analysis to identify potential synergies leading to business benefits and enhanced resource efficiency. These synergies can then be further explored to identify those that are commercially viable with existing capabilities and technologies.

The National Industrial Symbiosis Programme (NISP) is the world's first national IS project. The capacity of this innovative business-led initiative to identify profitable synergies and promote solutions to barriers is growing substantially. Companies are finding greater benefit from this form of facilitated collaborative action as compared with existing purchasing and disposal processes.

In addition to improvements in regional resource productivity, early results from NISP demonstrate that by working together businesses will benefit from cost reductions and new product opportunities with associated safeguarding of jobs and job creation.

CASE STUDY 1

NISP members The Sims Group and Marley Building Materials are dedicated to offering innovative recycling solutions and both aim to develop a 100% recycling option across each of their processes. These companies have identified a synergy by which a waste stream from a Sims Group reprocessing activity can be utilised as a raw material in Marley's manufacturing process.

The Sims Groups dismantles and reprocesses used washing machines. Washing machine manufacturers incorporate a large block of concrete into the design as ballast to ensure the machine is stable during the spinning and drying cycles. While most of the waste components from these machines can be recycled, the concrete is more problematic and is currently sent to landfill. The Group has developed a variety of innovative and sustainable methods to recycle most of the materials from this process, but has been seeking and optimal solution for this particular waste stream.

Marley Building Materials has developed a range of novel recycling solutions to utilise waste concrete materials in their manufacturing processes and has worked very closely with Sims Group through the NISP programme. This project will reduce the cost base for both businesses and help to safeguard jobs both now and in the long run. Additionally the synergy will d ivert o ver 3 000 t onnes f rom l andfill a nd t he t echnology s hould h ave a pplication for additional building material streams.

Having successfully implemented one synergy, the partners and other programme members are also exploring other aggregate replacements such as recycled glass and construction demolition wastes utilising their mutual network of local authority, industrial and commercial contacts.

CASE STUDY 2

Following a conversation at the Scottish Construction and Aggregates Partnership Marketplace, Encore recycled 1,500 tonnes of inert waste tipped on a Council site. The waste would otherwise have been transported by lorry to a landfill site 11 miles away. This equates to some 75 journeys at 22 miles each, or 1,650 road miles.

700 tonnes of high-grade aggregates were produced, and will be used by Falkirk Council in construction of a road as an alternative to virgin aggregates. 300 tonnes of fine soil were also produced, which will be used by the Falkirk Council Grounds Maintenance Department.

If the waste had been land filled, and virgin aggregates and soil purchased by the Council, both of these processes would require a certain amount of road miles. Through this project, the landfilling road miles have been saved, while the transportation of recycled and virgin materials will be roughly equivalent.

This project is regarded as a pilot for a potential permanent arrangement between the two parties which could see a spoil tip remediated releasing a further 700,000 Te's of aggregates and fine soil.

CONCLUSIONS

- The National Industrial Symbiosis Programme is a business led initiative which offers significant potential for improved resource efficiency and waste minimisation.

- The programme has already achieved an impressive level of return and offers participating businesses innovative pathways to new opportunities for sourcing both recycled materials and ideas.

ACKNOWLEDGEMENTS

The author would like to acknowledge the support provided to the programme by its' funding organisations and members.

CONFERENCE CLOSING PAPERS

RECYCLING POWER STATIONS BY-PRODUCTS-
A LONG HISTORY OF USE?

L K A Sear

United Kingdom Quality Ash Association
United Kingdom

ABSTRACT. The UK coal fired power industry has been recycling ash for over 50 years. Pulverised Fuel Ash (PFA) and Furnace Bottom Ash (FBA) have been used for a wide variety of applications with in excess of 50 million tonnes being utilised. As well as the environmental benefits of using PFA and FBA, there are often considerable technical benefits compared to naturally occurring materials. These markets have been successfully developed, but they are now under threat.

European Directives, which have existed for some time, are beginning to negatively impact on recycling industries. Recent European and National court rulings, whilst seeking to clarify the situation, have highlighted weaknesses in the legislation. These judgements have also been subject to misinterpretation. This has led to materials now being considered by the regulatory authorities as being waste. While being labelled a 'waste' in some applications is enough to put off many construction companies from using PFA and FBA, the bureaucracy and requirements of the Waste Management Licensing are such as to put the 'nail in the coffin' of ash recycling.
This paper will review the current situation, the problems with the Waste Management Licensing system and to try to predict the future for the PFA and FBA recycling industry.

Keywords: Pulverised Fuel Ash, PFA, Fly Ash, Environment, Recycling, Waste Management Licensing, European Legislation.

Lindon K A Sear is the Technical Officer of the United Kingdom Quality Ash Association representing the interests of the UK coal fired power station operators. He represents the members of the UKQAA on a number of British and European Standard Committees. He is involved in the steering committees of many research projects ranging from the environmental aspects of PFA/fly ash through to the thaumasite form of sulfate attack. In addition, Lindon has been closely involved with developments within the environmental field in recent years and has observed the increasing bureaucracy and regulation that is in fact harming the environment.

INTRODUCTION

Pulverised fuel ash (PFA) and Furnace Bottom Ash (FBA) have been produced in significant quantities since World War II, with the introduction of modern steam raising plant. Following research carried out in the late 1940's it was discovered that PFA could be used in cementitious applications as well as being used for fill applications. The first UK use in concrete was in the construction of the Breadblane Hydro electric scheme in 1954. Similarly, in 1952 PFA was first used as a fill material. FBA found its market was within the production of lightweight aggregate concrete blocks. The markets for the products developed gradually, culminating in the preparation of British Standards and incorporation within National specifications as being suitable for many applications.

Since the UK joined the European Community, various pieces of legislation have been enacted that affect industry. Environmental legislation has increasingly come to the fore as having the greatest effect on the construction industry, in particular. This legislation was introduced through directives and subsequent court cases that are legally binding on the UK regulators. It would appear in recent months that the interpretation of the directives and court cases has become an industry in its own right, though with a consequential threat to the by-products and recycling industries. The following is a review of the situation on power station by-products and the environmental legislation.

THE HISTORY OF PFA AND FBA

The first reference to the idea of utilising coal fly ash within modern day concrete was by McMillan & Powers[1] in 1934 and research in the United States of America [2] indicated that fly ash had a role in concrete in 1935. Later research also in the US [3] reported that fly ash was a possible artificial pozzolana. Trial applications and continuing research promoted the idea that introduction of a proportion of fly ash as replacement of cement would limit shrinkage cracking in mass concrete by reducing internal hydration temperatures.

The introduction of pulverised coal steam raising plant in the UK, particularly after the 1939-1945 war, resulted in the production of fly ash and it was the late 1940's that saw research into the use of the material. In particular the example of using fly ash in mass concrete dams was considered, and, following research at the University of Glasgow [4], the practice was adopted for construction of the Lednock[5], Clatworthy and Lubreoch Dams, see Figure 1. In 1954, construction commenced on Lednock dam, which was the first structure built with some 62,500 m^3 of concrete containing PFA being used, replacing some 3,000 tonnes of Portland cement. The dams formed part of the Scottish Hydro-Electric Board's Breadblane scheme and the subsequent durability of the structures has been found excellent.

There followed in the period 1954-58 numerous examples of the use of fly ash as cement replacement in structural concrete at the Fleet Telephone Exchange, Newman Spinney Power Station [6] and the High Marnham sub-station [7]. By the mid 1970's fly ash was regularly being used in concrete as an addition at the concrete plant within many power company structures and some notable public works [8] being constructed. Such usage was always on a basis of close monitoring by the site and within large construction projects. Within the UK, fly ash from coal combustion became known as Pulverised Fuel Ash (PFA) around this time to differentiate it from ashes derived from other processes.

Figure 1 – Placing concrete in Lednock Dam, ~1956

Although the use of PFA in concrete was accepted by British Standards it was not until 1965 when the first edition of BS3892 [9] was published that there was a standard for PFA for use in concrete. PFA was treated as a fine aggregate having three classes of fineness based on the specific surface area. During this period acceptance in the routine readymixed concrete supply market was not being achieved. During the 1970's readymixed concrete suppliers were producing ever more technically demanding concretes of higher strength and lower water cement ratios. It was perceived that the variability in quality and the supply problems of PFA when taken directly from the power station were unacceptable.

Within the UK one solution was found to the fineness variability problems when in 1975 Pozzolan Ltd. introduced the concept of supplying controlled fineness material. Controlling the PFA to a tightly controlled fineness involved either classifying the ash, to remove the coarse fractions, or selection of the finer material by continual monitoring of the ash production. In general, finer PFA enhances the pozzolanicity whilst reducing the water demand. An Agrément Board Certificate [10] was obtained for controlled fineness PFA in 1975. These changes were reflected within BS3892 [11] in 1982 with the various parts of the standard indicating the uses and quality of PFA. Controlled fineness PFA to BS3892: Part 1 was accepted as counting fully towards the cement content of a mix, whereas 'run of station' ashes were at the discretion of the site engineer. The latter were usually considered as inert fillers and are covered by BS3892: Part 2: 1984.

In 1985 two British Standards were published for cements containing pulverised fuel ash.
- BS6588 for Portland pulverised fuel ash cements permitted an ash level between 15 and 35% by mass of cement and;
- BS6610 for pozzolanic pulverised fuel ash cement permitted an ash level of 35 to 50% by mass of cement.

Before 1985, interground Portland PFA cement had been produced by Blue Circle in the North of England, under an Agrément Certificate. Classified PFA was increasingly accepted for use within concrete both on technical and economic grounds. Currently the use of classified PFA is widespread within the UK readymixed and precast concrete industries. Some 25% of the readymixed concrete produced in the UK contains a binder that consists of, typically, 30% PFA and 70% Portland cement. Currently some 500,000 tonnes per annum of classified PFA are used in readymixed and precast concrete. With European harmonisation, a new standard for fly ash, BS EN450 [12] 1995, was introduced. With the exception of the UK and Ireland, no other European countries routinely classify PFA for use in concrete. EN450 reflects this differing approach and allows a wider range of fineness for use in concrete than

BS3892 Part 1[13]. The enabling standard for EN450 fly ash, EN 206 [14], has taken a number of years to finalise and consequently the use of EN450 fly ash is somewhat restricted. However, the complementary standard to BS EN206, BS 8500 [15], reflects the better performance of finer PFA and selected/classified PFA to BS3892 Part 1 will remain permitted under BS8500 for the foreseeable future.

Figure 2 – PFA used as a fill material A45, Packington nr Coventry

As stated above, PFA's first recorded use as a fill material was in 1952. It has been extensively used in highway construction following considerable research during the 1950's and 1960's into its properties. It was incorporated into the Specification for Highway works and widely used as embankment fill during the period the motorway network was being built. It was found to be a very stable, lightweight material that settles very little with increasing strength with age, see Figure 2. It has also been used successfully on reinforced earth structures.

More recently PFA has been used in numerous other applications, e.g. grouting, sub-base layers in road construction, in brick manufacture, for soil stabilisation and remediation, etc. FBA has been used in block making for many years. Many still refer to 'breeze blocks', with breeze being the fore-runner to FBA. Because FBA is flushed from the furnace by high pressure water jets, it is a far superior product to 'breeze', which ceased to be available anyway as the older furnaces were replaced. Subsequently FBA became the preferred product by the block manufacturers as a lightweight aggregate. Currently 99% of UK FBA production is used in lightweight aggregate block manufacture.

Since its introduction well in excess of 50 million tonnes of power station ash products have been used in a wide variety of applications. To our knowledge there has been no recorded incident of significant environmental pollution due to use of these products, and yet they are under threat on environmental legislative grounds.

ENVIRONMENTAL LEGISLATION

The legislation that potentially affects PFA and numerous other by-product materials stems from the EU "The Waste Framework Directive (75/442/EEC as amended by 91/156/EEC)". This directive was enforced in England and Wales by the Environmental Protection Act of 1990. A series of exemptions were permitted under resulting legislation, The Waste Management License (WML) regulations, which were produced in 1994 and are interpreted using DOE Circular 11/94 [16].

In recent years the Environment Agency (EA) has stated that there is doubt as to the validity of DOE 11/94 resulting from various EU court cases. This resulted in the EA regarding the document as no longer valid and this being stated within parliament. However, even though this statement was made, the document was not withdrawn from the public domain. A revision of DOE 11/94 by the Department of the Environment, Food and Rural Affairs (DEFRA) has recently taken place. Additionally, the permitted exemptions have been revised and these should be published by the time of this conference. It is understood the main changes to the exemptions system are:

1. Applications for exemption are no longer automatic and a minimum of 21 days notice must be given.

2. The EA may object to an exemption being granted within the 21 day period.

3. A charge for the application for exemption will be made.

4. There will be no more significant exemptions permitted from the 1994 categories. There will be more restrictions to some exemptions, e.g. a maximum thickness limit on land restoration of 2m.

Having said all the above the legislation only relates to materials that are classified as 'wastes'. Obviously, domestic refuse being disposed of in a landfill operation is a waste and requires Waste Management Licensing (WML). What is less clear is when a material is a waste and when does it cease to be a waste, say after processing or use by another industry. So does a by-product, like PFA, legally become a waste? PFA is occasionally disposed in landfills that have WML's. However, the majority of the material is sold, therefore not discarded and sold for beneficial use to a construction company, i.e. it has a market, surely this is not waste? It is for this reason there are important questions such as;

- When is a material 'recovered'?

- When does it cease to be a waste – when it's recovered?

- What processing is considered necessary for 'recovery' to have occurred?

THE DEFINITIONS, UK and EU LAW AND COURT CASES

The overarching aims of the EU Waste Framework Directive are as follows:

- ***Waste Management Hierarchy***. Waste management strategies must aim primarily to prevent the generation of waste and to reduce its harmfulness. Where this is not possible, waste materials should be reused, recycled or recovered, or used as a source of energy. As a final resort, waste should be disposed of safely (e.g. by incineration or in landfill sites).

- **Self-Sufficiency at Community and, if possible, at Member State level.** Member states need to establish, in co-operation with other Member States, an integrated and adequate network of waste disposal facilities.

- **Best Available Technique Not Entailing Excessive Cost (BATNEEC).** Emissions from installations to the environment should be reduced as much as possible and in the most economically efficient way.

- **Proximity.** Wastes should be disposed of as close to the source as possible.

- **Producer Responsibility.** Economic operators, and particularly manufacturers of products, have to be involved in the objective to close the life cycle of substances, components and products from their production throughout their useful life until they become a waste.

These aims of the directive are very laudable and fully supported by most. However, the same directive does seem to have created a situation that will in fact reduce '*re-use, recycling and recovery*' simply because these terms are ill defined both in the mind of the producers, end users and regulators. While various bodies have asked for clarification, to date little has happened that gives the producer or user any confidence the aims are achievable. In fact the directive is in danger of having the opposite effect – to reduce the use of by-products, recovery and recycling. The problem is outlined as follows.

The definition of waste used in the EU Waste Framework Directive is:

- 'Waste' means any substance or object which the holder disposes of or is required to dispose of.

- 'disposal' means :

  o The collection, sorting, transport and treatment of waste as well as its storage and tipping above or below ground.

  o The transformation operations necessary for its re-use, recovery or recycling.

DOE 11/94 gives guidance on the concept of waste. It states in paragraph 2.28 that if a substance can be used in its present form in the same way as any other raw material without being subjected to a specialised recovery operation this would provide reasonable indication that the substance or object has NOT been discarded and therefore is not a waste. In 2.32 there is a specific comment on by-products, stating that useful by-products that change hands for value and remain part of the normal commercial cycle should not be regarded as being discarded. As PFA and FBA are both products that are used as alternatives to other raw materials, such as sand for fill and grouting and cement in cementitious applications, this would suggest they are not wastes.

Subsequent European Case Law has confused the interpretations of DOE 11/94. The following cases asked some specific questions:

- **Euro Tombesi and Adino Tombesi etc**[17]: Is a material that has economic value ever a waste? Answer: Waste is not to be understood as excluding substances and objects that have economic value.

- **Inter-Environnement Wallonie ASBL and Region Wallonne** [18]: Can a material that forms part of an integrated industrial process be regarded as a waste? Answer: Yes

- **Mayer Parry Recycling Ltd v Environment Agency** [19]: Under what circumstance is scrap a waste? Answer: Recovered material is not a waste. Difficulty arises from the definition of a recovery process and the point at which recovery is regarded as complete. Clarity was not given on when "recovery/recycling" was regarded as complete.

- **ARCO Chemie Nederland Ltd and Minister van Volkshuisvesting, Ruimtelijke Ordening en Milieubeheer** [20]: Is a residue used as a fuel a waste? Answer: Unclear answer – depends on circumstances. The concept of waste cannot be interpreted restrictively.

- **Mayer Parry Recycling Ltd v The Environment Agency** [21]: When has waste been recycled and therefore ceased to be a waste? This relates to the previous Mayer Parry case and the Packaging Regulations. Answer: When recovery is complete. This has been referred to the European Court of Justice for further consideration.

- **Castle Cement v The Environment Agency** [22]: Is waste derived fuel a waste? Answer: Yes

- **Palin Granit Oy v Korkein hallinto-oikeus** [23]: Should left over stone that is no threat to the environment awaiting suitable markets for sale be classified as a waste? Answer: The holder of leftover stone resulting from stone quarrying, which is stored for an indefinite length of time to await possible use, discards or intends to discard that leftover stone, which is accordingly to be classified as waste.

- **AvestaPolarit Chrome v Korkein hallinto-oikeus (Supreme Administrative Court, Finland)** [24]; Is leftover rock resulting from the extraction of ore and/or ore-dressing sand resulting from the dressing of ore in mining operations to be regarded as waste? Answer: As in Palin Granit leftover rock resulting from the extraction of ore and/or ore-dressing sand resulting from the dressing of ore in mining operations which is stored for an indefinite length of time to await possible use are to be classified as waste

- **Tribunale di Gela (Italy) PET coke ruling.**[25] Does petroleum coke fall within the meaning of waste? Answer: Petroleum coke which is produced intentionally or in the course of producing other petroleum fuels in an oil refinery and is certain to be used as fuel to meet the energy needs of the refinery and those of other industries does not constitute waste.

The EA interpretation of these cases, in particular the Palin Granit and AvestPolarit cases, leads to them to use a three point test for whether a product is a waste:

1. Re-use is not a mere possibility, but a certainty;

2. There must be no further processing required prior to re-use, and

3. The re-use must be an integral part of the production process.

In respect of PFA and FBA produced for sale the Electricity Supply Industry (ESI) would comment of these requirements as follows:

Re-use is a certainty: The design of the coal milling plant and the boiler are optimised to produce PFA and FBA of saleable quality. In addition the PFA is stored in both dry and conditioned forms for sale to the various construction markets, e.g. concrete, fill, grouting, block making, etc. There are dedicated sales teams for these materials that are active within the construction markets. In addition, the ESI has invested in a considerable amount of research over many years to develop the various applications for the products. It has been so successful that 99% of FBA produced is used in block making and 55% of PFA produced is used in a variety of applications. While it is not the intention of the power station to discard any PFA, any material that cannot be sold is then consigned to lagoons or landfill sites, see Figure 3. Here this material is regulated as a waste and covered by Integrated Prevention of Pollution and Control (IPPC) licensing.

Figure 3 - An ash disposal site - Barlow mound at Drax power station

The ESI feels it can clearly demonstrate that PFA and FBA have a certainty of use through the markets it has developed.

There must be no further processing required prior to re-use: PFA and FBA for re-use does not undergo any significant processing. For concrete applications PFA is normally supplied dry in cement tankers. It may be processed to enhance its properties, but this does not preclude from use without any processing. For fill and grout applications, water will be added to PFA to prevent dust problems. Usually, it will then be delivered in normal sheeted tipping vehicles, allowing it to be used as an alternative material to naturally occurring sand. FBA may be screened and crushed to appropriate sizes for the use, otherwise there is no processing. Any processing that does take place is substantially no different than would be applicable to naturally occurring materials, e.g. sand and aggregates. Additionally, the markets for PFA and FBA are in direct substitution for those of either the cementitious, fine and coarse aggregate markets.

Considering the above the ESI feels it demonstrates that no processing or recovery is required before it can be used in production processes.

The re-use must be an integral part of the production process: This test is disputed by the ESI. It results from the Palin Granit (paragraph 36) court ruling as follows:

> *However, having regard to the obligation, recalled at paragraph 23 of this judgment, to interpret the concept of waste widely in order to limit its inherent risks and pollution, the reasoning applicable to by-products should be confined to situations in which the reuse of the goods, materials or raw materials is not a mere possibility but a certainty, without any further processing prior to reuse and <u>as an integral part of the production process</u>.*

This paragraph seems to be contradicted by other paragraphs in the ruling and by paragraph 37 that states:

> *It therefore appears that, in addition to the criterion of whether a substance constitutes a production residue, a second relevant criterion for determining whether or not that substance is waste for the purposes of Directive 75/442 is the degree of likelihood that that substance will be reused, without any further processing prior to its reuse. If, in addition to the mere possibility of reusing the substance, there is also a financial advantage to the holder in so doing, the likelihood of reuse is high. In such circumstances, the substance in question must no longer be regarded as a burden which its holder seeks to 'discard, but as a genuine product.*

The wording in paragraph 36 relating to 'part of the production process' has been taken to mean by the EA that the re-use must occur within the power station. However, it could be interpreted that the by-product is an integral part of the production process and it requires no further processing prior to re-use, e.g. it has markets and is a genuine product. To an extent this view is supported by the PET coke ruling.

The ESI takes the latter view and concludes that:

- PFA and FBA are NOT wastes;
- They have certainty of re-use;
- They require no further processing;
- They are marketed as genuine products for use in the cementitious, fine and coarse aggregate markets.

THE EFFECTS ON MARKETS

The EA is responsible for the enforcement of waste management controls. It states in its Internal Note 1 [26] that explains their Definition of Waste that "*It is the responsibility of the person who produces or holds a substance or object to determine whether what they are holding is waste. People who are unclear as to their obligations should consider taking their own legal advice as ultimately the interpretation of the law is a matter of the courts.*" However, this does not stop the EA from giving its opinion when asked. Rather than make the above statement to an enquirer, the EA advice at the time of writing this paper is that PFA is a waste and will, depending on the application, either require an exemption or a Waste Management License. Clearly the ESI view is contrary to that of the EA's, i.e. PFA is NOT a waste. Additionally, the EA should not be giving such advice, based on its own internal documents that states this decision shall be made by the producer.

This complex legal position has been subject to review by a top barrister, who concurs with the ESI view. In addition other producers of by-products of a similar nature to power station ash have also sort legal review and been advised similarly, their products are NOT wastes. As a result this has resulted in a legal impasse that may only be resolved by yet another court case. It would appear the EU Waste Framework Directive has a number of flaws:

1. The definition of 'waste' is clear. However;

2. There is a lack of clear definitions of terms such as 'recovery', 'by-product', 'product', etc to support the definition of waste.

3. It was created with disposal of waste into landfill and the illegal dumping of wastes in mind – it did not consider the industrial by-product and recycling markets that were already in existence and how the Directive would impinge on such operations.

These problems are compounded by the National regulators who are concerned about correctly interpreting the directives and various court cases. As any EU country who fails to enact an EU Directive can be subject to very large fines, coupled with the UK's passion for exactly following the letter of the law, this leads to regulators who are not prepared to interpret or who take the most pessimistic view of the situation. The EA view is that '*the interpretation of the law is a matter for the courts*', e.g. that all decisions about a product will be made by the courts and not the regulator.

The result of this less than satisfactory situation is the regulator, if asked, will advise the end user of a product that by-products like PFA are classified as wastes and require either waste management licenses or exemption certificates. Of course the end user, not wishing to get embroiled in the bureaucracy of such legislation, opts for the simpler and totally unregulated natural aggregate option. Conversely if the end user deliberately decides that the material is NOT a waste and not subject to waste management or is ignorant of the situation, he accepts a small risk that he may end up defending this decision in court. Obviously the ESI would support any end user in such a battle, but many are put off such a decision. The result of this situation is:

1. While ignorance of the law is no excuse, badly drafted laws and directives lead to situations that can put the end user at risk of not complying.

2. The onus is placed on the producer and end user to decide whether a material is a waste and yet the regulator is able to advise to the contrary - with the only recourse being through the courts.

3. The markets for by-product materials are gradually eroded in favour of using natural materials, natural materials having less bureaucracy and risk associated with them.

4. The actual environmental impacts of the by-product or the natural material used in such construction applications have no bearing on the situation, which is solely one of legal interpretations.

CONCLUSIONS

The bureaucratic burden on the producers and users of by-product and recycled materials and the level of uncertainty over EU Directives, the UK interpretation of the directives and the inability of the regulators to make rulings other than to refer back to the courts is detrimental to both the environment, industry and ultimately the government. Such regulation is actively discouraging the use of by-products and recycled materials.

The solution is for the government and its departments, the regulators and the EU Commission to listen to industry and to work with industry. In this manner sensible, workable solutions will be forthcoming, rather than ad hoc legislation being created that is just counter productive. If this approach is not adopted, less recycling and recovery will occur, to the overall detriment of the environment. Eventually, because of increasing costs, industry will transfer to the developing countries where the environmental and bureaucratic burdens are less.

REFERENCES

[1] McMillan F R and Powers T C, A method of evaluating admixtures, Proceedings American Concrete Institute, Vol. 30, pp. 325-344, March-April 1934.

[2] Davis R E, Kelly J W, Troxell G E and Davis H E, 1935, Proportions of Mortars and Concretes containing Portland-Pozzolan Cements, ACI, Vol. 32, 80-114

[3] Davis R E, Carlston R W, Kelly J W and Davis H E, 1937, Properties of Cements and Concretes containing Fly ash, A.C.I., Journal 33, 577-612

[4] Fulton, A A and Marshall W T, 1956, The use of fly ash and similar materials in concrete, Proc. Inst. Civ. Engrs., Part 1, Vol. 5, 714-730

[5] Allen AC, Features of Lednock Dam, including the use of fly ash. Paper No. 6326 Proceedings of the Institute of Civil Engineers 1958:13 August: 179-196

[6] Richardson, L and Bailey J C, 1965, Design, construction and testing of pulverised fuel ash concrete structures at Newman Spinney Power Station, Parts I, II, III, CEGB Research and Development Report

[7] Howell, L H, 1958, Report on pulverised fuel ash as a partial replacement for cement in normal works c oncrete, C EGB, East Midlands D ivision. (also Ashtech '84, L ondon, 1984)

[8] Central Electricity Generating Board Technical Bulletins Nos. 1 to 48

[9] BS3892: Pulverised fuel ash for use in concrete, 1965.

[10] The Agrèment Board Certificate No. 75/283. Pozzolan - a selected fly ash for use in concrete.

[11] BS3892 Part 1, Pulverised fuel ash for use as a cementitious component in structural concrete, 1982, BSI.

[12] BS EN 450 Fly ash for concrete - Definitions, requirements and quality control, 1995, BSI, ISBN 0 580 24612 4.

[13] BS 3892: Part 1, S pecification for pulverised-fuel ash for use with Portland cement, 1997, BSI, ISBN 0 580 26785 7.

14   BS EN206: 2001, Concrete – Part 1: Specification, performance, production and conformity, BSI, London.

15   BS8500 (Draft for public comment): Concrete – complementary British Standard to BS EN206-1, Feb., 2001.

16   DoE Circular 11/94 Environmental Protection Act 1990: Part II Waste Management Licensing.

17   uro Tombesi and Adino Tombesi etc Joined Cases C-304/94, C-330/94, C-342/94 and C-224/95 – European Court of Justice – Advocate General F G Jacobs. *25 June 1997*

18   Inter-Environnement Wallonie ASBL and Region Wallonne – Case C-129/96 – European Court of Justice – Advocate General F G Jacobs. *18 December 1997*

19   Mayer Parry Recycling Ltd v Environment Agency – Chancery Division – Carnworth J - *9 November 1998*

20   ARCO Chemie Nederland Ltd and Minister van Volkshuisvesting, Ruimtelijke Ordening en Milieubeheer – Joined Cases C-418/97 (ARCO) and C-419/97 (Epon) – European Court of Justice – Advocate General S Alber - 15 June 2000

21   Mayer Parry Recycling Ltd v The Environment Agency – CO/512/00 – High Court of Justice – Mr Justice Collins - *8 September 2000*

22   Castle Cement v The Environment Agency - High Court of Justice – Justice Stanley Burnton - 22 March 2001

23   In Case C-9/00, Palin Granit Oy and Vehmassalon kansanterveystyön kuntayhtymän hallitus, JUDGMENT OF THE COURT (Sixth Chamber), 18 April 2002

24   Case C-114/01 AvestaPolarit Chrome Oy vKorkein hallinto-oikeus (Supreme Administrative Court, Finland), 10 April 2003

25   Case C-235/02, 15 January 2004, Article 104(3) of the Rules of Procedure - Directives 75/442/EEC and 91/156/EEC - Waste management - Definition of waste - Petroleum coke, REFERENCE to the Court under Article 234 EC by the Giudice per le indagini preliminari of the Tribunale di Gela (Italy) for a preliminary ruling in the criminal proceedings before that court against Mario Antonio Saetti and Andrea Frediani.

26   Environment Agency, Definition of Waste - Internal Note 1, UNDERSTANDING THE DEFINITION OF WASTE, September 2003.

MAKING RECYLING ECONOMIC – THE SUSTAINABILITY OF MATERIALS FOR THE BUILT ENVIRONMENT

A J W Harrison

TecEco Pty Ltd.

Australia

ABSTRACT. Our interaction with our planet earth is described as the techno-process which is the process of taking resources, manipulating them as required (modifying substances), making something with them, using what is made and then throwing what has been made that no longer has utility[i] away. The techno-process of take, manipulate, make, use and waste is discussed particularly in relation to reducing, reusing and recycling materials, earth systems, the web of life and materials in the built environment which, at over 70% of all materials flows, comprise the major component of the flow through the techno process. The most significant material flow is that of concrete as over two tonnes are produced per person on the planet per annum. The use of new calcium magnesium blended cements (TecEco cements) invented b y t he a uthor t hat u tilise w astes a nd s equester c arbon d ioxide i s d iscussed a s a n example of the significant impact changes in materials could make.

Reducing, re-using and recycling materials would reduce the impact of the techno process on the planet, but is currently undertaken more for "feel good" political reasons than sound economic reasons in many instances. Some unique practical solutions are offered to resolve this dilemma including a unique electronic identification system for materials and the use of new generation TecEco cement composite materials as potential repositories of wastes including CO_2.

Keywords: Abatement, Sustainable, Sustainability, Sequestration, CO_2, Brucite ($Mg(OH)_2$), Durability, Reactive magnesium oxide, Materials, Magnesian, Magnesia, Reactive magnesia (MgO), Magnesite ($MgCO_3$), Hydromagnesite ($Mg_5(CO_3)4(OH)_2 \cdot 4H_2O$), Fly ash, Pozzolan, Hydraulic cement, P ortland c ement, Concrete, P rocess energy, Embodied e nergy, Lifetime energy, Durability, Shrinkage, Cracking, Extraction, Permeability, Rheology, emissions, Matter, Materials, Substances, Wastes, Reduce, Reducing, Reuse, Re-using, recycle, Recycling, Manipulate, Utility, Digital, Silicon chip.

A John W Harrison, *B.Sc. B.Ec. FCPA.* John Harrison is managing director and chairman of TecEco Pty. L td. and best known for the invention of tec, eco and enviro-cements known mainly for their impact on sustainability. John is an authority on sustainable materials for the built environment and was a founder of the Association for the Advancement of Sustainable Materials in Construction.

THE IMPACT OF POPULATION GROWTH, TECHNOLOGY
AND CONSUMPTION

There is not the room in this paper to go into the detail of our legacy on this planet. Suffice to say we have taken over, we control the world and our impact has been very detrimental to virtually all earth systems. Our detrimental linkages with the environment have grown due to population increases and partly due to shifts in the technological basis of the techno – process whereby we take resources from the environment, manipulate the molecules, make something, use it and then when we have finished it waste it.

According to the American Association for the Advancement of Science Population and Environment Atlas[ii], "Consumption and technology impact on the environment by way of two major types of human activity. First, we use resources. We occupy or pre-empt the use of space, and so modify or remove entirely the habitats of many wild species. We extract or take resources -- growing food, catching fish, mining minerals, pumping groundwater or oil. This affects the stock of resources available for humans and for other species in the future.

Second, we dump wastes -- not just those that consumers throw away, but all the waste solids, liquids and gases that are generated from raw material to final product. These affect the state of land, groundwater, rivers, oceans, atmosphere and climate." If we want to survive the next millennia as a race we will need to take more responsibility for the environment.

THE TECHNO-PROCESS

Our interaction with the geosphere-biosphere[iiiiv] is described as the techno-process which is the process of extracting resources, manipulating them as required (modifying substances), making something with them, using what is made and then throwing what has been made that no longer has utility[v] away. The techno-function (Take→Manipulate→Make→Use→Waste) describes this techno – process.

The impact of the techno - function on the planet is significant. Resources are not unlimited and the planet does not have an infinite capacity to reabsorb wastes. To reduce linkages with the environment and for our own long term sustainable survival the take and waste need to be reduced and preferably eliminated to what is renewable and preferably biodegradable. To do this we must adopt the philosophy of Reducing, Re-using, Recycling and Recovering. Resources extracted from the geosphere-biosphere can be classified broadly into several non exclusive types. Resources are renewable or non renewable, short use or longer use materials.

- *Renewable Resources*
  A renewable resource is any natural resource (such as wood or solar energy) that can be replenished naturally[vi] with the passage of time. Can be either short use (renewable energy) or long use (wood). Depend on the health of the geosphere-biosphere for natural replenishment.
- *Non-Renewable Resources*
  The use of non-renewable resources as materials or energy sources leads to depletion of the Earth's reserves. Non renewable resources are characterised as non-renewable in human relevant periods.

For sustainability the extraction of resources and production of waste need to be reduced and preferably limited to the extraction of resources that are renewable and production of wastes that are biodegradable

Figure 1 –The Techno - Process.

- **Short Use Resources**
  Are renewable (food) or non renewable (fossil fuels). Have short use, are generally extracted modified and consumed, may (food, air, fuels) or may not (water) change chemically but are generally altered or contaminated on return back to the geosphere-biosphere (e.g. food consumed ends up as sewerage, water used is contaminated on return.)

- **Long Term Use Resources or Materials**
  Materials are "the substance or substances out of which a thing is or can be made[vii]." Alternatively they could be viewed as "the substance of which a thing is made or composed, component or constituent matter.[viii]"
  Some materials are renewable (wood), however most are not renewable unless totally recycled (metals, most plastics etc.) Materials generally have a longer cycle from extraction to return, remaining in the techno-sphere[ix] whilst being used and before eventually being wasted. Materials may (plastics) or may not (wood) be chemically altered and are further divided into organic (e.g. wood & paper) and inorganic (e.g. metals minerals etc.)

Other classifications are possible such as surface or sub surface resources, etc. but not relevant to my arguments in this paper. The techno-process is very inefficient in that large quantities of renewable and non renewable resources are extracted to produce small quantities of materials which themselves are used to produce even smaller volumes of things actually used, many of which do not retain utility (value to us) very long before they are in turn discarded.

"Studies show that between a half and three-fourths of the materials used in our industrial economy are generated and treated as waste before ever entering the economy. They are not seen or treated as commodities and aren't valued as such[x]". It is essential that we find more uses for these wastes for linkages to the biosphere-geosphere to be reduced. The inventive is that economically we would be better off.

EARTH SYSTEMS

The Earth has well-connected systems. Carbon dioxide emitted in one country is rapidly mixed throughout the atmosphere, and pollutants released into the ocean in one location are transported to distant parts of the planet. Local and regional emissions create global environmental problems. It is difficult to completely understand let alone put numbers to the complex flows and balances that go on around the planet. We do know however that pollution from wastes of various kinds has affected atmospheric composition, land cover, marine ecosystems, coastal zones, freshwater systems, global biological diversity and many other global systems.

Due to our pervading interference the constituent components of matter which are molecules are no longer produced or used in equilibrium. Of particular concern and therefore the most studied is the carbon cycle which is out of balance. Complex carbon based molecules put together by living matter over many millions of years are being used as if there was no tomorrow. The level of carbon dioxide waste from this process is rising too rapidly for conversion by photosynthesis or utilization by organisms for skeletons and shells. As a consequence the level of CO_2 in the air has risen from 280 parts per million in pre-industrial times to around 370 parts per million today. Methane concentrations have also risen by 145 percent over the same period. Before the industrial revolution gaseous chlorines did not exist in the atmosphere. By 1996 there were 2 731 parts per trillion, most of these produced in the 20th century[xi].

Rising atmospheric carbon dioxide concentration and its potential impact on future climate is an issue of global economic and political significance. Of possible even greater importance is the depletion of oxygen in the atmosphere. A study of all the other earth systems including for example the hydrosphere of which freshwater systems are a part and in relation to which water quality, salinity and supply are major issues or sea fisheries which are in major decline is beyond the scope of this paper. The Affect of the Techno-Process on the Web of Life

Figure 2 – Most Pollution Eventually Makes it to our Waterways Killing Fish and other Life[xii]

What distinguishes our planet from any other we have yet discovered is that there exists life, comprising carbon based molecules which have evolved in a delicate balance with the rest of the atoms and molecules that make up planet earth. Living matter is different from dead matter in that it contains genetic coding and has the ability to take atoms and molecules from the environment to build new replicates of itself for the future. This delicate balance is characterized by the flow of substances from the dead world to living matter and in reverse and has gone on for billions of years. Since the dawn of mankind and in particular the industrial revolution our ability and willingness to manipulate everything around us has however has wreaked havoc with living life forms.

For example, according to some estimates, seventy percent of Earth's coral reefs will cease to exist in the next forty years if the current trend continues. Every year, around one thousand (1,000) species are driven to extinction. The loss of tropical forests is especially alarming as they are the home habitat of over fifty percent of the world's species of plants and animals [xiii]. There is no doubt that we have seriously disturbing the "web of life" and evidence is mounting of worse future problems unless we can reduce our impact.

We take atoms and molecules from the dead and living matter around us in what is the techno-process and then when finished throw them "away". There is no such place as "away". Many of these transformed atoms and molecules enter the global commons and return as part of the flow negatively affecting our wellbeing and that of other living organisms on the planet. Persistent chemicals are not confined to the area in which they are created. Gases like Chloro fluoro hydrocarbons (CFC's) which is a waste enter what is the global commons. CFC's were developed during the 1930s and found widespread application after World War II. They are halogenated hydrocarbons, mostly trichlorofluoromethane and dichlorodifluoromethane and were used extensively as aerosol-spray propellants, refrigerants, solvents, and foam-blowing agents because they were non-toxic and non-flammable and readily converted from a liquid to a gas and vice versa.

CFCs were however found to pose a serious environmental threat because in the stratosphere they broke down under ultraviolet light releasing chlorine which destroyed ozone. A hole over Antarctica developed in the ozone layer which shields living organisms on earth from the harmful effects of the sun's ultraviolet radiation. "Because of a growing concern over stratospheric ozone depletion and its attendant dangers, a ban was imposed on the use of CFCs in aerosol-spray dispensers in the late 1970s by the United States, Canada, and the Scandinavian countries. In 1990, 93 nations agreed to end production of ozone-depleting chemicals by the end of the century, and in 1992 most of those same countries agreed to end their production of CFCs by 1996[xiv]."

Heavy metals have been of particular concern. The term "heavy metal" refers to metals that are relatively high in density and toxic or poisonous even at low concentrations. Examples include mercury (Hg), cadmium (Cd), arsenic (As), chromium (Cr), thallium (Tl), and lead (Pb). Heavy metals are natural components of the Earth's crust and cannot be degraded or destroyed. They are particularly dangerous when released as wastes because they tend to bio-accumulate up the food chain. For example, marine organisms can consume a particularly dangerous form of mercury called methylmercury. When fish eat these organisms, the methylmercury is not excreted, but retained in bodily tissues. The older the fish and the more organisms further down the food chain they have consumed, the greater the amount of methylmercury in their tissues. The accumulated methylmercury is concentrated as it is passed up the food chain and any organism at the top such as us faces a serious risk of mercury poisoning by eating such fish.

Heavy metals can enter a water supply from industrial and consumer waste, or even from acidic rain breaking down soils and releasing heavy metals into streams, lakes, rivers, and groundwater. We are constantly being alerted to the fact that living organisms in the far reaches of the globe contain significant traces of organic and metallic pollutants and that the deepest marine sediments, remotest glaciers and icecaps are contaminated. The list of contaminants is frightening and long. Pollution from waste is everybody's problem.

The Environmental Affects of Resource Extraction in the Techno – function

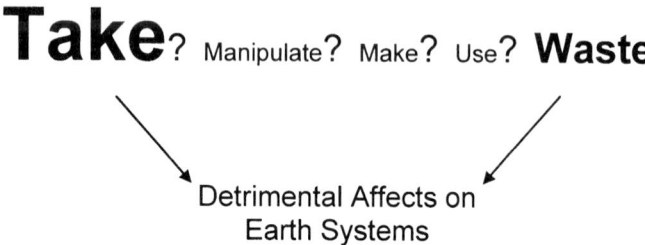

In the past the main cause of concern was that resources would not be sufficient to sustain the human race let alone the techno-process (even if it was not called that) "Frequent warnings were issued that we faced massive famines, or that we would "run out" of essential fuels and minerals.[xv]" Renewable resources such as water, fish stocks, even the air we breathe are today of much greater concern because they are now understood to be much more fragile and influenced strongly by overuse and pollution

According to the recent report "Global Change and the Earth System - A Planet under Pressure" IGBP SCIENCE No. 4[xvi], funded largely by the Swedish Government, our planet is changing quickly. In recent decades many environmental indicators have moved outside the range in which they have varied for the past half-million years. We are altering our life support system and potentially pushing the planet into a far less hospitable state. It is not population growth per say that is the problem; it is the increase in flows through the techno-process previously defined associated with increasing levels of the technology factor and associated consumption.

Exponential Growth

There is an explosion of substances through the techno-process many of which have damaging wastes associated with them. The worst are fossil fuels and cement production. The Industrial Revolution was the beginning of the transformation of societies into the energy-intensive economies of today. The consumption of fossil fuels has had a big impact on atmospheric composition. Another major source of emission is the production of cement. In 2003 the world produced 1.86 billion tons of cement[xvii] -- about a quarter of a ton for every man, woman and child on Earth. Production is now probably well over 2 billion tonnes. Over 13 billion tonnes of concrete are the result which is over two tonnes per person per annum on the planet.

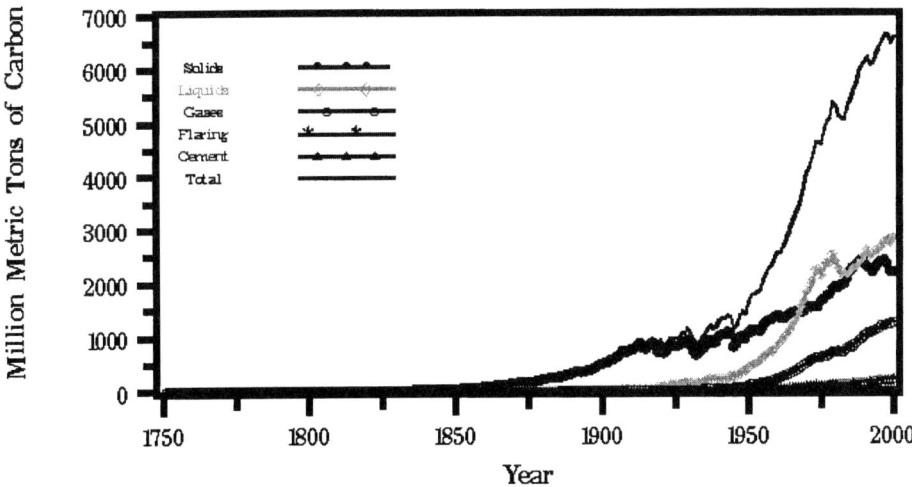

Figure 3 - Global CO_2 Emissions from Fossil Fuel Burning, Cement Production, and Gas Flaring for 1751-2000.[xviii]

The growth of energy consumption is closely correlated with the increases in gross national product which is a measure of economic development. The current consumption patterns of fossil fuels, as well as contributing to emissions, is not sustainable and neither is the production of cement (See Reducing the Impact of the Techno Process on page 382.) The emissions from the burning of fossil fuels and production of cement are shown in Figure 3.

The Environmental Affects of Wastes in the Techno-function

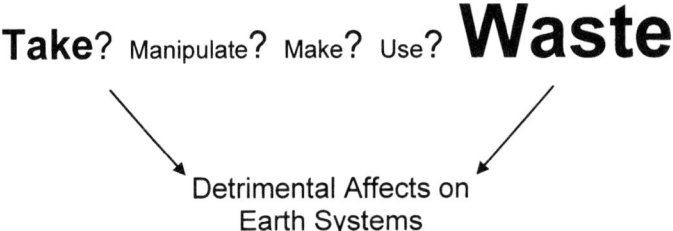

Of major concern is the problem of wastes which are the output of the techno function according to which wastes are created in producing and consuming resources. Huge volumes are produced. In the mid-1990s for example countries belonging to the OECD produced 1.5 billion tons of industrial waste and 579 million tons of municipal waste -- an annual total of almost 2 tons of waste for every person[ii]. The figures for building materials in waste streams vary around the globe and one of the problems is that the method in which audits are conducted also varies making it hard to obtain comparative statistics.

According to Maria Atkinson of the Green Building Council of Australia the figure nationally of waste going to landfill from construction and deconstruction activities (predominantly the churn of refurbishments) was around 40%[xix]. The flow of unwanted or waste materials is affecting our planet. The solid wastes that are not incinerated generally go to landfill and pollute water courses and the local area. Liquid and gaseous pollutants are more insidious and spread invisibly in the global commons.

Landfill is the technical term for filling large holes in the ground with waste. These holes may be specially excavated for the purpose, may be old quarries, mine shafts and even railway cuttings. More recently the mountains or islands made of waste are being created. Apart from wasting what are potentially resources landfill sites produce climate changing gases such as methane which is some 25 times more powerful than CO_2 as a greenhouse gas but only remains in the atmosphere for about ten years and so looses it's greenhouse effect quickly compared to CO_2 which remains in the atmosphere significantly longer.

The current atmospheric concentration of methane is 1.8 ppm or $25 \times 1.8 = 45$ ppm CO_2 equivalent. This is 12% of CO_2 concentration and its growing 2.5 times as fast[xx]. The current concentration of CO_2 on the other hand is around 370 ppm. Landfill can cause ill health in the area, lead to the contamination of land, underground water, streams and coastal waters and gives rise to various nuisances including increased traffic, noise, odours, smoke, dust, litter and pests.

According to the EPA "Resources that simply become waste are not available for future generations and extraction and harvesting of additional resources can have long term environmental impacts. Even as we implement protective waste management programs, toxic chemicals still can find their way into the environment throughout the life cycle of materials in extraction, production, transportation, use, and reuse. Persistent, bio accumulative and toxic chemicals released into the environment can present long term risks to human health and the environment, even when released in small quantities.[xxi]"

Reducing the Impact of the Techno Process

It is essential that the human race, with all the power it has over the environment moves rapidly towards reducing the linkages between the techno-sphere and the geosphere-biosphere before it is too late. To do this the inputs and outputs of the techno-function need to be reduced. We need to change the techno function to:

Reuse

Take less? Manipulate? Make? Use? Waste less

Manipulate ← Recycle

And more desirably to:

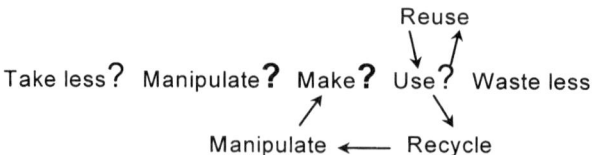

Reuse

Take only renewables ? Manipulate? Make? Use? Waste only what is biodegradable or can be re-assimilated

Manipulate ← Recycle

To make a material difference to the quantum of resources extracted and consequent and subsequent w aste w e n eed t o t ackle t he b ig p roblems f irst a nd t he b iggest p roblem i s t he weather. So far "everybody is talking about the weather but nobody does anything about it[xxii]" Now is the time for focussed action to modify the principal techno sub processes that pollute with carbon dioxide.

Fossil Fuels

Energy consumption results in the most gaseous releases because of our dependence on fossil fuels. There is a strong need to kick the fossil fuel habit however this is unlikely to happen unless alternative sources of energy become more economical. "This may be sooner than we think as "just under half of the world's total endowment of oil and gas has been extracted already, and that output will begin to decline within the next five years, pushing prices up sharply.[xxiii]" Most geologist however concur that thirty rather than five years is more likely. Even if we do kick the fossil fuel habit before it kicks us, it will take centuries to bring the carbon balance back down to levels in the 50's. Abatement and sequestration on a massive scale are essential. "Complementary to traditional areas of energy research, such as improving energy efficiency or shifting to renewable or nuclear energy sources, carbon sequestration will allow continued use of fossil energy, buying decades of time needed for transitioning into less carbon-intensive and more energy-efficient methods for generating energy in the future.[xxiv]" It is encouraging that there is a slow shift to renewable energy as Figure 4 shows.

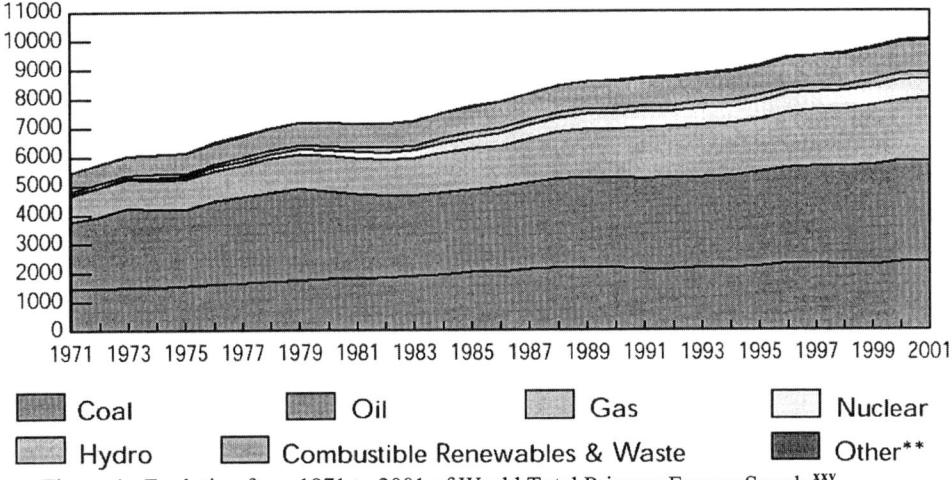

Figure 4 - Evolution from 1971 to 2001 of World Total Primary Energy Supply[xxv]

Cement and Concrete

The main material used for buildings and infrastructure is concrete. Concrete is made by utilising a cement such as Portland cement to bond stone and sand together. Ordinary Portland Cement (OPC) is the most common cement used and the concrete made with it is an ideal construction material, as it is generally economic, durable, easily handled and readily available. Contrary to lay understanding Portland cement concretes have low embodied energies compared to other building materials such as aluminium and steel, have relatively high thermal capacity and are therefore relatively environmentally friendly.

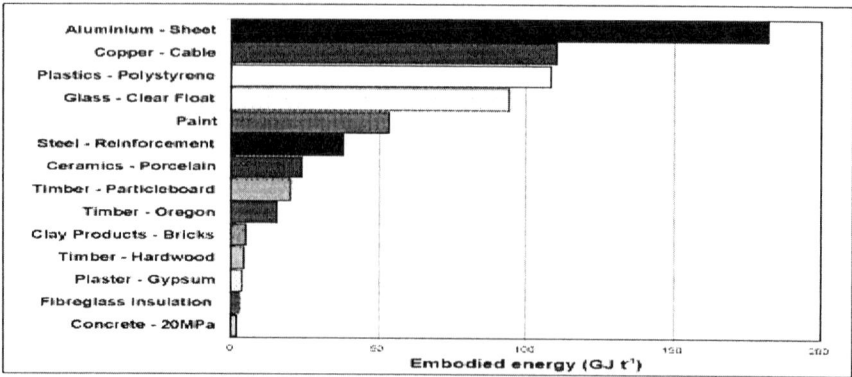

Figure 5 - Embodied Energy of Building Materials[xxvi]

However concrete, based mainly on Portland cement clinker, is the most widely used material on Earth. Production for the year ended 30 June 2003 was 1.86 billion tonnes[xvii], enough to produce o ver 7 c ubic k m o f c oncrete p er year or o ver t wo t onnes o r o ne c ubic m etre p er person on the planet resulting in significant global emissions.

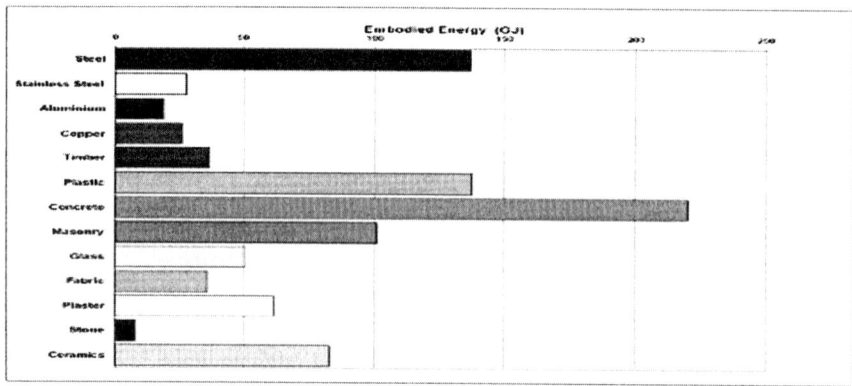

Figure 6 - Embodied Energy in Buildings[xxvi]

As a consequence of the huge volume of Portland cement manufactured, considerable energy is consumed (See Figure 6 - Embodied Energy in Buildingsxxvi on page 384) resulting in carbon dioxide emissions. Carbon dioxide is also released chemically from the calcination of limestone used in the manufacturing process. Various figures are given in the literature for the intensity of carbon emission with Portland cement production and these range from 0.74 tonnes coal/ tonne cement [xxvii] to as high as 1.24 tonne [xxviii] and 1.30 tonne.[xxix] The figure of one tonne of carbon dioxide for every tonne of Portland cement manufactured[xxx] given by New Scientist Magazine is generally accepted.

The production of cement has increased significantly since the end of World War II.

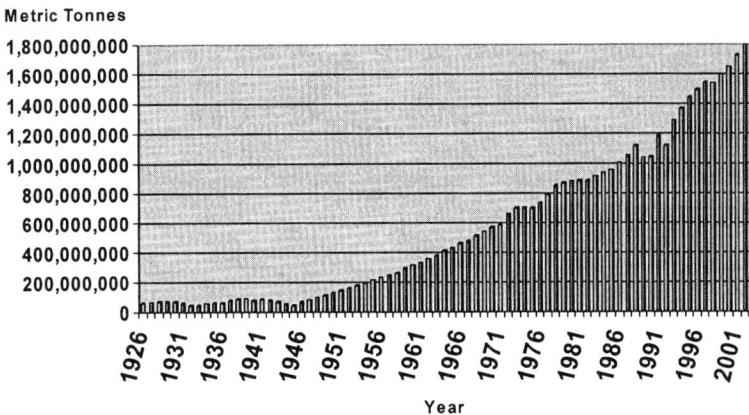

Metric Tonnes

Year

Table 1 - Cement Production = Carbon Dioxide Emissions from Cement Production 1926-2002[xxxi]

These releases are due to:

- The burning of fossil fuels in the kilns used;
- Emissions associated with electricity used during the manufacturing process, and;
- The chemical release of CO_2 from calcining limestone.

As of 2004 some 2.00 billion tonnes of Portland Cement (OPC) were produced globally[xxxii]. This accounts for more embodied energy than any other material in the construction sector[xxxiii]. The manufacture of OPC is one the biggest single contributors to the greenhouse effect, accounting for between 5%[xxxiv] and 10%[xxxv,xxxvi] of global anthropogenic[xxxvii] CO_2 emissions. Global production of cement is likely to increase significantly over the coming decades as:

- Global population grows;
- GDP grows;
- Urban development continues; and,
- Through increasing industrialisation.

A direct consequence of such huge usage and growing demand is the associated enormous potential for environmental benefits and improvements in sustainability.

Summary

The key to more rapid moves toward sustainability is to change the technology paradigm and this is discussed in more detail later in this paper (See Making Reducing, Re-using and Recycling Work in the Long Term without Subsidies or Taxes On page 390). If solar cells for example suddenly became very cheap to make a far higher proportion of electricity would be generated using them.

THE BUILT ENVIRONMENT - A FOCUS FOR SUSTAINABILITY EFFORTS

Our understanding of the flows and interactions in the global commons is very inadequate. The widely held view is that sustainable management strategies are complex to devise and politically difficult to introduce. What if they were economic? The obvious place that seems to have been missed by just about everybody to focus sustainability efforts is the built environment. It is our footprint on the globe. Given the size of the built environment there are huge opportunities for changing the techno function.

The dominant proportion of what we take, manipulate and make that we do not consume immediately goes into the materials with which we build the techno-sphere. Buildings and infrastructure are our footprint on the globe and probably account for around 70% of all materials flows and of this "Buildings account for 40 percent of the materials and about a third of the energy consumed by the world economy. Combined with eco-city design principles, green building technologies therefore have the potential to make an enormous contribution to a required 50% reduction in the energy and material intensity of consumption in the post-modern world.[xxxviii]"

"In 1999, construction activities contributed over 35% of total global CO_2 emissions - more than any other industrial activity. Mitigating and reducing the impacts contributed by these activities is a significant challenge for urban planners, designers, architects and the construction industry, especially in the context of population and urban growth, and the associated requirement for houses, offices, shops, factories and roads[xxxix]." According to the Human Settlements Theme Report, State of the Environment Australia 2001[xl], "Carbon dioxide (CO_2) emissions are highly correlated with the energy consumed in manufacturing building materials. "On average, 0.098 tonnes of CO_2 are produced per gigajoule of embodied energy of materials used in construction. The energy embodied in the existing building stock in Australia is equivalent to approximately 10 years of the total energy consumption for the entire nation. Choices of materials and design principles have a significant impact on the energy required to construct a building. However, this energy content of materials has been little considered in design until recently, despite such impacts being recognized for over 20 years."

To date the main practical emphasis has been on designing building with low lifetime energies. Little effort has been made to reduce the impact of materials on lifetime energies, embodied energies and emissions.

WHAT GOVERNMENTS ARE DOING TO ENCOURAGE SUSTAINABILITY

For sustainability the take and waste components of the techno-process need to be reduced and preferably eliminated to what is renewable and preferably biodegradable.
Governments generally view sustainability as a desirable goal and various policy options are being experimented with such as:

1. Research and Development Funding Priorities.

2. Procurement policies.
   Government in Australia is more than 1/3 of the economy and can strongly influence change through:

- Life cycle purchasing policy.
- Funding of public projects and housing linked to sustainability such as **recycling**.
3. Intervention Policies.
    - Building codes including mandatory adoption of performance specification.
    - Requiring the recognition and **accounting for externalities**
    - **Extended producer responsibility (EPR) legislation**
    - Mandatory use of minimum standard materials that are more sustainable
    - Mandatory eco-labelling
4. Taxation and Incentive Policies
    - Direct o r i ndirect t axes, b onuses o r r ebates t o d iscourage/encourage s ustainable construction etc.
    - A national system of carbon taxes.
    - An international system of carbon trading.
5. Sustainability Education

Consider building codes, research and development funding and policies of encouraging recycling and legislating for extended user responsibility as they are most relevant to the materials theme of this paper.

Building Codes

The emphasis has been on lifetime rather than embodied energies because potential lifetime energy reductions through good design are significant. Most OECD countries have set up energy efficiency standards for new dwellings and service sector buildings: this includes all European countries, Australia, Canada, the USA, Japan, Korea and New Zealand. Some non-OECD countries outside Europe have also established mandatory or voluntary standards for service buildings and Singapore and the Philippines were among the first. To date building codes have not encouraged the use of more energy efficient materials, in spite of the huge impact of materials not only on embodied energies but lifetime energies as well.

Research and Development Funding Priorities

The early North American and to some extent European approach was to "prime the pump" and research global warming. The early Australian approach to sustainability was much cruder and involved the outright purchase of abatement with little research. To date materials science, which the paper demonstrates a s fundamental has however r eceived little funding globally. As of early 2004 European priorities were genomics and biotechnology for health, information society technologies, nanotechnologies, intelligent materials and new production processes, aeronautics and space, food safety and health risks, sustainable development and global change, citizens and governance in the European knowledge-based society. In the USA Federal funding priorities include nanotechnology, defense, and aeronautics.

Geological sequestration has also been a priority in many countries as it is associated with the petroleum industry and doubts have been raised as to the transparency of funding[xli] To date Australian research priorities have not included materials science however the Australian research funding priorities for 2004 – 2005 may well lead the way globally as included under the heading "Frontier Technologies for Building and Transforming Australia" are advanced materials. "Advanced materials for applications in construction, communications, transport, agriculture and medicine (examples include ceramics, organics, biomaterials, smart material and fabrics, composites, polymers and light metals).[xlii]" The Killer Application – TecEco Cements on page 393 are material composites that include wastes and sequester CO_2.

Recycling

Recycling involves a series of activities by which wastes are collected, sorted, processed and converted into raw materials and used in the production of new products. Recycling is carried out by individuals, volunteers, businesses and governments. For high value waste recycling is profitable and undertaken by business, usually by buying back wastes, but not so as the value declines. Governments generally have recognized the importance of recycling but have gone about the introduction of recycling through councils and local authorities in completely the wrong way. As the hazards of discarded wastes do not correlate with their value, many wastes are recycled by the authority of legislation or power of producer organisations.

Instead of being forced upon us using good taxpayer dollars for "feel good" reasons, much more effort should have been put into the development of technologies to make the process economic. With market forces driving recycling much more would occur much more efficiently. The problem is to make the process of recycling less "feel good" and more economic so it is driven by market forces.

Accounting for Externalities - the True Cost of the Techno-process

"...the exponential growth curve of cost associated with negative impacts or "externalities" such as climate change, salinity, acid sulphate soils, river system degradation, or general pollution, has up until now, been a legacy for future generations to deal with. For decades the cash p rice o f g oods a nd s ervices h as b een a rtificially d eflated, w ith m uch o f t he r eal c ost being outsourced on to the environment. The costs, however, that are backing up on us - bush fires, dust storms, floods, soil erosion, salinity, changes in disease patterns, hurricanes and cyclones - can often be attributed, at least in part, to climate change. An integral part of the dilemma we have is the denial that anything truly threatening is happening"[xliii].

Accounting systems that recognise the value of natural capital[xliv] are required so that the true costs of sub techno-processes that extract and waste are born by those who gain the benefit of doing so.

Extended producer responsibility (EPR)

EPR incorporates negative externalities from product use and end-of-life in product prices Producers are made responsible for environmental effects over the entire product life cycle so that the cost of compliance cannot be shifted to a third party and must therefore be incorporated i nto p roduct p rices. E xamples o f E PR r egulations i nclude e missions a nd f uel economy standards (use stage) and product take back requirements (end of life) such as deposit legislation, and mandatory returns policies which tend to force design with disassembly in mind.

Producers are made responsible for collecting and recycling end-of-life products. Waste-management costs are shifted to those most capable of reducing disposal costs by changing designs for recyclability, longevity, reduced toxicity, and limited volume of waste generated. Disposal costs are reflected in product prices so consumers can make more informed decisions. The above solutions all involve a cost. What if benefits could be incorporated?

What Governments Should be Doing and Why

Much as been written about the role of governments and the need to bring about common good.

1. The first requirement is that the people in power realise their urgent responsibility to promote sustainability. The democratic system has a fatal flaw in that the outlook of politicians and therefore governments is usually not much beyond the next election. As a consequence policy is generally extremely short sighted and too directly connected to the needs of the here and now rather than mindful of the future. Change is occurring, sustainability issues are becoming recognised as important but all too slowly.
2. The second requirement is that governments all over the world co-operate to bring about sustainability

Problems on a global scale are not just the concern of one or two countries but all people on the planet. World federalists believe we need a system of democratic global governance on top of (not instead of) national governments. Such a system would provide enforceable legal mechanisms for resolving conflicts and safeguarding the environment. Perhaps they have a point.

In spite of the two UN 'Habitat' conferences on urban prospects,[xlv] and their huge impact, cities have not been given serious attention in the mainstream sustainability debate until very recently. For example the World Conservation Strategy of 1980, which first used the term "sustainable development," paid little attention to accelerating urbanization. The Brundtland report[xlvi] did discuss the issue, but the main emphasis was on the "urban crisis in developing countries." The role of the built environment, particularly in rich cities has been neglected; however this is difficult to reconcile with physical reality.

"The world population reached 6 billion in 1999.....At the current rate the world will have 7 billion people soon after the year 2010. The overwhelming share of world population growth is taking place in developing countries (...95.2% in 1990-2000; 97.6% in 2000-10; and 98.4% in 2010-20). The population of developing countries has more than doubled in 35 years, growing from 1.89 billion in 1955 to 4.13 billion in 1990. Significant proportions of population increases in the developing countries have been and will be absorbed by urban areas (...71.8% in 1990-2000; 83.4% in 2000-10; and 93.4 in 2010-20). Urban settlements in the developing countries are, at present, growing five times as fast as those in the developed countries. Cities in the developing countries are already faced by enormous backlogs in shelter, infrastructure and services and confronted with increasingly overcrowded transportation systems, insufficient water supply, deteriorating sanitation and environmental pollution.[xlvii]"

Since the wealthiest 25 percent of the human population consume 80 percent of the world's economic output[xlviii], approximately 64 % of the world's economic production/consumption and pollution is associated with cities in *rich* countries. Only 12 percent is tied to cities in the developing world[xlix]. In short, "half the people and three-quarters of the world's environmental problems reside in cities, and rich cities, mainly in the developed North, impose by far the greater load on the ecosphere and global commons[l]". It is time for governments to take an active role, to recognize their responsibility to seek sustainability as a cornerstone to all government expenditure and policy and facilitate economic systems that encourage sustainability such as carbon trading and EPR. The problem is to achieve "common good" without a disproportionate impact on taxation.

Making Reducing, Re-using and Recycling Work in the Long Term without Subsidies or Taxes

Even though governments through policy can introduce change it is important that technologies that seek sustainability are also fundamentally economic otherwise they are inefficient and not viable in the long run.

Economic viability attracts investment, and insufficient investment is finding its way into sustainability. Natural capital is undervalued.

Consider the techno-function: Take→Manipulate→Make→Use→Waste

The function has an input and output rate and volume dimension. If the impacts of the techno-function on the geosphere-biosphere are to be reduced then the rate and volume of flows through the function need to be reduced and they are therefore of vital interest. A large proportion of what passes through the techno-process that is not renewable or easily reassimilated ends up as the materials with which we build our techno-sphere. The rate and volume of these flows can be reduced by:

- Reducing the input to produce the same output
- Recycling so that fewer resources are required to be extracted.

By reducing, re-using and recycling the function becomes:

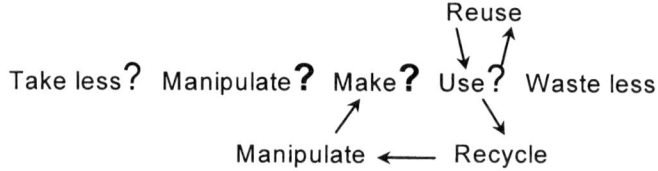

And more desirably:

Reuse

Take only   →Manipulate→Make→Use→   Waste only what is
renewables   biodegradable or can be re-
                                              assimilated
                    Manipulate ←—— Recycle

Reducing the "Take" & "Waste"

The biggest factor in increasing economic growth and raising living standards over time has been the economy's ability to produce more out of less, i.e. to become more productive. Productivity is good economics and driven by positive market forces. Productive companies reduce inputs for the same output quality, volume and cost. An increase in productivity would put less demand on natural resources which is something we all would agree is desirable.

Arguably on a personal v alues scale in the future productivity will become the domain of robots and the challenge will be what we do rather than how e fficiently we do it. This is especially true of knowledge workers who in western countries comprise probably three quarters of the work force[li].

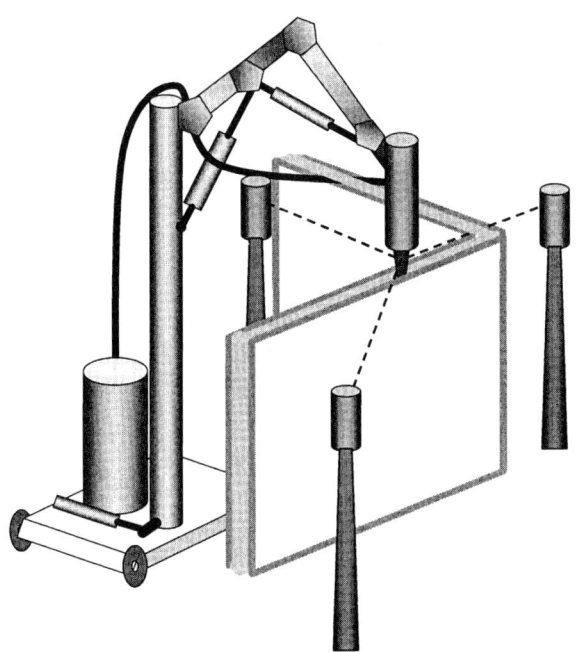

Figure 7 - Robot Construction in the Future

For robotics to be used in construction it will be necessary to squeeze self hardening materials out of a nozzle so they just sit without deformation until they gain full strength. Various materials from structural containing fibres for reinforcement to void filling and insulating will be required and like a colour inkjet printer will be selected as specified by the design. Accuracy will be far greater than currently possible, wonderful architectural shapes as yet unthought of will be used and fibres will provide reinforcing. Walls will most likely have a low strength foamed insulating cementitious material between the faces making services easy to add at a later time. Conduits could also be provided by design. The use of robots in construction will reduce the waste of new construction materials immensely. Just like an inkjet printer only uses the right amount of ink, only the exact amount of material will be used. The introduction of robots to construction will also mean more wastes can be utilised for building materials. More self hardening materials will be required, not less and mineral binders like TecEco's new cements have the obvious advantage of being able to utilise a large quantity of wastes (See The Killer Application – TecEco Cements on page 393).

There is no doubt that on a global scale, reducing the rate and volume of inputs required to satisfy the needs of the techno-sphere and output of waste materials that no longer satisfy needs will reduce our detrimental linkages on the environment. The use of Robots is but one practical idea as to how this could be done.

Re-using and Recycling

Before re-using and recycling can become economic the main economic hurdles to overcome are the laws of supply and demand and economies of scale. To do this we need to change the technical paradigm. Henry Ford leveraged his success selling cars to devise more efficient methods of production. As a consequence he was able to sell his cars more cheaply, increasing sales, providing more money for innovation, which reduced costs even further and so on. Ford was able to sell more at lower prices and yet make more money by achieving economies of scale. The laws of economics rely on positive feedback loops. Industrial economies of scale tend to increase value linearly, while the laws of supply and demand would dictate that exponentially more is sold or used the lower the price for the same quality. It can however take years before these laws kick in.

For example during the first 10 years, Microsoft's profits were negligible. They started to rise in 1985 and then exploded. The experience of Federal Express was similar. The same applies to fax machines and the internet which similarly festered for some time before becoming ubiquitous. The question is whether the world can wait for an explosion in the recycling business to take place. There is a desperate need to achieve sustainability quickly. What factor or factors are missing? What will make it happen as a matter of profitable economics rather than policy?

The trouble is that right now it just costs too much to reuse and recycle for these processes to be driven by economics alone. As a consequence government intervention in the form of regulation (Germany, some other countries and some states in some countries e.g. South Australia in Australia to some extent) and subsidies (most of the rest of the developed world) are required for what is no doubt a desirable social outcome. How can re-using and recycling move beyond the desirable subsidized by tax dollars to the preferred pushed and dragged by sound economics?

Currently it is more expensive to reuse and recycle than to use newly extracted resources. There would be a rapid turnaround in the sustainability industry if this hurdle could be overcome so that it was cheaper to reuse or recycle. There are two main costs involved in re-using and recycling. The costs of sorting waste streams and then transporting sorted recyclable materials back to a location in which they can be reused. The second law of thermodynamics (the law of entropy) was formulated in the middle of the last century by Clausius and Thomson. Like most natural processes, waste streams tend to follow this law in that wastes at the point of elimination from the techno-process tend to be all mixed up. Disorder is prevalent for two main reasons; things are made with mixed materials and the waste collection process tends to mix them up even more.

The current technical paradigm for the techno-process generally requires separate inputs. Costs are incurred and waste generated in separating what is required from the balance of material as nature itself rarely concentrates. As mentioned earlier, one study found that around 93 percent of materials used in production do not end up in saleable products but in waste. Re-using and recycling is even more uneconomic because the cost of un-mixing even more complex waste streams is prohibitive. After recycling is completed there is a cost of returning the materials back to manufacturers who can use them. Simultaneously dealing with the disassembly/sorting constraints, cost, material problems and transport issues during recycling are critical challenges. I once had the pleasure of a long discussion with Edward de Bono, the inventor of the words "lateral thinking" about a new technology I have invented that you will hear about later in this paper. He said that what was needed for market success was a "killer" application, an application that just could not help but succeed.

To get over the laws of increasing returns and economies of scale and to make the sorting of wastes economic so that wastes become low cost inputs for the techno-process new paradigms are required. The way forward involves at least:
- A new killer technology in the form of a method for sorting wastes
- A killer application for unsorted wastes

The "Killer" Technology - Silicon Identification of Materials

The means to very efficiently sort wastes may just lie in the silicon chip. The cost and size of intelligent silicon with embedded thought are both falling exponentially. Silicon chips already have a diverse range of uses For example they are being used in paint by car manufacturers for identification purposes and one was recently put in the ear of my dog for the same reason. Silicon chips will one day be as plentiful as what they could be embedded in. They will tell us the cost at the check-out, the manufacturer, warranty details who the owner is and what waste stream a robot should put them into when eventually wasted. Remember the impact bar-coding had in supermarkets? Silicon embedded in products will do the same thing for the cost of sorting waste streams. Robots will efficiently and productively be able to distinguish different types of plastic, glass, metals ceramics and so on.

The only economic hurdle that would remain would be the efficient transportation back to manufacturing points of these waste streams. This could be overcome with a ubiquitous killer application. We are progressing towards a silcon intelligence defined flow of materials from producer to consumer. Wal-Mart, one of the biggest US retailers has initiated the introduction of smart tagging of all goods that it will sell in the store. So too have the US defense department for all provisions. Everything will have a tag that indicates what order it was on, delivery details, price, disposal etc. The US packaging giant, Smurfit-Stone want to eliminate the smart label and implant chips directly into packaging. TecEco want to go a stage further and implant intelligent silicon that will do everything everybody wants it to do, directly into the materials out of which things are made.

The Killer Application – TecEco Cements

What if wastes could be utilized depending on their class of properties rather than specific properties? What if an application could be found that could utilize vast volumes of materials that offered useful broadly defined properties such as light weight, tensile strength, insulating capacity, strength or thermal capacity? After all, the best thing to do with wastes is to use them if at all possible.

There are many wastes that are just too costly to further sort into very specific wastes streams such as many plastics. There are also waste streams such as mine tailings, furnace sands, quarry dusts and the like for which no particular use could otherwise be found. Glasses tend to share in common a lot of properties as do plastics, wood, ceramics and so on. Glasses are brittle, tough and abrasion resistant. Plastics are generally light, insulating and have tensile strength. What if it did not matter if they were mixed up together?

The solution is to use these materials in composites for their properties rather than for their composition. The problem then becomes one of finding a potentially cheap, un-reactive but strong binder with the right rheology for use by for example robots of the futureReducing the "Take" & "Waste" on page 390. Eliminate plastics, epoxies and other inorganic binders that are just too expensive and the choice for durability and cost is a mineral binder. Ordinary Portland cement concretes are a good start; over two tonnes are produced per person on the planet per annum. Unfortunately they are too reactive to use with a wide range of fillers. The breakthrough has been the development of a wide range of blended calcium-magnesium binders with a low long term pH and that are internally much drier.

Sustainability by regulation subsidy or taxes is not itself economically sustainable. The new TecEco binder technologies are a breakthrough in that they change the technology paradigm. "...it is technology that defines what is a resource as well as what our effective supply of that resource is.[lii]" Wastes can be used and in large quantities, there is wealth in waste and significant overall improvements in sustainability are achievable with economic benefits. A major advantage of the TecEco technology over all other sequestration and abatement proposals it that the technology itself is viable even without a value being placed on abatement and sequestration. Another advantage of the magnesian tec, eco and enviro-cement technologies is that they define improvements in the material properties of concrete.

The TecEco technology is but one very important example of where materials can make a big difference to sustainability by providing composites that utilise wastes. There are other ways improvements in materials will make a big difference.

MATERIALS – THE KEY TO SUSTAINABILITY

Materials are the lasting substances that flow through the techno-process. They are the link between the biosphere-geosphere and techno-sphere. The use of more sustainable materials is fundamental to our survival on the planet. The choice of materials that we use to construct our techno-sphere ultimately controls emissions, lifetime and embodied energies, maintenance of utility, recyclability and the properties of wastes returned to the biosphere - geosphere.

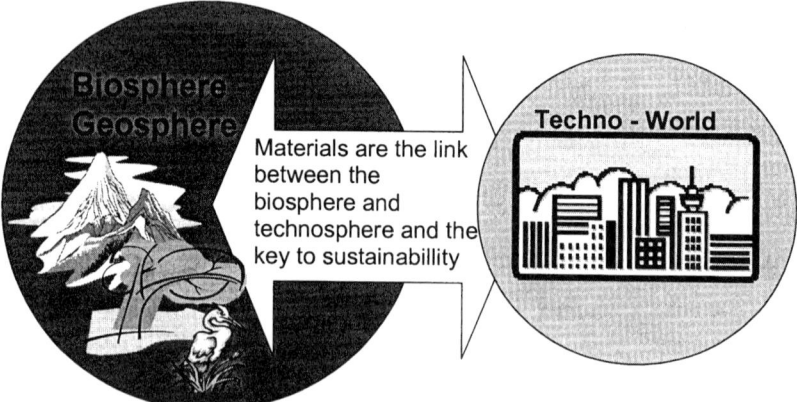

Figure 8 - Materials - The Link between the Biosphere - geosphere and Techno-sphere

Fundamental c hanges a re necessary to achieve real sustainability and if t hese are to occur without economic disruption, as the materials we use control the sustainability of the systems we proliferate, the materials paradigm we live in will also have to change.

- Materials science will become of great importance as the race to develop materials for the future gets underway

- The properties of many materials are too focussed. It should be possible in the future to develop new materials with more than one property currently considered as conflicting.

- For example materials that are good insulators do not generally have a high heat capacity. Combining insulating and heat capacity has huge potential for reducing the lifetime energies of buildings. Another example is that materials that are light in weight are not generally strong.

- The embodied energy of most materials is too high. In the future it may be possible to develop ways of making existing materials or new materials with the same functionality and properties but with lower embodied energies.

  The new TecEco kiln for example combines grinding which is only 1-2% efficient with heating to reduce overall energy inputs by some 30%

- Materials will be required that are either biodegradable or easily recycled within the techno-sphere techno – process are required.

  An example of a biodegradable replacement for a non biodegradable product includes rice paper takeaway eating containers and utensils instead of plastic. An example of easily recycled materials would be printer cartridges which are economic to return for refilling or polling booths in Australia which are all returned for recycling as paper under the watchful eye of the electorate commission.

- Materials that remain useful for longer are required.

  Things were once and still are to some extent built to break down after the period of their warranty. Henry Ford started this by reducing the quality of the items that lasted the longest in his cars. There is a way forward however. Imagine if functionality and service were purchased instead of energy and things. An example given in the book "natural Capitalism" was that instead of buying electricity one would purchase heating and lighting[liii]. I recently had the pleasure of entertaining in Tasmania the representative of a Brazilian company, Magnesita S.A. He told me that his company no longer just sold refractory bricks. They were paid on the basis of downtime experienced by their steel producing clients. With better quality bricks and less downtime they were paid more.

The Importance of Sustainable Materials for the Built Environment

Urbanization has serious negative implications for global sustainability yet the impact and the associated opportunities for improvement have been given little emphasis.
Given the enormous materials flows involved, the obvious place to improve sustainability is the built environment. The materials used determine net emissions, the impact of extraction,

how they can be reused and the effects on earth systems of wastage. To reduce the impact of the techno - function that describes the flow of these materials from take to waste it is fundamental that we think about the materials we use to construct our built environment and the molecules they are made of. With the right materials technology, because of its sheer size the built environment could reduce the take from the geosphere-biosphere and utilise many different wastes including carbon dioxide.

Materials used to construct the built environment should, as well as the required properties have low embodied energies, low lifetime energies, and low greenhouse gas emissions when considered on a whole of life cycle basis. They should also be preferably made from renewable resources and either easily recycled or reassimilated by the geosphere-biosphere

The Lesson for Governments

As Paul Zane Pilzer says "Technology is the major determinant of wealth as it determines the nature and supply of physical resources[lii]." Why is it then that so little government research funding is to change the technical paradigm for reducing, re-using and recycling materials? Materials are after all a major part of the flow of resources in the techno function and fundamental for sustainability. Instead of for example like the Australian Greenhouse Gas Abatement program which had little for science and financially supported projects with significant abatement including converting from coal to natural gas, governments need to focus on fundamental research that change the technology factor. As Pilzer's first law or alchemy states "By enabling us to make productive use of particular raw materials, technology determines what constitutes a physical resource[lii]." Pilzer goes on further to explain that definitional technologies are those that enable us to make use of particular resources. Wastes are potentially a huge resource. Improvements in recovery and utilisation technologies will one day make them of significant value.

Fortunately some governments such as the EU are starting to research how we could live more sustainably on the planet. As of today however I am not aware of any country or group of countries that prioritise the development research into materials as a way of reducing the take and waste in the techno - function, maximising utility and making re-use and recycling more profitable.

Concrete - The Material for the Future

As previously discussed under the heading THE BUILT ENVIRONMENT - A FOCUS FOR SUSTAINABILITY EFFORTS on page 386, concrete is the single biggest material flow on the planet by a big margin. There is tremendous scope to add strength and improve sustainability and other properties of concrete as a material through the addition of other substances including wastes, many of which would add tensile strength, insulating capacity or reduce weight. New composites made in this way will be the high performance materials of the future.

Materials such as the new magnesian tec, eco and enviro-cements will have a role in the development of these new materials as they not only absorb carbon dioxide in bricks, blocks, pavers, mortars and porous pavement, but also improve properties and allow the incorporation of a wider range of materials.

CONCLUSIONS

The way forward is clear, technology can help us change the techno process so that we take and waste less. By doing so the process becomes more economic and thus self propelled with less government intervention. Finding 3 under the heading Transport and Urban Design of the recent ISOS conference in Australia applies globally. It stated in part. "…The Federal Government should promote Australian building innovations (e.g. eco-cement) that contribute global solutions towards sustainability; provide more sustainable city innovation R&D funds; and re-direct some housing and transport funds towards sustainable cities demonstration projects[liv]."

Technology can make it possible to achieve a far greater measure of sustainability, to economically reduce, re-use and recycle. The potential multipliers from spending on research and development are huge.

With the development of definitional materials technologies as a result of appropriate research and development funding as Amory Lovins of the Rocky Mountain Institute puts it sustainability *"will happen, and happen rapidly – because it's profitable[lv]"*.

FOOTNOTES AND REFERENCES

[i] Utility is an economic term for value to the user.

[ii] The American Association for the Advancement of Science Population and Environment Atlas at http://atlas.aaas.org/index.php?part=1&sec=waste valid as at 24/04/04

[iii] For the purposes of this paper the geosphere is defined as the solid earth including the continental and oceanic crust as well as the various layers of the earth's interior

[iv] For the purposes of this paper the biosphere is defined as "living organisms and the part of the earth and its atmosphere in which living organisms exist or that is capable of supporting life"

[v] Utility is an economic term for value to the user.

[vi] Ed. Given a healthy environment

[vii] dictionary.com at http://dictionary.reference.com, valid as at 24/04/04

[viii] The Collins Dictionary and Thesaurus in One Volume, Harper Collins, 1992

[ix] The term techno-sphere refers to our footprint on the globe, our technical world of cars, buildings, infrastructure etc.

[x] David Schaller, Sustainable Development Coordinator, US EPA Region 8 - Denver, Beyond Sustainability: From Scarcity to Abundance, BioInspire Newsletter, 13 February 2004

[xi] WRI, World Resources 1998-99 Database Diskette, 1998.

[xii] Microsoft Office 2003 clipart.

[xiii] Powerof10 web site at http://www.powersof10.com/p10_day/planet.pdf

[xiv] http://www.c-f-c.com/supportdocs/cfcs.htm

[xv] The American Association for the Advancement of Science Population and Environment Atlas at http://atlas.aaas.org/index.php?part=1&sec=waste valid as at 24/04/04

[xvi] "*Global Change and the Earth System - A Planet under Pressure*" IGBP SCIENCE No. 4http://www.igbp.kva.se/cgi-bin/php/publications_books.show.php?section_id=48&article_id=105&onearticle=, download the summary from http://www.igbp.kva.se/cgi-bin/php/frameset.php)

[xvii] US Government Survey, Mineral Commodities Summary, Cement 2004 at http://minerals.usgs.gov/minerals/pubs/commodity/cement/cemenmcs04.pdf valid 27/04/04

[xviii] From Marland, G., T.A. Boden, and R. J. Andres. "*Global, Regional, and National CO2 Emissions. In Trends: A Compendium of Data on Global Change.*" 2003, Carbon Dioxide Information Analysis Center, Oak Ridge National Laboratory, U.S. Department of Energy, Oak Ridge, Tenn., U.S.A. at http://cdiac.esd.ornl.gov/trends/emis/glo.htm valid 27/04/04

[xix] Atkinson, Maria, Speech, Energy Efficiency Conference 14 November 2003, Green Building Council of Australia, at http://www.gbcaus.org/gbc.asp?sectionid=16&docid=116 as at 24/04/04

[xx] University of Oregon, Electronic Universe Educational Server http://zebu.uoregon.edu/1998/es202/l14.html valid 06/05/04

[xxi] Office of Solid Waste Strategic Planning Document, 2003 - 2008, 9/23/03

[xxii] Mark Twain

[xxiii] According to Dr. Colin Campbell, a petro-geologist in and article titled *How Long Can Oil the Last*, The Sunday Business Post 27th October 2002

[xxiv] National Energy Technology Laboratory (Numerous Authors) *Chemical And Geologic Sequestration Of Carbon Dioxide*, at http://www.netl.doe.gov/products/r&d/annual_reports/2001/cgscdfy01.pdf , page 7 valid 28/12/03.

[xxv] International Energy Agency, Key World Energy Statistics 2003 Edition downloaded from http://www.iea.org/bookshop/add.aspx?id=144, 27/04/04

[xxvi] URL: www.dbce.csiro.au/ind-serv/brochures/embodied/embodied.htm (last accessed 07 March 2000)

[xxvii] New Scientist, 19 July 1997, page 14.

[xxviii] According to the article *Cement and Concrete: Environmental Considerations* in Environmental Building News volume 2, No 2 – March/April 1993 researchers at the Oak Ridge National Laboratories (USA) put the figure at 1.24 tonnes of CO2 for every tonne of Portland cement.

[xxix] Pers comm. Dr Selwyn Tucker, CSIRO Department of Building Construction and Engineering, Melbourne

[xxx] Pearce, F., "The Concrete Jungle Overheats", New Scientist, 19 July, No 2097, 1997 (page 14).

[xxxi] Cement production = Carbon dioxide emissions at 1 tonne cement= 1 tonne CO_2 Source of data USGS cement XLS file. Data collected by Van Oss, Hendrik G. and Kelly, Thomas D. Last modification: April 15, 2004 Available at http://minerals.usgs.gov/minerals/pubs/of01-006/ as at 29/04/04

[xxxii] USGS figures extrapolated to 2004

[xxxiii] Dr Selwyn Tucker, CSIRO on line brochure at http://www.dbce.csiro.au/ind-serv/brochures/embodied/embodied.htm valid 05/08/2000

[xxxiv] Hendriks C.A., Worrell E, de Jager D., Blok K., and Riemer P. Emission Reductions of Greenhouse Gases from the Cement Industry. International Energy Agency Conference Paper at www.ieagreen.org.uk.

[xxxv] Davidovits, J A Practical Way to Reduce Global Warming The Geopolymer Institute info@geopolymer.org, http://www.geopolymer.org/

xxxvi Pearce, F., "The Concrete Jungle Overheats", New Scientist, 19 July, No 2097, 1997 (page 14).

xxxvii Anthropogenic – human produced

xxxviii *The Built Environment and the Ecosphere: A Global Perspective*, Rees, William E

Professor and Director, The University of British Columbia, School of Community and Regional Planning

xxxix *Energy and Cities: Sustainable Building and Construction Summary of Main Issues* IETC Side Event at UNEP Governing Council, 6 February, 2001 - Nairobi, Kenya

xl CSIRO (numerous authors), "Human Settlements Theme Report, State of the Environment Australia 2001", Australian Government Department of Environment and Heritage. See http://www.deh.gov.au/soe/2001/settlements/settlements02-5c.html, valid 29/04/04

xli Wilson, Nigel *"Call for public input to carbon plan"*, The Weekend Australian, May 1-2, 2004,'Resources',p 7.

xlii Descriptions of Designated National Research Priorities and associated Priority Goals, http://www.arc.gov.au/pdf/2004_designated_national_research_priorities_&_associate.pdf valid 06/05/04

xliii Fiona Wain, CEO, Environment Business Australia, Canberra Times, 5 February 2004

xliv Hawken Paul, Lovins Amory, Lovins L. Hunter, *Natural Capitalism: Creating the Next Industrial Revolution,* Earthscan Publications Pty. Ltd. *2000*

xlv Vancouver in 1976 and Istanbul in 1996.

xlvi In 1983 the United Nations appointed an international commission to propose strategies for "sustainable development" - ways to improve human well-being in the short term without threatening the local and global environment in the long term.The Commission was chaired by Norwegian Prime-Minister Gro Harlem Brundtland, and it's report "Our Common Future", published in 1987 was widely known as "The Brundtland Report."

xlvii http://www.unchs.org/habrdd/global.html, valid as at 22/02/04

xlviii WCED 1987

xlix Rees,W. E. (1999) *The Built Environment and the Ecosphere: A Global Perspective*, Building Research and Information 27: (4/5): 206-220

l Rees, W. E. (1997) Is "sustainable city" an oxymoron? *Local Environment* 2, 303-310.

li For an excellent discussion on the affects of knowledge on politics, business and society see Peter F Drucker, *Post Capitalist Society*, ButterworthHeinemann, 1993

lii Pilzer, Paul Zane, *Unlimited Wealth, The Theory and Practice of Economic Alchemy*, Crown Publishers Inc. New York.1990

liii Hawken Paul, Lovins Amory, Lovins L. Hunter, *Natural Capitalism: Creating the Next Industrial Revolution,* Earthscan Publications Pty. Ltd. *2000*

liv ISOS Conference, 14th November Canberra, ACT, Australia communique downloadable from http://www.isosconference.org.au/entry.html.

lv The Bulletin, April 25, 2000.

INDEX OF AUTHORS

SUBJECT INDEX

This index has been compiled from the keywords assigned to the papers, edited and extended as appropriate. Number refers to the first page of the relevant paper